Art Lew, Holger Mauch

Dynamic Programming

Studies in Computational Intelligence, Volume 38

Editor-in-chief
Prof. Janusz Kacprzyk
Systems Research Institute
Polish Academy of Sciences
ul. Newelska 6
01-447 Warsaw
Poland
E-mail: kacprzyk@ibspan.waw.pl

Further volumes of this series can be found on our homepage: springer.com

Vol. 22. N. Nedjah, E. Alba, L. de Macedo Mourelle (Eds.)
Parallel Evolutionary Computations, 2006
ISBN 3-540-32837-8

Vol. 23. M. Last, Z. Volkovich, A. Kandel (Eds.)
Algorithmic Techniques for Data Mining, 2006
ISBN 3-540-33880-2

Vol. 24. Alakananda Bhattacharya, Amit Konar, Ajit K. Mandal
Parallel and Distributed Logic Programming, 2006
ISBN 3-540-33458-0

Vol. 25. Zoltán Ésik, Carlos Martín-Vide, Victor Mitrana (Eds.)
Recent Advances in Formal Languages and Applications, 2006
ISBN 3-540-33460-2

Vol. 26. Nadia Nedjah, Luiza de Macedo Mourelle (Eds.)
Swarm Intelligent Systems, 2006
ISBN 3-540-33868-3

Vol. 27. Vassilis G. Kaburlasos
Towards a Unified Modeling and Knowledge-Representation based on Lattice Theory, 2006
ISBN 3-540-34169-2

Vol. 28. Brahim Chaib-draa, Jörg P. Müller (Eds.)
Multiagent based Supply Chain Management, 2006
ISBN 3-540-33875-6

Vol. 29. Sai Sumathi, S.N. Sivanandam
Introduction to Data Mining and its Applications, 2006
ISBN 3-540-34689-9

Vol. 30. Yukio Ohsawa, Shusaku Tsumoto (Eds.)
Chance Discoveries in Real World Decision Making, 2006
ISBN 3-540-34352-0

Vol. 31. Ajith Abraham, Crina Grosan, Vitorino Ramos (Eds.)
Stigmergic Optimization, 2006
ISBN 3-540-34689-9

Vol. 32. Akira Hirose
Complex-Valued Neural Networks, 2006
ISBN 3-540-33456-4

Vol. 33. Martin Pelikan, Kumara Sastry, Erick Cantú-Paz (Eds.)
Scalable Optimization via Probabilistic Modeling, 2006
ISBN 3-540-34953-7

Vol. 34. Ajith Abraham, Crina Grosan, Vitorino Ramos (Eds.)
Swarm Intelligence in Data Mining, 2006
ISBN 3-540-34955-3

Vol. 35. Ke Chen, Lipo Wang (Eds.)
Trends in Neural Computation, 2006
ISBN 3-540-36121-9

Vol. 36. Ildar Batyrshin, Janusz Kacprzyk, Leonid Sheremetor, Lotfi A. Zadeh (Eds.)
Perception-based Data Mining and Decision Making in Economics and Finance, 2006
ISBN 3-540-36244-4

Vol. 37. Jie Lu, Da Ruan, Guangquan Zhang (Eds.)
E-Service Intelligence, 2006
ISBN 3-540-37015-3

Vol. 38. Art Lew, Holger Mauch
Dynamic Programming, 2007
ISBN 3-540-37013-7

Art Lew
Holger Mauch

Dynamic Programming

A Computational Tool

With 55 Figures and 5 Tables

Prof. Art Lew
Department of Information and Computer Sciences
University of Hawaii at Manoa
1680 East-West Road
Honolulu, HI 96822
USA
E-mail: artlew@hawaii.edu

Dr. Holger Mauch
Department of Computer Science
Natural Sciences Collegium
Eckerd College
4200, 54th Ave. S.
Saint Petersburg, FL 33711
USA
E-mail: mauchh@eckerd.edu

ISSN print edition: 1860-949X
ISSN electronic edition: 1860-9503
ISBN 978-3-642-07200-0 e-ISBN 978-3-540-37014-7

Springer is a part of Springer Science+Business Media
springer.com

Softcover reprint of the hardcover 1st edition 2007

Cover design: deblik, Berlin

To the Bellman Continuum, in memory of Richard Bellman. A.L.

To my family. H.M.

Preface

Dynamic programming has long been applied to numerous areas in mathematics, science, engineering, business, medicine, information systems, biomathematics, artificial intelligence, among others. Applications of dynamic programming have increased as recent advances have been made in areas such as neural networks, data mining, soft computing, and other areas of computational intelligence. The value of dynamic programming formulations and means to obtain their computational solutions has never been greater.

This book describes the use of dynamic programming as a computational tool to solve discrete optimization problems.

(1) We first formulate large classes of discrete optimization problems in dynamic programming terms, specifically by deriving the dynamic programming functional equations (DPFEs) that solve these problems. A text-based language, *gDPS*, for expressing these DPFEs is introduced. gDPS may be regarded as a high-level specification language, not a conventional procedural computer programming language, but which can be used to obtain numerical solutions.

(2) We then define and examine properties of Bellman nets, a class of Petri nets that serves both as a formal theoretical model of dynamic programming problems, and as an internal computer data structure representation of the DPFEs that solve these problems.

(3) We also describe the design, implementation, and use of a software tool, called *DP2PN2Solver*, for solving DPFEs. DP2PN2Solver may be regarded as a program generator, whose input is a DPFE, expressed in the input specification language gDPS and internally represented as a Bellman net, and whose output is its numerical solution that is produced indirectly by the generation of "solver" code, which when executed yields the desired solution.

This book should be of value to different classes of readers: students, instructors, practitioners, and researchers. We first provide a tutorial introduction to dynamic programming and to Petri nets. For those interested in dynamic programming, we provide a useful software tool that allows them to obtain numerical solutions. For researchers having an interest in the fields of dynamic programming and Petri nets, unlike most past work which applies dynamic programming to solve Petri net problems, we suggest ways to apply Petri nets to solve dynamic programming problems.

For students and instructors of courses in which dynamic programming is taught, usually as one of many other problem-solving methods, this book provides a wealth of examples that show how discrete optimization problems can be formulated in dynamic programming terms. Dynamic programming has been and continues to be taught as an "art", where how to use it must be learned by example, there being no mechanical way to apply knowledge of the general principles (e.g., the principle of optimality) to new unfamiliar problems. Experience has shown that the greater the number and variety of problems presented, the easier it is for students to apply general concepts. Thus, one objective of this book is to include many and more diverse examples. A further distinguishing feature of this book is that, for all of these examples, we not only formulate the DP equations but also show their computational solutions, exhibiting computer programs (in our specification language) as well as providing as output numerical answers (as produced by the automatically generated solver code).

In addition, we provide students and instructors with a software tool (DP2PN2Solver) that enables them to obtain numerical solutions of dynamic programming problems without requiring them to have much computer programming knowledge and experience. This software tool can be downloaded from either of the following websites:

http://natsci.eckerd.edu/~mauchh/Research/DP2PN2Solver
http://www2.hawaii.edu/~icl/DP2PN2Solver

Further information is given in Appendix B. Having such software support allows them to focus on dynamic programming rather than on computer programming. Since many problems can be solved by different dynamic programming formulations, the availability of such a computational tool, that makes it easier for readers to experiment with their own formulations, is a useful aid to learning.

The DP2PN2Solver tool also enables practitioners to obtain numerical solutions of dynamic programming problems of interest to them without requiring them to write conventional computer programs. Their time, of course, is better spent on problem formulation and analysis than on program design and debugging. This tool allows them to verify that their formulations are correct, and to revise them as may be necessary in their problem solving efforts. The main limitation of this (and any) dynamic programming tool for many practical problems is the size of the state space. Even in this event, the tool may prove useful in the formulation stage to initially test ideas on simplified scaled-down problems.

As a program generator, DP2PN2Solver is flexible, permitting alternate front-ends and back-ends. Inputs other than in the gDPS language are possible. Alternative DPFE specifications can be translated into gDPS or directly into Bellman nets. Output solver code (i.e., the program that numerically solves a given DPFE) may be in alternative languages. The solver code emphasized in this book is Java code, largely because it is universally and freely available on practically every platform. We also discuss solver codes for spreadsheet systems and Petri net simulators. By default, the automatically generated solver code is hidden from the average user, but it can be inspected and modified directly by users if they wish.

Furthermore, this book describes research into connections between dynamic programming and Petri nets. It was our early research into such connections that ultimately lead to the concept of Bellman nets, upon which the development of our DP2PN2Solver tool is based. We explain here the underlying ideas associated with Bellman nets. Researchers interested in dynamic programming or Petri nets will find many open questions related to this work that suggest avenues of future research. For example, additional research might very likely result in improvements in the DP2PN2Solver tool, such as to address the state-space size issue or to increase its diagnostic capabilities. Every other aspect of this work may benefit from additional research.

Thus, we expect the DP2PN2Solver tool described in this book to undergo revisions from time to time. In fact, the tool was designed modularly to make it relatively easy to modify. As one example, changes to the gDPS specification language syntax can be made by simply revising its BNF definition since we use a compiler-compiler rather than a compiler to process it. Furthermore, alternate input languages (other than gDPS) and solver codes (other than Java) can be added as optional modules, without changing the existing modules. We welcome suggestions from readers on how the tool (or its description) can be improved. We may be contacted at artlew@hawaii.edu or mauchh@eckerd.edu. Updates to the software and to this book, including errata, will be placed on the aforementioned websites.

Acknowledgements. The authors wish to thank Janusz Kacprzyk for including this monograph in his fine series of books. His encouragement has been very much appreciated.

Honolulu, June 2006, *Art Lew*
St. Petersburg, June 2006, *Holger Mauch*

Contents

Part I Dynamic Programming

Part I

Dynamic Programming

1

Introduction to Dynamic Programming

This book concerns the use of a method known as *dynamic programming* (DP) to solve large classes of optimization problems. We will focus on discrete optimization problems for which a set or sequence of decisions must be made to optimize (minimize or maximize) some function of the decisions. There are of course numerous methods to solve discrete optimization problems, many of which are collectively known as mathematical programming methods. Our objective here is not to compare these other mathematical programming methods with dynamic programming. Each has advantages and disadvantages, as discussed in many other places. However, we will note that the most prominent of these other methods is *linear programming*. As its name suggests, it has limitations associated with its linearity assumptions whereas many problems are nonlinear. Nevertheless, linear programming and its variants and extensions (some that allow nonlinearities) have been used to solve many real world problems, in part because very early in its development software tools (based on the simplex method) were made available to solve linear programming problems. On the other hand, no such tools have been available for the much more general method of dynamic programming, largely due to its very generality. One of the objectives of this book is to describe a software tool for solving dynamic programming problems that is general, practical, and easy to use, certainly relative to any of the other tools that have appeared from time to time.

One reason that simplex-based tools for solving linear programming problems have been successful is that, by the nature of linear programming, problem specification is relatively easy. A basic LP problem can be specified essentially as a system or matrix of equations with a finite set of numerical variables as unknowns. That is, the input to an LP software tool can be provided in a tabular form, known as a *tableaux*. This also makes it easy to formulate LP problems as a spreadsheet. This led to spreadsheet system providers to include in their product an LP solver, as is the case with Excel.

A software tool for solving dynamic programming problems is much more difficult to design, in part because the problem specification task in itself

A. Lew and H. Mauch: *Introduction to Dynamic Programming*, Studies in Computational Intelligence (SCI) **38**, 3–43 (2007)
www.springerlink.com

presents difficulties. A DP problem specification is usually in the form of a complex (nonlinear) *recursive* equation, called the *dynamic programming functional equation* (DPFE), where the DPFE often involves nonnumerical variables that may include sets or strings. Thus, the input to a DP tool must necessarily be general enough to allow for complex DPFEs, at the expense therefore of the simplicity of a simple table. The DP tool described in this book assumes that the input DPFE is provided in a text-based specification language that does not rely on mathematical symbols. This decision conforms to that made for other mathematical programming languages, such as AMPL and LINGO.

In this introductory chapter, we first discuss the basic principles underlying the use of dynamic programming to solve discrete optimization problems. The key task is to formulate the problem in terms of an equation, the DPFE, such that the solution of the DPFE is the solution of the given optimization problem. We then illustrate the computational solution of the DPFE for a specific problem (for linear search), either by use of a computer program written in a conventional programming language, or by use of a spreadsheet system. It is not easy to generalize these examples to solve DP problems that do not resemble linear search. Thus, for numerous dissimilar DP problems, a significant amount of additional effort is required to obtain their computational solutions. One of the purposes of this book is to reduce this effort.

In Chap. 2, we show by example numerous types of optimization problems that can be solved using DP. These examples are given, first to demonstrate the general utility of DP as a problem solving methodology. Other books are more specialized in the kinds of applications discussed, often focusing on applications of interest mainly to operations research or to computer science. Our coverage is much more comprehensive. Another important reason for providing numerous examples is that it is often difficult for new students of the field to see from a relatively small sample of problems how DP can be applied to other problems. How to apply DP to new problems is often learned by example; the more examples learned, the easier it is to generalize. Each of the sample problems presented in Chap. 2 was computationally solved using our DP tool. This demonstrates the generality, flexibility, and practicality of the tool.

In Part II of this book, we show how each of the DPFEs given in Chap. 2 can be expressed in a text-based specification language, and then show how these DPFEs can be formally modeled by a class of Petri nets, called Bellman nets. Bellman nets serve as the theoretical underpinnings for the DP tool we later describe, and we describe our research into this subject area.

In Part III of this book, we describe the design and implementation of our DP tool. This tool inputs DPFEs, as given in Part II, and produces numerical solutions, as given in Part IV.

In Part IV of this book, we present computational results. Specifically, we give the numerical solutions to each of the problems discussed in Chap. 2, as provided by our DP tool.

Appendix A of this book provides program listings for key portions of our DP tool. Appendix B of this book is a User/Reference Manual for our DP tool.

This book serves several purposes.

1. It provides a practical introduction to how to solve problems using DP. From the numerous and varied examples we present in Chap. 2, we expect readers to more easily be able to solve new problems by DP. Many other books provide far fewer or less diverse examples, hoping that readers can generalize from their small sample. The larger sample provided here should assist the reader in this process.
2. It provides a software tool that can be and has been used to solve all of the Chap. 2 problems. This tool can be used by readers in practice, certainly to solve academic problems if this book is used in coursework, and to solve many real-world problems, especially those of limited size (where the state space is not excessive).
3. This book is also a research monograph that describes an important application of Petri net theory. More research into Petri nets may well result in improvements in our tool.

1.1 Principles of Dynamic Programming

Dynamic programming is a method that in general solves optimization problems that involve making a sequence of decisions by determining, for each decision, subproblems that can be solved in like fashion, such that an optimal solution of the original problem can be found from optimal solutions of subproblems. This method is based on Bellman's Principle of Optimality, which he phrased as follows [1, p.83].

> An optimal policy has the property that whatever the initial state and initial decision are, the remaining decisions must constitute an optimal policy with regard to the state resulting from the first decision.

More succinctly, this principle asserts that "optimal policies have optimal subpolicies." That the principle is valid follows from the observation that, if a policy has a subpolicy that is not optimal, then replacement of the subpolicy by an optimal subpolicy would improve the original policy. The principle of optimality is also known as the "optimal substructure" property in the literature. In this book, we are primarily concerned with the computational solution of problems for which the principle of optimality is given to hold. For DP to be computationally efficient (especially relative to evaluating all possible sequences of decisions), there should be common subproblems such that subproblems of one are subproblems of another. In this event, a solution to a subproblem need only be found once and reused as often as necessary; however, we do not incorporate this requirement as part of our definition of DP.

In this section, we will first elaborate on the nature of sequential decision processes and on the importance of being able to separate the costs for each of the individual decisions. This will lead to the development of a general equation, the *dynamic programming functional equation* (DPFE), that formalizes the principle of optimality. The methodology of dynamic programming requires deriving a special case of this general DPFE for each specific optimization problem we wish to solve. Numerous examples of such derivations will be presented in this book. We will then focus on how to numerically solve DPFEs, and will later describe a software tool we have developed for this purpose.

1.1.1 Sequential Decision Processes

For an optimization problem of the form $\text{opt}_{d \in \Delta}\{H(d)\}$, d is called the *decision*, which is chosen from a set of eligible decisions Δ, the optimand H is called the *objective function*, and $H^* = H(d^*)$ is called the *optimum*, where d^* is that value of $d \in \Delta$ for which $H(d)$ has the optimal (minimum or maximum) value. We also say that d^* *optimizes* H, and write $d^* = \arg\text{opt}_d\{H(d)\}$. Many optimization problems consist of finding a set of decisions $\{d_1, d_2, \ldots, d_n\}$, that taken together yield the optimum H^* of an objective function $h(d_1, d_2, \ldots, d_n)$. Solution of such problems by *enumeration*, i.e., by evaluating $h(d_1, d_2, \ldots, d_n)$ concurrently, for all possible combinations of values of its decision arguments, is called the "brute force" approach; this approach is manifestly inefficient. Rather than making decisions concurrently, we assume the decisions may be made in some specified sequence, say $(d_1, d_2, \ldots, d_n)$, i.e., such that

$$\begin{aligned} H^* &= \text{opt}_{(d_1,d_2,\ldots,d_n)\in\Delta}\{h(d_1, d_2, \ldots, d_n)\} \\ &= \text{opt}_{d_1\in D_1}\{\text{opt}_{d_2\in D_2}\{\ldots\{\text{opt}_{d_n\in D_n}\{h(d_1, d_2, \ldots, d_n)\}\}\ldots\}\}, \end{aligned} \tag{1.1}$$

in what are known as *sequential decision processes*, where the ordered set $(d_1, d_2, \ldots, d_n)$ belongs to some decision space $\Delta = D_1 \times D_2 \times \ldots \times D_n$, for $d_i \in D_i$. Examples of decision spaces include: $\Delta = B^n$, the special case of Boolean decisions, where each decision set D_i equals $B = \{0, 1\}$; and $\Delta = \Pi(\mathbf{D})$, a permutation of a set of eligible decisions $\mathbf{D}$. The latter illustrates the common situation where decisions d_i are interrelated, e.g., where they satisfy constraints such as $d_i \neq d_j$ or $d_i + d_j \leq M$. In general, each decision set D_i depends on the decisions $(d_1, d_2, \ldots, d_{i-1})$ that are earlier in the specified sequence, i.e., $d_i \in D_i(d_1, d_2, \ldots, d_{i-1})$. Thus, to show this dependence explicitly, we rewrite (1.1) in the form

$$\begin{aligned} H^* &= \text{opt}_{(d_1,d_2,\ldots,d_n)\in\Delta}\{h(d_1, d_2, \ldots, d_n)\} \\ &= \text{opt}_{d_1\in D_1}\{\text{opt}_{d_2\in D_2(d_1)}\{\ldots\{\text{opt}_{d_n\in D_n(d_1,\ldots,d_{n-1})}\{h(d_1, \ldots, d_n)\}\}\ldots\}\}. \end{aligned} \tag{1.2}$$

This nested set of optimization operations is to be performed from inside-out (right-to-left), the innermost optimization yielding the optimal choice for d_n as a function of the possible choices for $d_1, \ldots, d_{n-1}$, denoted $d_n^*(d_1, \ldots, d_{n-1})$, and the outermost optimization $\text{opt}_{d1 \in D1}\{h(d_1, d_2^*, \ldots, d_n^*)\}$ yielding the optimal choice for d_1, denoted d_1^*. Note that while the initial or "first" decision d_1 in the specified sequence is the outermost, the optimizations are performed inside-out, each depending upon outer decisions. Furthermore, while the optimal solution may be the same for any sequencing of decisions, e.g.,

$$\begin{aligned} &\text{opt}_{d_1 \in D_1}\{\text{opt}_{d_2 \in D_2(d_1)}\{\cdots\{\text{opt}_{d_n \in D_n(d_1,\ldots,d_{n-1})}\{h(d_1, \ldots, d_n)\}\}\ldots\}\} \\ &= \text{opt}_{d_n \in D_n}\{\text{opt}_{d_{n-1} \in D_{n-1}(d_n)}\{\cdots\{\text{opt}_{d_1 \in D_1(d_2,\ldots,d_n)}\{h(d_1, \ldots, d_n)\}\}\ldots\}\} \end{aligned} \tag{1.3}$$

the decision sets D_i may differ since they depend on different outer decisions. Thus, efficiency may depend upon the order in which decisions are made.

Referring to the foregoing equation, for a given sequencing of decisions, if the outermost decision is "tentatively" made initially, whether or not it is optimal depends upon the ultimate choices d_i^* that are made for subsequent decisions d_i; i.e.,

$$\begin{aligned} H^* &= \text{opt}_{d_1 \in D_1}\{\text{opt}_{d_2 \in D_2(d_1)}\{\cdots\{\text{opt}_{d_n \in D_n(d_1,\ldots,d_{n-1})}\{h(d_1, \ldots, d_n)\}\}\ldots\}\} \\ &= \text{opt}_{d1 \in D1}\{h(d_1, d_2^*(d_1), \ldots, d_n^*(d_1))\} \end{aligned} \tag{1.4}$$

where each of the choices $d_i^*(d_1)$ for $i = 2, \ldots, n$ is constrained by — i.e., is a function of — the choice for d_1. Note that determining the optimal choice $d_1^* = \arg \text{opt}_{d_1 \in D_1}\{h(d_1, d_2^*(d_1), \ldots, d_n^*(d_1))\}$ requires evaluating h for all possible choices of d_1 unless there is some reason that certain choices can be excluded from consideration based upon a priori (given or derivable) knowledge that they cannot be optimal. One such class of algorithms would choose $d_1 \in D_1$ independently of (but still constrain) the choices for $d_2, \ldots, d_n$, i.e., by finding the solution of a problem of the form $\text{opt}_{d_1 \in D_1}\{H'(d_1)\}$ for a function H' of d_1 that is myopic in the sense that it does not depend on other choices d_i. Such an algorithm is optimal if the locally optimal solution of $\text{opt}_{d_1}\{H'(d_1)\}$ yields the globally optimal solution H^*.

Suppose that the objective function h is *(strongly) separable* in the sense that

$$h(d_1, \ldots, d_n) = C_1(d_1) \circ C_2(d_2) \circ \ldots \circ C_n(d_n) \tag{1.5}$$

where the decision-cost functions C_i represent the costs (or profits) associated with the individual decisions d_i, and where $\circ$ is an associative binary operation, usually addition or multiplication, where $\text{opt}_d\{a \circ C(d)\} = a \circ \text{opt}_d\{C(d)\}$ for any a that does not depend upon d. In the context of sequential decision processes, the cost C_n of making decision d_n may be a function not only of the decision itself, but also of the state $(d_1, d_2, \ldots, d_{n-1})$ in which the decision is made. To emphasize this, we will rewrite (1.5) as

$$h(d_1,\ldots,d_n) = C_1(d_1|\emptyset) \circ C_2(d_2|d_1) \circ \ldots \circ C_n(d_n|d_1,\ldots,d_{n-1}). \quad (1.6)$$

We now define h as *(weakly) separable* if

$$h(d_1,\ldots,d_n) = C_1(d_1) \circ C_2(d_1,d_2) \circ \ldots \circ C_n(d_1,\ldots,d_n). \quad (1.7)$$

(Strong separability is, of course, a special case of weak separability.) If h is (weakly) separable, we then have

$$\begin{aligned}
&\text{opt}_{d_1\in D_1}\{\text{opt}_{d_2\in D_2(d_1)}\{\ldots\{\text{opt}_{d_n\in D_n(d_1,\ldots,d_{n-1})}\{h(d_1,\ldots,d_n)\}\}\ldots\}\} \\
&= \text{opt}_{d_1\in D_1}\{\text{opt}_{d_2\in D_2(d_1)}\{\ldots\{\text{opt}_{d_n\in D_n(d_1,\ldots,d_{n-1})}\{C_1(d_1|\emptyset) \circ C_2(d_2|d_1) \circ \ldots \\
&\quad \ldots \circ C_n(d_n|d_1,\ldots,d_{n-1})\}\}\ldots\}\} \\
&= \text{opt}_{d_1\in D_1}\{C_1(d_1|\emptyset) \circ \text{opt}_{d_2\in D_2(d_1)}\{C_2(d_2|d_1) \circ \ldots \\
&\quad \ldots \circ \text{opt}_{d_n\in D_n(d_1,\ldots,d_{n-1})}\{C_n(d_n|d_1,\ldots,d_{n-1})\}\ldots\}\}. \quad (1.8)
\end{aligned}$$

Let the function $f(d_1,\ldots,d_{i-1})$ be defined as the optimal solution of the sequential decision process where the decisions $d_1,\ldots,d_{i-1}$ have been made and the decisions $d_i,\ldots,d_n$ remain to be made; i.e.,

$$\begin{aligned}
f(d_1,\ldots,d_{i-1}) = \text{opt}_{d_i}\{&\text{opt}_{d_{i+1}}\{\ldots\{\text{opt}_{d_n}\{C_i(d_i|d_1,\ldots,d_{i-1}) \circ \\
&C_{i+1}(d_{i+1}|d_1,\ldots,d_i) \circ \ldots \circ C_n(d_n|d_1,\ldots,d_{n-1})\}\}\ldots\}\}.
\end{aligned} \quad (1.9)$$

Explicit mentions of the decision sets D_i are omitted here for convenience. We have then

$$\begin{aligned}
f(\emptyset) &= \text{opt}_{d_1}\{\text{opt}_{d_2}\{\ldots\{\text{opt}_{d_n}\{C_1(d_1|\emptyset) \circ C_2(d_2|d_1) \circ \ldots \\
&\quad \ldots \circ C_n(d_n|d_1,\ldots,d_{n-1})\}\}\ldots\}\} \\
&= \text{opt}_{d_1}\{C_1(d_1|\emptyset) \circ \text{opt}_{d_2}\{C_2(d_2|d_1) \circ \ldots \\
&\quad \ldots \circ \text{opt}_{d_n}\{C_n(d_n|d_1,\ldots,d_{n-1})\}\ldots\}\} \\
&= \text{opt}_{d_1}\{C_1(d_1|\emptyset) \circ f(d_1)\}. \quad (1.10)
\end{aligned}$$

Generalizing, we conclude that

$$f(d_1,\ldots,d_{i-1}) = \text{opt}_{d_i\in D_i(d_1,\ldots,d_{i-1})}\{C_i(d_i|d_1,\ldots,d_{i-1}) \circ f(d_1,\ldots,d_i)\}. \quad (1.11)$$

Equation (1.11) is a recursive functional equation; we call it a *functional equation* since the unknown in the equation is a function f, and it is *recursive* since f is defined in terms of f (but having different arguments). It is the *dynamic programming functional equation* (DPFE) for the given optimization problem. In this book, we assume that we are given DPFEs that are properly formulated, i.e., that their solutions exist; we address only issues of how to obtain these solutions.

1.1.2 Dynamic Programming Functional Equations

The problem of solving the DPFE for $f(d_1,\ldots,d_{i-1})$ depends upon the subproblem of solving for $f(d_1,\ldots,d_i)$. If we define the state $S=(d_1,\ldots,d_{i-1})$ as the sequence of the first $i-1$ decisions, where $i=|S|+1=|\{d_1,\ldots,d_{i-1}\}|+1$, we may rewrite the DPFE in the form

$$f(S)=\mathrm{opt}_{d_i\in D_i(S)}\{C_i(d_i|S)\circ f(S')\}, \tag{1.12}$$

where S is a state in a set $\mathcal{S}$ of possible states, $S'=(d_1,\ldots,d_i)$ is a next-state, and $\emptyset$ is the initial state. Since the DPFE is recursive, to terminate the recursion, its solution requires *base cases* (or "boundary" conditions), such as $f(S_0)=b$ when $S_0\in\mathcal{S}_{base}$, where $\mathcal{S}_{base}\subset\mathcal{S}$. For a *base* (or *terminal*) state S_0, $f(S_0)$ is not evaluated using the DPFE, but instead has a given numerical constant b as its value; this value b may depend upon the base state S_0.

It should be noted that the sequence of decisions need not be limited to a fixed length n, but may be of indefinite length, terminating when a base case is reached. Different classes of DP problems may be characterized by how the states S, and hence the next-states S', are defined. It is often convenient to define the state S, not as the sequence of decisions made so far, with the next decision d chosen from $D(S)$, but rather as the set from which the next decision can be chosen, so that $D(S)=$ or $d\in S$. We then have a DPFE of the form

$$f(S)=\mathrm{opt}_{d\in S}\{C(d|S)\circ f(S')\}. \tag{1.13}$$

We shall later show that, for some problems, there may be multiple next-states, so that the DPFE has the form

$$f(S)=\mathrm{opt}_{d\in S}\{C(d|S)\circ f(S')\circ f(S'')\} \tag{1.14}$$

where S' and S'' are both next-states. A DPFE is said to be *r-th order* (or *nonserial* if $r>1$) if there may be r next-states.

Simple serial DP formulations can be modeled by a state transition system or directed graph, where a state S corresponds to a node (or vertex) and a decision d that leads from state S to next-state S' is represented by a branch (or arc or edge) with label $C(d_i|S)$. $D(S)$ is the set of possible decisions when in state S, hence is associated with the successors of node S. More complex DP formulations require a more general graph model, such as that of a Petri net, which we discuss in Chap. 5.

Consider the directed graph whose nodes represent the states of the DPFE and whose branches represent possible transitions from states to next-states, each such transition reflecting a decision. The label of each branch, from S to S', denoted $b(S,S')$, is the cost $C(d|S)$ of the decision d, where $S'=T(S,d)$, where $T:\mathcal{S}\times D\rightarrow\mathcal{S}$ is a next-state transition or transformation function. The DPFE can then be rewritten in the form

$$f(S) = \text{opt}_{S'}\{b(S, S') + f(S')\}, \tag{1.15}$$

where $f(S)$ is the length of the shortest path from S to a terminal or *target* state S_0, and where each decision is to choose S' from among all (eligible) successors of S. (Different problems may have different eligibility constraints.) The base case is $f(S_0) = 0$.

For some problems, it is more convenient to use a DPFE of the "reverse" form

$$f'(S) = \text{opt}_{S'}\{f'(S') + b(S', S)\}, \tag{1.16}$$

where $f'(S)$ is the length of the shortest path from a designated state S_0 to S, and S' is a predecessor of S; S_0 is also known as the *source* state, and $f(S_0) = 0$ serves as the base case that terminates the recursion for this alternative DPFE. We call these *target-state* and *designated-source* DPFEs, respectively. We also say that, in the former case, we go "backward" from the target to the source, whereas, in the latter case, we go forward from the "source" to the target.

Different classes of DP formulations are distinguished by the nature of the decisions. Suppose each decision is a number chosen from a set $\{1, 2, \ldots, N\}$, and that each number must be chosen once and only once (so there are N decisions). Then if states correspond to possible permutations of the numbers, there are $O(N!)$ such states. Here we use the "big-O" notation ([10, 53]): we say $f(N)$ is $O(g(N))$ if, for a sufficiently large N, $f(N)$ is bounded by a constant multiple of $g(N)$. As another example, suppose each decision is a number chosen from a set $\{1, 2, \ldots, N\}$, but that not all numbers must be chosen (so there may be less than N decisions). Then if states correspond to subsets of the numbers, there are $O(2^N)$ such states. Fortuitously, there are many practical problems where a reduction in the number of relevant states is possible, such as when only the final decision d_{i-1} in a sequence $(d_1, \ldots, d_{i-1})$, together with the time or stage i at which the decision is made, is significant, so that there are $O(N^2)$ such states. We give numerous examples of the different classes in Chap. 2.

The solution of a DP problem generally involves more than only computing the value of $f(S)$ for the goal state S^*. We may also wish to determine the initial optimal decision, the optimal second decision that should be made in the next-state that results from the first decision, and so forth; that is, we may wish to determine the optimal sequence of decisions, also known as the optimal "policy" , by what is known as a reconstruction process. To reconstruct these optimal decisions, when evaluating $f(S) = \text{opt}_{d \in D(S)}\{C(d|S) \circ f(S')\}$ we may save the value of d, denoted d^*, that yields the optimal value of $f(S)$ at the time we compute this value, say, tabularly by entering the value $d^*(S)$ in a table for each S. The main alternative to using such a policy table is to reevaluate $f(S)$ as needed, as the sequence of next-states are determined; this is an example of a space versus time tradeoff.

1.1.3 The Elements of Dynamic Programming

The basic form of a dynamic programming functional equation is

$$f(S) = \text{opt}_{d \in D(S)} \{R(S, d) \circ f(T(S, d))\}, \tag{1.17}$$

where S is a *state* in some *state space* $\mathcal{S}$, d is a *decision* chosen from a *decision space* $D(S)$, $R(S, d)$ is a *reward function* (or *decision cost*, denoted $C(d|S)$ above), $T(S, d)$ is a next-state *transformation* (or *transition*) function, and $\circ$ is a binary operator. We will restrict ourselves to discrete DP, where the state space and decision space are both discrete sets. (Some problems with continuous states or decisions can be handled by discretization procedures, but we will not consider such problems in this book.) The elements of a DPFE have the following characteristics.

State The state S, in general, incorporates information about the sequence of decisions made so far. In some cases, the state may be the complete sequence, but in other cases only partial information is sufficient; for example, if the set of all states can be partitioned into equivalence classes, each represented by the last decision. In some simpler problems, the length of the sequence, also called the stage at which the next decision is to be made, suffices. The initial state, which reflects the situation in which no decision has yet been made, will be called the *goal state* and denoted S^*.

Decision Space The decision space $D(S)$ is the set of possible or "eligible" choices for the next decision d. It is a function of the state S in which the decision d is to be made. Constraints on possible next-state transformations from a state S can be imposed by suitably restricting $D(S)$. If $D(S) = \emptyset$, so that there are no eligible decisions in state S, then S is a terminal state.

Objective Function The objective function f, a function of S, is the optimal profit or cost resulting from making a sequence of decisions when in state S, i.e., after making the sequence of decisions associated with S. The goal of a DP problem is to find $f(S)$ for the goal state S^*.

Reward Function The reward function R, a function of S and d, is the profit or cost that can be attributed to the next decision d made in state S. The reward $R(S, d)$ must be separable from the profits or costs that are attributed to all other decisions. The value of the objective function for the goal state, $f(S^*)$, is the combination of the rewards for the complete optimal sequence of decisions starting from the goal state.

Transformation Function(s) The transformation (or transition) function T, a function of S and d, specifies the next-state that results from making a decision d in state S. As we shall later see, for nonserial DP problems, there may be more than one transformation function.

Operator The operator is a binary operation, usually addition or multiplication or minimization/maximization, that allows us to combine the returns of separate decisions. This operation must be associative if the returns of decisions are to be independent of the order in which they are made.

Base Condition Since the DPFE is recursive, base conditions must be specified to terminate the recursion. Thus, the DPFE applies for S in a state space $\mathcal{S}$, but

$$f(S_0) = b,$$

for S_0 in a set of base-states not in $\mathcal{S}$. Base-values b are frequently zero or infinity, the latter to reflect constraints. For some problems, setting $f(S_0) = \pm\infty$ is equivalent to imposing a constraint on decisions so as to disallow transitions to state S_0, or to indicate that $S_0 \notin \mathcal{S}$ is a state in which no decision is eligible.

To solve a problem using DP, we must define the foregoing elements to reflect the nature of the problem at hand. We give several examples below. We note first that some problems require certain generalizations. For example, some problems require a second-order DPFE having the form

$$f(S) = \mathrm{opt}_{d \in D(S)}\{R(S,d) \circ f(T_1(S,d)) \circ f(T_2(S,d))\}, \tag{1.18}$$

where T_1 and T_2 are both transformation functions to account for the situation in which more than one next-state can be entered, or

$$f(S) = \mathrm{opt}_{d \in D(S)}\{R(S,d) \circ p_1.f(T_1(S,d)) \circ p_2.f(T_2(S,d))\}, \tag{1.19}$$

where T_1 and T_2 are both transformation functions and p_1 and p_2 are multiplicative weights. In probabilistic DP problems, these weights are probabilities that reflect the probabilities associated with their respective state-transitions, only one of which can actually occur. In deterministic DP problems, these weights can serve other purposes, such as "discount factors" to reflect the time value of money.

1.1.4 Application: Linear Search

To illustrate the key concepts associated with DP that will prove useful in our later discussions, we examine a concrete example, the optimal "linear search" problem. This is the problem of permuting the data elements of an array A of size N, whose element x has probability p_x, so as to optimize the linear search process by minimizing the "cost" of a permutation, defined as the expected number of comparisons required. For example, let $A = \{a, b, c\}$ and $p_a = 0.2$, $p_b = 0.5$, and $p_c = 0.3$. There are six permutations, namely, *abc, acb, bac, bca, cab, cba*; the cost of the fourth permutation *bca* is 1.7, which can be calculated in several ways, such as

$$1p_b + 2p_c + 3p_a \text{ [using Method S]}$$

and

$$(p_a + p_b + p_c) + (p_a + p_c) + (p_a) \text{ [using Method W]}.$$

This optimal permutation problem can be regarded as a sequential decision process where three decisions must be made as to where the elements of A are to be placed in the final permuted array A'. The decisions are: which element is to be placed at the beginning of A', which element is to be placed in the middle of A', and which element is to be placed at the end of A'. The order in which these decisions are made does not necessarily matter, at least insofar as obtaining the correct answer is concerned; e.g., to obtain the permutation bca, our first decision may be to place element c in the middle of A'. Of course, some orderings of decisions may lead to greater efficiency than others. Moreover, the order in which decisions are made affects later choices; if c is chosen in the middle, it cannot be chosen again. That is, the decision set for any choice depends upon (is constrained by) earlier choices. In addition, the cost of each decision should be separable from other decisions. To obtain this separability, we must usually take into account the order in which decisions are made. For Method S, the cost of placing element x in the i-th location of A' equals ip_x regardless of when the decision is made. On the other hand, for Method W, the cost of a decision depends upon when the decision is made, more specifically upon its decision set. If the decisions are made in order from the beginning to the end of A', then the cost of deciding which member d_i of the respective decision set D_i to choose next equals $\sum_{x \in D_i} p_x$, the sum of the probabilities of the elements in $D_i = A - \{d_1, \ldots, d_{i-1}\}$. For example, let d_i denote the decision of which element of A to place in position i of A', and let D_i denote the corresponding decision set, where $d_i \in D_i$. If the decisions are made in the order $i = 1, 2, 3$ then $D_1 = A, D_2 = A - \{d_1\}, D_3 = A - \{d_1, d_2\}$. For Method S, if the objective function is written in the form $h(d_1, d_2, d_3) = 1p_{d_1} + 2p_{d_2} + 3p_{d_3}$, then

$$
\begin{aligned}
f(\emptyset) &= \min_{d_1 \in A} \{ \min_{d_2 \in A - \{d_1\}} \{ \min_{d_3 \in A - \{d_1, d_2\}} \{1p_{d_1} + 2p_{d_2} + 3p_{d_3}\}\}\} \\
&= \min_{d_1 \in A} \{1p_{d_1} + \min_{d_2 \in A - \{d_1\}} \{2p_{d_2} + \min_{d_3 \in A - \{d_1, d_2\}} \{3p_{d_3}\}\}\} \qquad (1.20)
\end{aligned}
$$

For Method W, if the objective function is written in the form $h(d_1, d_2, d_3) = \sum_{x \in A} p_x + \sum_{x \in A - \{d_1\}} p_x + \sum_{x \in A - \{d_1, d_2\}} p_x$, then

$$
\begin{aligned}
& f(\emptyset) \\
&= \min_{d_1 \in A} \{ \min_{d_2 \in A - \{d_1\}} \{ \min_{d_3 \in A - \{d_1, d_2\}} \{ \sum_{x \in A} p_x + \sum_{x \in A - \{d_1\}} p_x + \sum_{x \in A - \{d_1, d_2\}} p_x \}\}\} \\
&= \min_{d_1 \in A} \{ \sum_{x \in A} p_x + \min_{d_2 \in A - \{d_1\}} \{ \sum_{x \in A - \{d_1\}} p_x + \min_{d_3 \in A - \{d_1, d_2\}} \{ \sum_{x \in A - \{d_1, d_2\}} p_x \}\}\}.
\end{aligned}
\qquad (1.21)
$$

However, if the decisions are made in reverse order $i = 3, 2, 1$, then $D_3 = A, D_2 = A - \{d_3\}, D_1 = A - \{d_2, d_3\}$, and the above must be revised accordingly. It should also be noted that if $h(d_1, d_2, d_3) = 0 + 0 + (1p_{d_1} + 2p_{d_2} + 3p_{d_3})$, where all of the cost is associated with the final decision d_3, then

$$f(\emptyset) = \min_{d_1 \in A}\{0 + \min_{d_2 \in A-\{d_1\}}\{0 + \min_{d_3 \in A-\{d_1,d_2\}}\{1p_{d_1} + 2p_{d_2} + 3p_{d_3}\}\}\}, \tag{1.22}$$

which is equivalent to enumeration. We conclude from this example that care must be taken in defining decisions and their interrelationships, and how to attribute separable costs to these decisions.

1.1.5 Problem Formulation and Solution

The optimal linear search problem of permuting the elements of an array A of size N, whose element x has probability p_x, can be solved using DP in the following fashion. We first define the state S as the set of data elements from which to choose. We then are to make a sequence of decisions as to which element of A should be placed next in the resulting array. We thus arrive at a DPFE of the form

$$f(S) = \min_{x \in S}\{C(x|S) + f(S - \{x\})\}, \tag{1.23}$$

where the reward or cost function $C(x|S)$ is suitably defined. Note that $S \in 2^A$, where 2^A denotes the power set of A. Our goal is to solve for $f(A)$ given the base case $f(\emptyset) = 0$. (This is a target-state formulation, where $\emptyset$ is the target state.)

This DPFE can also be written in the complementary form

$$f(S) = \min_{x \notin S}\{C(x|S) + f(S \cup \{x\})\}, \tag{1.24}$$

for $S \in 2^A$, where our goal is to solve for $f(\emptyset)$ given the base case $f(A) = 0$.

One definition of $C(x|S)$, based upon Method W, is as follows:

$$C_W(x|S) = \sum_{y \in S} p_y.$$

This function depends only on S, not on the decision x. A second definition, based upon Method S, is the following:

$$C_S(x|S) = (N + 1 - |S|)p_x.$$

This function depends on both S and x. These two definitions assume that the first decision is to choose the element to be placed first in the array. The solution of the problem is 1.7 for the optimal permutation *bca*. (Note: If we assume instead that the decisions are made in reverse, where the first decision chooses the element to be placed last in the array, the same DPFE applies but with $C'_S(x|S) = |S|p_x$; we will call this the *inverted* linear search problem. The optimal permutation is *acb* for this inverted problem.) If we order S by descending probability, it can be shown that the first element x^* in this ordering of S (that has maximum probability) minimizes the set $\{C(x|S)+f(S-x)\}$. Use of this "heuristic", also known as a *greedy* policy, makes performing the minimization operation of the DPFE unnecessary; instead, we need only find the maximum of a set of probabilities $\{p_x\}$. There are many optimization problems solvable by DP for which there are also greedy policies that reduce the amount of work necessary to obtain their solutions; we discuss this further in Sec. 1.1.14.

The inverted linear search problem is equivalent to a related problem associated with ordering the elements of a set A, whose elements have specified lengths or weights w (corresponding to their individual retrieval or processing times), such that the sum of the "sequential access" retrieval times is minimized. This optimal *permutation* problem is also known as the "tape storage" problem [22, pp.229–232], and is equivalent to the "shortest processing time" scheduling (SPT) problem. For example, suppose $A = \{a, b, c\}$ and $w_a = 2$, $w_b = 5$, and $w_c = 3$. If the elements are arranged in the order acb, it takes 2 units of time to sequentially retrieve a, 5 units of time to retrieve c (assuming a must be retrieved before retrieving c), and 10 units of time to retrieve b (assuming a and c must be retrieved before retrieving b). The problem of finding the optimal permutation can be solved using a DPFE of the form

$$f(S) = \min_{x\in S}\{|S|w_x + f(S - \{x\})\}, \tag{1.25}$$

as for the inverted linear search problem. $C(x|S) = |S|w_x$ since choosing x contributes a cost of w_x to each of the $|S|$ decisions that are to be made.

Example 1.1. Consider the linear search example where $A = \{a, b, c\}$ and $p_a = 0.2$, $p_b = 0.5$, and $p_c = 0.3$. The target-state DPFE (1.23) may be evaluated as follows:

$$\begin{aligned}
f(\{a,b,c\}) &= \min\{C(a|\{a,b,c\}) + f(\{b,c\}), C(b|\{a,b,c\}) + f(\{a,c\}),\\
&\qquad C(c|\{a,b,c\}) + f(\{a,b\})\}\\
f(\{b,c\}) &= \min\{C(b|\{b,c\}) + f(\{c\}), C(c|\{b,c\}) + f(\{b\})\}\\
f(\{a,c\}) &= \min\{C(a|\{a,c\}) + f(\{c\}), C(c|\{a,c\}) + f(\{a\})\}\\
f(\{a,b\}) &= \min\{C(a|\{a,b\}) + f(\{b\}), C(b|\{a,b\}) + f(\{a\})\}\\
f(\{c\}) &= \min\{C(c|\{c\}) + f(\emptyset)\}\\
f(\{b\}) &= \min\{C(b|\{b\}) + f(\emptyset)\}\\
f(\{a\}) &= \min\{C(a|\{a\}) + f(\emptyset)\}\\
f(\emptyset) &= 0
\end{aligned}$$

For Method W, these equations reduce to the following:

$$\begin{aligned}
f(\{a,b,c\}) &= \min\{C_W(a|\{a,b,c\}) + f(\{b,c\}), C_W(b|\{a,b,c\}) + f(\{a,c\}),\\
&\qquad C_W(c|\{a,b,c\}) + f(\{a,b\})\}\\
&= \min\{1.0 + f(\{b,c\}), 1.0 + f(\{a,c\}), 1.0 + f(\{a,b\})\}
\end{aligned}$$

$$
\begin{aligned}
&= \min\{1.0 + 1.1, 1.0 + 0.7, 1.0 + 0.9\} = 1.7\\
f(\{b,c\}) &= \min\{C_W(b|\{b,c\}) + f(\{c\}), C_W(c|\{b,c\}) + f(\{b\})\}\\
&= \min\{0.8 + f(\{c\}), 0.8 + f(\{b\})\}\\
&= \min\{0.8 + 0.3, 0.8 + 0.5\} = 1.1\\
f(\{a,c\}) &= \min\{C_W(a|\{a,c\}) + f(\{c\}), C_W(c|\{a,c\}) + f(\{a\})\}\\
&= \min\{0.5 + f(\{c\}), 0.5 + f(\{a\})\}\\
&= \min\{0.5 + 0.3, 0.5 + 0.2\} = 0.7\\
f(\{a,b\}) &= \min\{C_W(a|\{a,b\}) + f(\{b\}), C_W(b|\{a,b\}) + f(\{a\})\}\\
&= \min\{0.7 + f(\{b\}), 0.7 + f(\{a\})\}\\
&= \min\{0.7 + 0.5, 0.7 + 0.2\} = 0.9\\
f(\{c\}) &= \min\{C_W(c|\{c\}) + f(\emptyset)\} = \min\{0.3 + f(\emptyset)\} = 0.3\\
f(\{b\}) &= \min\{C_W(b|\{b\}) + f(\emptyset)\} = \min\{0.5 + f(\emptyset)\} = 0.5\\
f(\{a\}) &= \min\{C_W(a|\{a\}) + f(\emptyset)\} = \min\{0.2 + f(\emptyset)\} = 0.2\\
f(\emptyset) &= 0
\end{aligned}
$$

For Method S, these equations reduce to the following:

$$
\begin{aligned}
f(\{a,b,c\}) &= \min\{C_S(a|\{a,b,c\}) + f(\{b,c\}), C_S(b|\{a,b,c\}) + f(\{a,c\}),\\
&\qquad C_S(c|\{a,b,c\}) + f(\{a,b\})\}\\
&= \min\{1 \times 0.2 + f(\{b,c\}), 1 \times 0.5 + f(\{a,c\}), 1 \times 0.3 + f(\{a,b\})\}\\
&= \min\{0.2 + 1.9, 0.5 + 1.2, 0.3 + 1.6\} = 1.7\\
f(\{b,c\}) &= \min\{C_S(b|\{b,c\}) + f(\{c\}), C_S(c|\{b,c\}) + f(\{b\})\}\\
&= \min\{2 \times 0.5 + f(\{c\}), 2 \times 0.3 + f(\{b\})\}\\
&= \min\{1.0 + 0.9, 0.6 + 1.5\} = 1.9\\
f(\{a,c\}) &= \min\{C_S(a|\{a,c\}) + f(\{c\}), C_S(c|\{a,c\}) + f(\{a\})\}\\
&= \min\{2 \times 0.2 + f(\{c\}), 2 \times 0.3 + f(\{a\})\}\\
&= \min\{0.4 + 0.9, 0.6 + 0.6\} = 1.2\\
f(\{a,b\}) &= \min\{C_S(a|\{a,b\}) + f(\{b\}), C_S(b|\{a,b\}) + f(\{a\})\}\\
&= \min\{2 \times 0.2 + f(\{b\}), 2 \times 0.5 + f(\{a\})\}\\
&= \min\{0.4 + 1.5, 1.0 + 0.6\} = 1.6\\
f(\{c\}) &= \min\{C_S(c|\{c\}) + f(\emptyset)\} = \min\{3 \times 0.3 + f(\emptyset)\} = 0.9\\
f(\{b\}) &= \min\{C_S(b|\{b\}) + f(\emptyset)\} = \min\{3 \times 0.5 + f(\emptyset)\} = 1.5\\
f(\{a\}) &= \min\{C_S(a|\{a\}) + f(\emptyset)\} = \min\{3 \times 0.2 + f(\emptyset)\} = 0.6\\
f(\emptyset) &= 0
\end{aligned}
$$

It should be emphasized that the foregoing equations are to be *evaluated* in reverse of the order they have been presented, starting from the base case $f(\emptyset)$ and ending with the goal $f(\{a,b,c\})$. This evaluation is said to be "bottom-up". The goal cannot be evaluated first since it refers to values not available initially. While it may not be evaluated first, it is convenient to start at the goal to systematically *generate* the other equations, in a "top-down" fashion, and then sort the equations as necessary to evaluate them. We discuss such a generation process in Sect. 1.2.2. An alternative to generating a sequence of equations is to recursively evaluate the DPFE, starting at the goal, as described in Sect. 1.2.1.

As indicated earlier, we are not only interested in the final answer ($f(A) = 1.7$), but also in "reconstructing" the sequence of decisions that yields that answer. This is one reason that it is generally preferable to evaluate DPFEs nonrecursively. Of the three possible initial decisions, to choose a, b, or c first in goal-state $\{a,b,c\}$, the optimal decision is to choose b. Decision b yields the minimum of the set $\{2.1, 1.7, 1.9\}$, at a cost of 1.0 for Method W or at a cost of 0.5 for Method S, and causes a transition to state $\{a,c\}$. For Method W, the minimum value of $f(\{a,c\})$ is 0.7, obtained by choosing c at a cost of 0.5, which yields the minimum of the set $\{0.8, 0.7\}$, and which causes a transition to state $\{a\}$; the minimum value of $f(\{a\})$ is 0.2, obtained by necessarily choosing a at a cost of 0.2, which yields the minimum of the set $\{0.2\}$, and which causes a transition to base-state $\emptyset$. Thus, the optimal policy is to choose b, then c, and finally a, at a total cost of $1.0+0.5+0.2 = 1.7$. For Method S, the minimum value of $f(\{a,c\})$ is 1.2, obtained by choosing c at a cost of 0.6, which yields the minimum of the set $\{1.3, 1.2\}$, and which causes a transition to state $\{a\}$; the minimum value of $f(\{a\})$ is 0.6, obtained by necessarily choosing a at a cost of 0.6, which yields the minimum of the set $\{0.6\}$, and which causes a transition to base-state $\emptyset$. Thus, the optimal policy is to choose b, then c, and finally a, at a total cost of $0.5+0.6+0.6 = 1.7$.

1.1.6 State Transition Graph Model

Recall the directed graph model of a DPFE discussed earlier. For any state S, $f(S)$ is the length of the shortest path from S to the target state $\emptyset$. For Method W, the shortest path overall has length $1.0 + 0.5 + 0.2 = 1.7$; for Method S, the shortest path overall has length $0.5 + 0.6 + 0.6 = 1.7$. The foregoing calculations obtain the answer 1.7 by adding the branches in the order $(1.0+(0.5+(0.2)))$ or $(0.5+(0.6+(0.6)))$, respectively. The answer can also be obtained by adding the branches in the reverse order $(((1.0)+0.5)+0.2)$ or $(((0.5)+0.6)+0.6)$. With respect to the graph, this reversal is equivalent to using the designated-source DPFE (1.16), or equivalently

$$f'(S) = \min_{S'}\{f'(S') + C(x|S')\}, \tag{1.26}$$

where S' is a predecessor of S in that some decision x leads to a transition from S' to S, and where $f'(S)$ is the length of the shortest path from the source state S^* to any state S, with goal $f'(\emptyset)$ and base state $S^* = \{a,b,c\}$.

Example 1.2. For the linear search example, the designated-source DPFE (1.26) may be evaluated as follows:

$$\begin{aligned}
f'(\{a,b,c\}) &= 0\\
f'(\{b,c\}) &= \min\{f'(\{a,b,c\}) + C(a|\{a,b,c\})\}\\
f'(\{a,c\}) &= \min\{f'(\{a,b,c\}) + C(b|\{a,b,c\})\}\\
f'(\{a,b\}) &= \min\{f'(\{a,b,c\}) + C(c|\{a,b,c\})\}\\
f'(\{c\}) &= \min\{f'(\{b,c\}) + C(b|\{b,c\}), f'(\{a,c\}) + C(a|\{a,c\})\}\\
f'(\{b\}) &= \min\{f'(\{b,c\}) + C(c|\{b,c\}), f'(\{a,b\}) + C(a|\{a,b\})\}\\
f'(\{a\}) &= \min\{f'(\{a,c\}) + C(c|\{a,c\}), f'(\{a,b\}) + C(b|\{a,b\})\}\\
f'(\emptyset) &= \min\{f'(\{a\}) + C(a|\{a\}), f'(\{b\}) + C(b|\{b\}),\\
&\qquad f'(\{c\}) + C(c|\{c\})\}
\end{aligned}$$

For Method W, these equations reduce to the following:

$$\begin{aligned}
f'(\{a,b,c\}) &= 0\\
f'(\{b,c\}) &= \min\{f'(\{a,b,c\}) + C_W(a|\{a,b,c\})\} = \min\{0+1.0\} = 1.0\\
f'(\{a,c\}) &= \min\{f'(\{a,b,c\}) + C_W(b|\{a,b,c\})\} = \min\{0+1.0\} = 1.0\\
f'(\{a,b\}) &= \min\{f'(\{a,b,c\}) + C_W(c|\{a,b,c\})\} = \min\{0+1.0\} = 1.0\\
f'(\{c\}) &= \min\{f'(\{b,c\}) + C_W(b|\{b,c\}), f'(\{a,c\}) + C_W(a|\{a,c\})\}\\
&= \min\{1.0+0.8, 1.0+0.5\} = 1.5\\
f'(\{b\}) &= \min\{f'(\{b,c\}) + C_W(c|\{b,c\}), f'(\{a,b\}) + C_W(a|\{a,b\})\}\\
&= \min\{1.0+0.8, 1.0+0.7\} = 1.7\\
f'(\{a\}) &= \min\{f'(\{a,c\}) + C_W(c|\{a,c\}), f'(\{a,b\}) + C_W(b|\{a,b\})\}\\
&= \min\{1.0+0.5, 1.0+0.7\} = 1.5\\
f'(\emptyset) &= \min\{f'(\{a\}) + C_W(a|\{a\}), f'(\{b\}) + C_W(b|\{b\}),\\
&\qquad f'(\{c\}) + C_W(c|\{c\})\}\\
&= \min\{1.5+0.2, 1.7+0.5, 1.5+0.3\} = 1.7
\end{aligned}$$

For Method S, these equations reduce to the following:

$$\begin{aligned}
f'(\{a,b,c\}) &= 0\\
f'(\{b,c\}) &= \min\{f'(\{a,b,c\}) + C_S(a|\{a,b,c\})\} = \min\{0+0.2\} = 0.2\\
f'(\{a,c\}) &= \min\{f'(\{a,b,c\}) + C_S(b|\{a,b,c\})\} = \min\{0+0.5\} = 0.5\\
f'(\{a,b\}) &= \min\{f'(\{a,b,c\}) + C_S(c|\{a,b,c\})\} = \min\{0+0.3\} = 0.3\\
f'(\{c\}) &= \min\{f'(\{b,c\}) + C_S(b|\{b,c\}), f'(\{a,c\}) + C_S(a|\{a,c\})\}\\
&= \min\{0.2+1.0, 0.5+0.4\} = 0.9\\
f'(\{b\}) &= \min\{f'(\{b,c\}) + C_S(c|\{b,c\}), f'(\{a,b\}) + C_S(a|\{a,b\})\}\\
&= \min\{0.2+0.6, 0.3+0.4\} = 0.7\\
f'(\{a\}) &= \min\{f'(\{a,c\}) + C_S(c|\{a,c\}), f'(\{a,b\}) + C_S(b|\{a,b\})\}\\
&= \min\{0.5+0.6, 0.3+1.0\} = 1.1\\
f'(\emptyset) &= \min\{f'(\{a\}) + C_S(a|\{a\}), f'(\{b\}) + C_S(b|\{b\}),
\end{aligned}$$

$$f'(\{c\}) + C_S(c|\{c\})\}$$
$$= \min\{1.1 + 0.6, 0.7 + 1.5, 0.9 + 0.9\} = 1.7$$

Here, we listed the equations in order of their (bottom-up) evaluation, with the base case $f'(\{a, b, c\})$ first and the goal $f'(\emptyset)$ last.

1.1.7 Staged Decisions

It is often convenient and sometimes necessary to incorporate stage numbers as a part of the definition of the state. For example, in the linear search problem there are N distinct decisions that must be made, and they are assumed to be made in a specified order. We assume that N, also called the *horizon*, is finite and known. The first decision, made at stage 1, is to decide which data item should be placed first in the array, the second decision, made at stage 2, is to decide which data item should be placed second in the array, etc. Thus, we may rewrite the original DPFE (1.23) as

$$f(k, S) = \min_{x \in S}\{C(x|k, S) + f(k + 1, S - \{x\})\}, \tag{1.27}$$

where the state now consists of a stage number k and a set S of items from which to choose. The goal is to find $f(1, A)$ with base condition $f(N+1, \emptyset) = 0$. Suppose we again define $C(x|k, S) = (N + 1 - |S|)p_x$. Since $k = N + 1 - |S|$, we have $C(x|k, S) = kp_x$. This cost function depends on the stage k and the decision x, but is independent of S.

For the inverted linear search (or optimal permutation) problem, where the first decision, made at stage 1, is to decide which data item should be placed last in the array, the second decision, made at stage 2, is to decide which data item should be placed next-to-last in the array, etc., the staged DPFE is the same as (1.27), but where $C(x|k, S) = kw_x$, which is also independent of S. While this simplification is only a modest one, it can be very significant for more complicated problems.

Incorporating stage numbers as part of the definition of the state may also be beneficial in defining base-state conditions. We may use the base condition $f(k, S) = 0$ when $k > N$ (for any S); the condition $S = \emptyset$ can be ignored. It is far easier to test whether the stage number exceeds some limit ($k > N$) than whether a set equals some base value ($S = \emptyset$). Computationally, this involves a comparison of integers rather than a comparison of sets.

Stage numbers may also be regarded as transition times, and DPFEs incorporating them are also called *fixed-time models*. Stage numbers need not be consecutive integers. We may define the stage or *virtual time* k to be some number that is associated with the k-th decision, where k is a sequence counter. For example, adding consecutive stage numbers to the DPFE (1.25) for the (inverted) linear search problem, we have

$$f(k, S) = \min_{x \in S}\{|S|w_x + f(k + 1, S - \{x\})\}, \tag{1.28}$$

where the goal is to find $f(1,A)$ with base-condition $f(k,S)=0$ when $k>N$. We have $C(x|S)=|S|w_x$ since choosing x contributes a length of w_x to each of the $|S|$ decisions that are to be made. Suppose we define the *virtual time* or stage k as the "length-so-far" when the next decision is to be made. Then

$$f(k,S)=\min_{x\in S}\{(k+w_x)+f(k+w_x,S-\{x\})\}, \tag{1.29}$$

where the goal is to find $f(0,A)$ with base-condition $f(k,S)=0$ when $k=\sum_{x\in A}w_x$ or $S=\emptyset$. The cost of a decision x in state (k,S), that is $C(x|k,S)=(k+w_x)$, is the length-so-far k plus the retrieval time w_x for the chosen item x, and in the next-state resulting from this decision the virtual time or stage k is also increased by w_x.

Example 1.3. For the linear search problem, the foregoing staged DPFE (1.28) may be evaluated as follows:

$$\begin{aligned}
f(1,\{a,b,c\}) &= \min\{C(a|1,\{a,b,c\})+f(2,\{b,c\}),\\
&\qquad C(b|1,\{a,b,c\})+f(2,\{a,c\}),C(c|1,\{a,b,c\})+f(2,\{a,b\})\}\\
&= \min\{6+11,15+7,9+9\}=17\\
f(2,\{b,c\}) &= \min\{C(b|2,\{b,c\})+f(3,\{c\}),C(c|2,\{b,c\})+f(3,\{b\})\}\\
&= \min\{10+3,6+5\}=11\\
f(2,\{a,c\}) &= \min\{C(a|2,\{a,c\})+f(3,\{c\}),C(c|2,\{a,c\})+f(3,\{a\})\}\\
&= \min\{4+3,6+2\}=7\\
f(2,\{a,b\}) &= \min\{C(a|2,\{a,b\})+f(3,\{b\}),C(b|2,\{a,b\})+f(3,\{a\})\}\\
&= \min\{4+5,10+2\}=9\\
f(3,\{c\}) &= \min\{C(c|3,\{c\})+f(4,\emptyset)\}=\min\{3+0\}=3\\
f(3,\{b\}) &= \min\{C(b|3,\{b\})+f(4,\emptyset)\}=\min\{5+0\}=5\\
f(3,\{a\}) &= \min\{C(a|3,\{a\})+f(4,\emptyset)\}=\min\{2+0\}=2\\
f(4,\emptyset) &= 0
\end{aligned}$$

Example 1.4. In contrast, the foregoing virtual-stage DPFE (1.29) may be evaluated as follows:

$$\begin{aligned}
f(0,\{a,b,c\}) &= \min\{(0+2)+f((0+2),\{b,c\}),(0+5)+f((0+5),\{a,c\}),\\
&\qquad (0+3)+f((0+3),\{a,b\})\}\\
&= \min\{2+15,5+17,3+15\}=17\\
f(2,\{b,c\}) &= \min\{(2+5)+f((2+5),\{c\}),(2+3)+f((2+3),\{b\})\}\\
&= \min\{7+10,5+10\}=15\\
f(5,\{a,c\}) &= \min\{(5+2)+f((5+2),\{c\}),(5+3)+f((5+3),\{a\})\}\\
&= \min\{7+10,8+10\}=17\\
f(3,\{a,b\}) &= \min\{(3+2)+f((3+2),\{b\}),(3+5)+f((3+5),\{a\})\}
\end{aligned}$$

$$
\begin{aligned}
&= \min\{5+10, 8+10\} = 15 \\
f(7,\{c\}) &= \min\{(7+3) + f((7+3),\emptyset)\} = 10+0 = 10 \\
f(5,\{b\}) &= \min\{(5+5) + f((5+5),\emptyset)\} = 10+0 = 10 \\
f(8,\{a\}) &= \min\{(8+2) + f((8+2),\emptyset)\} = 10+0 = 10 \\
f(10,\emptyset) &= 0
\end{aligned}
$$

1.1.8 Path-States

In a graph representation of a DPFE, we may let state S be defined as the ordered sequence of decisions $(d_1, \ldots, d_{i-1})$ made so far, and represent it by a node in the graph. Then each state S is associated with a path in this graph from the initial (goal) state $\emptyset$ to state S. The applicable path-state DPFE, which is of the form (1.24), is

$$f(S) = \min_{x \notin S}\{C(x|S) + f(S \cup \{x\})\}. \tag{1.30}$$

The goal is to solve for $f(\emptyset)$ given the base cases $f(S_0) = 0$, where each $S_0 \in \mathcal{S}_{base}$ is a terminal state in which no decision remains to be made.

Example 1.5. For the linear search example, the foregoing DPFE (1.30) may be evaluated as follows:

$$
\begin{aligned}
f(\emptyset) &= \min\{C(a|\emptyset) + f(a), C(b|\emptyset) + f(b), C(c|\emptyset) + f(c)\} \\
f(a) &= \min\{C(b|a) + f(ab), C(c|a) + f(ac)\} \\
f(b) &= \min\{C(a|b) + f(ba), C(c|b) + f(bc)\} \\
f(c) &= \min\{C(a|c) + f(ca), C(b|c) + f(cb)\} \\
f(ab) &= \min\{C(c|ab) + f(abc)\} \\
f(ac) &= \min\{C(b|ac) + f(acb)\} \\
f(ba) &= \min\{C(c|ba) + f(bac)\} \\
f(bc) &= \min\{C(a|bc) + f(bca)\} \\
f(ca) &= \min\{C(b|ca) + f(cab)\} \\
f(cb) &= \min\{C(a|cb) + f(cba)\} \\
f(abc) &= f(acb) = f(bac) = f(bca) = f(cab) = f(cba) = 0
\end{aligned}
$$

where $C(x|S)$ may be either the weak or strong versions. There are $N!$ individual bases cases, each corresponding to a permutation. However, the base-cases are equivalent to the single condition that $f(S) = 0$ when $|S| = N$.

For this problem, the information regarding the ordering of the decisions incorporated in the definition of the state is not necessary; we need only know the members of the decision sequence S so that the next decision d will be a different one (i.e., so that $d \notin S$). If the state is considered unordered, the complexity of the problem decreases from $O(N!)$ for permutations to $O(2^N)$ for subsets. For some problems, the state must *also* specify the most recent decision if it affects the choice or cost of the next decision. In other problems, the state need specify *only* the most recent decision.

We finally note that the equations of Example 1.5 can also be used to obtain the solution to the problem if we assume that $C(x|S) = 0$ (as would be the case when we cannot determine separable costs) and consequently the base cases must be defined by enumeration (instead of being set to zero), namely, $f(abc) = 2.1$, $f(acb) = 2.3$, $f(bac) = 1.8$, $f(bca) = 1.7$, $f(cab) = 2.2$, and $f(cba) = 1.9$.

1.1.9 Relaxation

The term *relaxation* is used in mathematics to refer to certain iterative methods of solving a problem by successively obtaining better approximations x_i to the solution x^*. (Examples of relaxation methods are the Gauss-Seidel method for solving systems of linear equations, and gradient-based methods for finding the minimum or maximum of a continuous function of n variables.)

In the context of discrete optimization problems, we observe that the minimum of a finite set $x^* = \min\{a_1, a_2, \ldots, a_N\}$ can be evaluated by a sequence of pairwise minimization operations

$$x^* = \min\{\min\{\ldots\{\min\{a_1, a_2\}, a_3\}, \ldots\}, a_N\}.$$

The sequence of partial minima, $x_1 = a_1$, $x_2 = \min\{x_1, a_2\}$, $x_3 = \min\{x_2, a_3\}$, $x_4 = \min\{x_3, a_4\}, \ldots$, is the solution of the recurrence relation $x_i = \min\{x_{i-1}, a_i\}$, for $i > 1$, with initial condition $x_1 = a_1$. (Note that $\min\{x_1, a_2\}$ will be called the "innermost min".) Instead of letting $x_1 = a_1$, we may let $x_1 = \min\{x_0, a_1\}$, where $x_0 = \infty$. We may regard the sequence $x_1, x_2, x_3, \ldots$ as "successive approximations" to the final answer x^*. Alternatively, the recursive equation $x = \min\{x, a_i\}$ can be solved using a successive approximations process that sets a "minimum-so-far" variable x to the minimum of its current value and some next value a_i, where x is initially ∞. Borrowing the terminology used for infinite sequences, we say the finite sequence x_i, or the "minimum-so-far" variable x, "converges" to x^*. {In this book, we restrict ourselves to finite processes for which N is fixed, so "asymptotic" convergence issues do not arise.} We will also borrow the term *relaxation* to characterize such successive approximations techniques.

One way in which the relaxation idea can be applied to the solution of dynamic programming problems is in evaluating the minimization operation of the DPFE

$$\begin{aligned} f(S) &= \min_{x \in S}\{C(x|S) + f(S'_x)\} \\ &= \min\{C(x_1|S) + f(S'_{x_1}), \\ &\qquad C(x_2|S) + f(S'_{x_2}), \ldots, \\ &\qquad C(x_m|S) + f(S'_{x_m})\}, \end{aligned} \tag{1.31}$$

where $S = \{x_1, x_2, \ldots x_m\}$ and S'_x is the next-state resulting from choosing x in state S. Rather than computing all the values $C(x|S) + f(S'_x)$, for each $x \in S$, before evaluating the minimum of the set, we may instead compute $f(S)$ by successive approximations as follows:

$$\begin{aligned} f(S) = \min\{&\min\{\ldots\{\min\{C(x_1|S) + f(S'_{x_1}),\\ &C(x_2|S) + f(S'_{x_2})\}, \ldots\},\\ &C(x_m|S) + f(S'_{x_m})\}. \end{aligned} \tag{1.32}$$

In using this equation to compute $f(S)$, the values of $f(S'_{x_i})$, as encountered in proceeding in a left-to-right (inner-to-outer) order, should all have been previously computed. To achieve this objective, it is common to order (topologically) the values of S for which $f(S)$ is to be computed, as in Example 1.6 of Sect. 1.1.10. An alternative is to use a staged formulation.

Consider the staged "fixed-time" DPFE of the form

$$f(k, S) = \min_x\{C(x|k, S) + f(k-1, S'_x)\}, \tag{1.33}$$

which, for each S, defines a sequence $f(0, S), f(1, S), f(2, S), f(3, S), \ldots$ of successive approximations to $f(S)$. The minimum member of the sequence is the desired answer, i.e., $f(S) = \min_k\{f(k, S)\}$. {Here, we adopt the Java "overloading" convention that f with one argument differs from f with two arguments.} Note that $f(k, S)$ is a function not of $f(k-1, S)$, but of $f(k-1, S'_x)$, where S'_x is the next-state; e.g., $f(1, S)$ depends not on $f(0, S)$ but on $f(0, S')$. Since the sequence of values $f(k, S)$ is not necessarily monotonic, we define a new sequence $F(k, S)$ by the "relaxation" DPFE

$$F(k, S) = \min\{F(k-1, S), \min_x\{C(x|k, S) + F(k-1, S'_x)\}\}. \tag{1.34}$$

In Example 1.9 of Sect. 1.1.10, we will see that this new sequence $F(0, S)$, $F(1, S), F(2, S), F(3, S), \ldots$ is monotonic, and converges to $f(S)$.

1.1.10 Shortest Path Problems

In the solution to the linear search problem we gave earlier, we used a state transition graph model and noted that solving the linear search problem was equivalent to finding the shortest path in a graph. There are a myriad of other problems that can be formulated and solved as graph optimization problems, so such problems are of special importance. Some of the problems are more complex, however, such as when the graph is cyclic.

For acyclic graphs, the shortest path from a source node s to a target node t can be found using a DPFE of a (target-state) form similar to (1.15):

$$f(p) = \min_q\{b(p, q) + f(q)\}, \tag{1.35}$$

where $b(p,q)$ is the distance from p to q, and $f(p)$ is the length of the shortest path from node p to node t. We may either restrict node $q \in \text{succ}(p)$ to be a successor of node p, or let $b(p,q) = \infty$ if $q \notin \text{succ}(p)$. {For acyclic graphs, we may also assume $b(p,p) = \infty$ for all p.} Our goal is to find $f(s)$ with base condition $f(t) = 0$. In this formulation, the state p is defined as the node in the graph at which we make a decision to go some next node q before continuing ultimately to the designated target.

Example 1.6 (SPA). As a numerical example, consider the graph in Fig. 1.1 with nodes $\{s, x, y, t\}$ and branches $\{(s,x), (s,y), (x,y), (x,t), (y,t)\}$ with branch distances $\{3, 5, 1, 8, 5\}$, respectively. For illustrative purposes, we add a "dummy" branch (s,t) having distance $b(s,t) = \infty$.

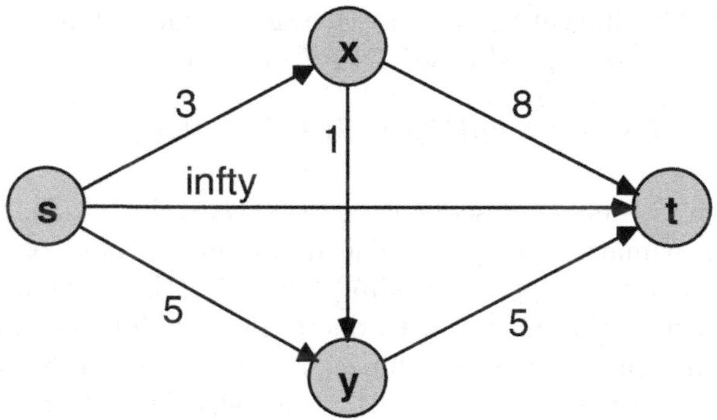

Fig. 1.1. Shortest Path Problem in an Acyclic Graph

The DPFE for the shortest path from s to t yields the following equations:

$$
\begin{aligned}
f(s) &= \min\{b(s,x) + f(x), b(s,y) + f(y), b(s,t) + f(t)\}, \\
f(x) &= \min\{b(x,y) + f(y), b(x,t) + f(t)\}, \\
f(y) &= \min\{b(y,t) + f(t)\}, \\
f(t) &= 0.
\end{aligned}
$$

Consider evaluating the minimizing value of $f(s)$ by relaxation, i.e.,

$$
\begin{aligned}
f(s) &= \min\{b(s,x) + f(x), b(s,y) + f(y), b(s,t) + f(t)\}, \\
&= \min\{\min\{b(s,x) + f(x), b(s,y) + f(y)\}, b(s,t) + f(t)\}.
\end{aligned}
$$

This is of the same form as (1.32). $f(x)$ and $f(y)$ in the innermost min should be evaluated before $f(t)$, but in fact both $f(x)$ and $f(y)$ depend upon $f(t)$. Thus, $f(t)$ should be evaluated first.

Substituting the above given branch distances into the foregoing equations, we have

$$\begin{aligned}
f(s) &= \min\{3 + f(x), 5 + f(y), \infty + f(t)\},\\
f(x) &= \min\{1 + f(y), 8 + f(t)\},\\
f(y) &= \min\{5 + f(t)\},\\
f(t) &= 0.
\end{aligned}$$

If these equations are evaluated in "bottom-up" order, then we have $f(t) = 0$, $f(y) = 5$, $f(x) = 6$, and $f(s) = 9$.

If the graph is acyclic, then the graph can be topologically sorted and the DPFE can be evaluated for p in this order such that evaluation of $f(p)$ will always be in terms of previously calculated values of $f(q)$. On the other hand, if the graph is cyclic, so that for example p and q are successors of each other, then $f(p)$ may be defined in terms of $f(q)$, and $f(q)$ may be defined in terms of $f(p)$. This circular definition presents difficulties that require special handling.

For convenience, we will assume that cyclic graphs do not contain *self-loops*, i.e., branches from a node p to itself. For a graph having such a branch, if $b(p,p)$ is positive, that branch cannot be in the shortest path since its omission would lead to a shorter path, hence we may omit the branch. On the other hand, if $b(p,p)$ is negative, the problem is not well-defined since there is no shortest path at all. If $b(p,p) = 0$, the problem is also not well-defined since a shortest path may have an infinite number of branches, hence we may also omit the branch.

One way to handle cyclic graphs (having no self-loops), where $f(q)$ may depend on $f(p)$, is to use relaxation to solve the DPFE

$$f(p) = \min_{q}\{b(p,q) + f(q)\}, \tag{1.36}$$

where $f(p) = \infty$ for $p \neq t$ initially, and $f(t) = 0$.

Example 1.7 (SPC—successive approximations). To the preceding example, suppose we add a branch (y,x) with distance $b(y,x) = 2$ (see Fig. 1.2).

The graph is then cyclic, and the equations obtained from the DPFE are as follows:

$$\begin{aligned}
f(s) &= \min\{b(s,x) + f(x), b(s,y) + f(y), b(s,t) + f(t)\\
&= \min\{3 + f(x), 5 + f(y), \infty + f(t)\},\\
f(x) &= \min\{b(x,y) + f(y), b(x,t) + f(t)\} = \min\{1 + f(y), 8 + f(t)\},\\
f(y) &= \min\{b(y,x) + f(x), b(y,t) + f(t)\} = \min\{2 + f(x), 5 + f(t)\},\\
f(t) &= 0.
\end{aligned}$$

We note that $f(x)$ depends on $f(y)$, and $f(y)$ depends on $f(x)$.

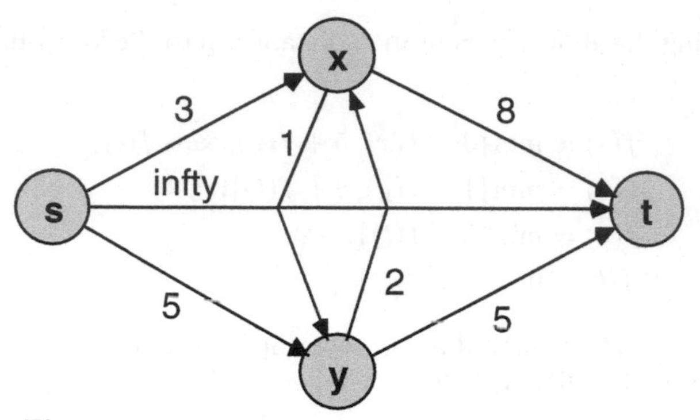

Fig. 1.2. Shortest Path Problem in an Cyclic Graph

Solving these equations by relaxation, assuming $f(s) = f(x) = f(y) = \infty$ and $f(t) = 0$ initially, we have, as the first successive approximation

$$\begin{aligned} f(s) &= \min\{3+\infty, 5+\infty, \infty+0\} = \infty, \\ f(x) &= \min\{1+\infty, 8+0\} = 8, \\ f(y) &= \min\{2+\infty, 5+0\} = 5, \\ f(t) &= 0, \end{aligned}$$

as the second successive approximation

$$\begin{aligned} f(s) &= \min\{3+8, 5+5, \infty+0\} = 10, \\ f(x) &= \min\{1+5, 8+0\} = 6, \\ f(y) &= \min\{2+8, 5+0\} = 5, \\ f(t) &= 0, \end{aligned}$$

and as the third successive approximation

$$\begin{aligned} f(s) &= \min\{3+6, 5+5, \infty+0\} = 9, \\ f(x) &= \min\{1+5, 8+0\} = 6, \\ f(y) &= \min\{2+6, 5+0\} = 5, \\ f(t) &= 0. \end{aligned}$$

Continuing this process to compute additional successive approximations will result in no changes. In this event, we say that the relaxation process has converged to the final solution, which for this example is $f(s) = 9$ (and $f(x) = 6$, $f(y) = 5$, $f(t) = 0$).

Another way to handle cyclic graphs is to introduce staged decisions, using a DPFE of the same form as Eq. (1.33).

$$f(k,p) = \min_q\{b(p,q) + f(k-1,q)\}, \tag{1.37}$$

where $f(k,p)$ is the length of the shortest path from p to t having *exactly* k branches. k, the number of branches in a path, ranges from 0 to $N-1$, where N is the number of nodes in the graph. We may disregard paths having N or more branches that are necessarily cyclic hence cannot be shortest (assuming the graph has no negative or zero-length cycle, since otherwise the shortest path problem is not well-defined). The base conditions are $f(0,t) = 0$, $f(0,p) = \infty$ for $p \neq t$, and $f(k,t) = \infty$ for $k > 0$. We must evaluate $\min_k\{f(k,s)\}$ for $k = 0, \ldots, N-1$.

Example 1.8 (SPC-fixed time). For the same example as above, the staged fixed-time DPFE (1.37) can be solved as follows:

$$\begin{aligned}
f(k,s) &= \min\{b(s,x) + f(k-1,x), b(s,y) + f(k-1,y), b(s,t) + f(k-1,t)\},\\
f(k,x) &= \min\{b(x,y) + f(k-1,y), b(x,t) + f(k-1,t)\},\\
f(k,y) &= \min\{b(y,x) + f(k-1,x), b(y,t) + f(k-1,t)\},\\
f(k,t) &= 0.
\end{aligned}$$

Its solution differs from the foregoing relaxation solution in that $f(0,t) = 0$ initially, but $f(k,t) = \infty$ for $k > 0$. In this case, we have, as the first successive approximation

$$\begin{aligned}
f(1,s) &= \min\{3 + f(0,x), 5 + f(0,y), \infty + f(0,t)\} = \infty,\\
f(1,x) &= \min\{1 + f(0,y), 8 + f(0,t)\} = 8,\\
f(1,y) &= \min\{2 + f(0,x), 5 + f(0,t)\} = 5,\\
f(1,t) &= \infty,
\end{aligned}$$

as the second successive approximation

$$\begin{aligned}
f(2,s) &= \min\{3 + f(1,x), 5 + f(1,y), \infty + f(1,t)\} = 10,\\
f(2,x) &= \min\{1 + f(1,y), 8 + f(1,t)\} = 6,\\
f(2,y) &= \min\{2 + f(1,x), 5 + f(1,t)\} = 10,\\
f(2,t) &= \infty,
\end{aligned}$$

and as the third and final successive approximation

$$\begin{aligned}
f(3,s) &= \min\{3 + f(2,x), 5 + f(2,y), \infty + f(2,t)\} = 9,\\
f(3,x) &= \min\{1 + f(2,y), 8 + f(2,t)\} = 11,\\
f(3,y) &= \min\{2 + f(2,x), 5 + f(2,t)\} = 8,\\
f(3,t) &= \infty.
\end{aligned}$$

Unlike the situation in the preceding example, the values of $f(k,p)$ do not converge. Instead, $f(p) = \min_k\{f(k,p)\}$, i.e.,

$$\begin{aligned}
f(s) &= \min\{\infty, \infty, 10, 9\} = 9, \\
f(x) &= \min\{\infty, 8, 6, 11\} = 6, \\
f(y) &= \min\{\infty, 5, 10, 8\} = 5, \\
f(t) &= \min\{0, \infty, \infty, \infty\} = 0.
\end{aligned}$$

We emphasize that the matrix $f(k,p)$ must be evaluated rowwise (varying p for a fixed k) rather than columnwise.

Example 1.9 (SPC-relaxation). Suppose we define $F(k,p)$ as the length of the shortest path from p to t having k *or fewer* branches. F satisfies a DPFE of the same form as f,

$$F(k,p) = \min_q\{b(p,q) + F(k-1,q)\}. \tag{1.38}$$

The goal is to compute $F(N-1,s)$. The base conditions are $F(0,t) = 0$, $F(0,p) = \infty$ for $p \neq t$, but where $F(k,t) = 0$ for $k > 0$. In this formulation, the sequence of values $F(k,p)$, for $k = 0, \ldots, N-1$, are successive approximations to the length of the shortest path from p to t having at most $N-1$ branches, hence the goal can be found by finding $F(k,s)$, for $k = 0, \ldots, N-1$. We observe that the values of $F(k,p)$ are equal to the values obtained as the k-th successive approximation for $f(p)$ in Example 1.7.

For a fixed k, if p has m successors $q_1, \ldots, q_m$, then

$$\begin{aligned}
F(k,p) &= \min_q\{b(p,q) + F(k-1,q)\} \\
&= \min\{b(p,q_1) + F(k-1,q_1), b(p,q_2) + F(k-1,q_2), \ldots, \\
&\qquad b(p,q_m) + F(k-1,q_m)\},
\end{aligned}$$

which can be evaluated by relaxation (as discussed in Sect. 1.1.9) by computing

$$\begin{aligned}
F(k,p) = \min\{&\min\{\ldots\{\min\{b(p,q_1) + F(k-1,q_1), b(p,q_2) + F(k-1,q_2)\}, \\
&\ldots\}, b(p,q_m) + F(k-1,q_m)\}.
\end{aligned} \tag{1.39}$$

For each p, $F(k,p)$ is updated once for each successor $q_i \in \mathrm{succ}(p)$ of p, assuming $F(k-1,q)$ has previously been evaluated for all q. We emphasize that computations are staged, for $k = 1, \ldots, N-1$.

It should be noted that, given k, in solving (1.39) for $F(k,p)$ as a function of $\{F(k-1,q) | q \in \mathrm{succ}(p)\}$, if $F(k,q)$ has been previously calculated, then it may be used instead of $F(k-1,q)$. This variation is the basis of the Bellman-Ford algorithm [10, 13], for which the sequence $F(k,p)$, for $k = 0, \ldots, N-1$, may converge more rapidly to the desired solution $F(N-1,s)$.

We also may use the path-state approach to find the shortest path in a cyclic graph, using a DPFE of the form

$$f(p_1, \ldots, p_i) = \min_{q \notin \{p_1, \ldots, p_i\}} \{b(p_i, q) + f(p_1, \ldots, p_i, q)\}, \tag{1.40}$$

where the state is the sequence of nodes $p_1, \ldots, p_i$ in a path S, and q is a successor of p_i that does not appear earlier in the path S. The next-state S' appends q to the path S. The goal is $f(s)$ (where $i = 1$) and the base condition is $f(p_1, \ldots, p_i) = 0$ if $p_i = t$. In Chap. 2, we show how this approach can be used to find longest simple (acyclic) paths (LSP) and to find shortest Hamiltonian paths, the latter also known as the "traveling salesman" problem (TSP).

In the foregoing, we made no assumption on branch distances. If we restrict branch distances to be positive, the shortest path problem can be solved more efficiently using some variations of dynamic programming, such as Dijkstra's algorithm (see [10, 59]). If we allow negative branch distances, but not negative cycles (otherwise the shortest path may be infinitely negative or undefined), Dijkstra's algorithm may no longer find the shortest path. However, other dynamic programming approaches can still be used. For example, for a graph having negative cycles, the path-state approach can be used to find the shortest *acyclic* path.

1.1.11 All-Pairs Shortest Paths

There are applications where we are interested in finding the shortest path from any source node s to any target node t, i.e., where s and t is an arbitrary pair of the N nodes in a graph having no negative or zero-length cycles and, for the reasons given in the preceding section, having no self-loops. Of course, the procedures discussed in Sect. 1.1.10 can be used to solve this "all-pairs" shortest path problem by treating s and t as variable parameters. In practice, we would want to perform the calculations in a fashion so as to avoid recalculations as much as possible.

Relaxation. Using a staged formulation, let $F(k, p, q)$ be defined as the length of the shortest path from p to q having k or fewer branches. Then, applying the relaxation idea of (1.34), the DPFE for the target-state formulation is

$$F(k, p, q) = \min\{F(k-1, p, q), \min_r\{b(p, r) + F(k-1, r, q)\}\}, \tag{1.41}$$

for $k > 0$, with $F(0, p, q) = 0$ if $p = q$ and $F(0, p, q) = \infty$ if $p \neq q$. The analogous DPFE for the designated-source formulation is

$$F'(k, p, q) = \min\{F'(k-1, p, q), \min_r\{F'(k-1, p, r) + b(r, q)\}\}. \tag{1.42}$$

If we artificially let $b(p, p) = 0$ for all p (recall that we assumed there are no self-loops), then the former (1.41) reduces to

$$F(k, p, q) = \min_r\{b(p, r) + F(k-1, r, q)\}, \tag{1.43}$$

where r may now include p, and analogously for the latter (1.42). The Bellman-Ford variation, that uses $F(k,r,q)$ or $F'(k,p,r)$ instead of $F(k-1,r,q)$ or $F'(k-1,p,r)$, respectively, applies to the solution of these equations. If the number of branches in the shortest path from p to q, denoted k^*, is less than $N-1$, then the goal $F(N-1,p,q) = F(k^*,p,q)$ will generally be found without evaluating the sequence $F(k,p,q)$ for all k; when $F(k+1,p,q) = F(k,p,q)$ (for all p and q), the successive approximations process has converged to the desired solution.

Floyd-Warshall. The foregoing DPFE (1.41) is associated with a divide-and-conquer process where a path from p to q (having at most k branches) is divided into subpaths from p to r (having 1 branch) and from r to q (having at most $k-1$ branches). An alternative is to divide a path from p to q into subpaths from p to r and from r to t that are restricted not by the number of branches they contain but by the set of intermediate nodes r that the paths may traverse. Let $F(k,p,q)$ denote the length of the shortest path from p to q that traverses (passes through) intermediate nodes only in the ordered set $\{1,\ldots,k\}$. Then the appropriate DPFE is

$$F(k,p,q) = \min\{F(k-1,p,q), F(k-1,p,k) + F(k-1,k,q)\}, \quad (1.44)$$

for $k > 0$, with $F(0,p,q) = 0$ if $p = q$ and $F(0,p,q) = b(p,q)$ if $p \neq q$. This is known as the *Floyd-Warshall* all-pairs shortest path algorithm. Unlike the former DPFE (1.41), where r may have up to $N-1$ values (if p has every other node as a successor), the minimization operation in the latter DPFE (1.44) is over only two values.

We note that the DPFEs for the above two algorithms may be regarded as matrix equations, which define matrices F^k in terms of matrices F^{k-1}, where p and q are row and column subscripts; since p, q, r, and k are all $O(N)$, the two algorithms are $O(N^4)$ and $O(N^3)$, respectively.

1.1.12 State Space Generation

The numerical solution of a DPFE requires that a function $f(S)$ be evaluated for all states in some state space S. This requires that these states be generated systematically. State space generation is discussed in, e.g., [12]. Since not all states may be reachable from the goal S^*, it is generally preferable to generate only those states reachable from the source S^*, and that this be done in a breadth-first fashion. For example, in Example 1.1, the generated states, from the source-state or goal to the target-state or base, are (in the order they are generated):

$$\{a,b,c\}, \{b,c\}, \{a,c\}, \{a,b\}, \{c\}, \{b\}, \{a\}, \emptyset.$$

In Example 1.2, these same states are generated in the same order although, since its DPFE is of the reverse designated-source form, the first state is the base and the last state is the goal. In Example 1.3, the generated states, from the source (goal) to the target (base), are:

$$(1, \{a, b, c\}), (2, \{b, c\}), (2, \{a, c\}), (2, \{a, b\}), (3, \{c\}), (3, \{b\}), (3, \{a\}), (4, \emptyset).$$

In Example 1.4, the generated states, from the source (goal) to the target (base), are:

$$(0, \{a, b, c\}), (2, \{b, c\}), (5, \{a, c\}), (3, \{a, b\}), (8, \{c\}), (5, \{b\}), (7, \{a\}), (10, \emptyset).$$

In Example 1.5, the generated states, from the source (goal) to the targets (bases), are:

$$\emptyset, a, b, c, ab, ac, ba, bc, ca, cb, abc, acb, bac, bca, cab, cba.$$

We note that this state space generation process can be automated for a given DPFE, say, of the form (1.17),

$$f(S) = \mathrm{opt}_{d \in D(S)}\{R(S, d) \circ f(T(S, d))\}, \tag{1.45}$$

using $T(S, d)$ to generate next-states starting from S^*, subject to constraints $D(S)$ on d, and terminating when S is a base state. This is discussed further in Sect. 1.2.2.

1.1.13 Complexity

The complexity of a DP algorithm is very problem-dependent, but in general it depends on the exact nature of the DPFE, which is of the general nonserial form

$$f(S) = \mathrm{opt}_{d \in D(S)}\{R(S, d) \circ p_1.f(T_1(S, d)) \circ p_2.f(T_2(S, d)) \circ \ldots\}. \tag{1.46}$$

A foremost consideration is the size or dimension of the state space, because that is a measure of how many optimization operations are required to solve the DPFE. In addition, we must take into account the number of possible decisions for each state. For example, for the shortest path problem, assuming an acyclic graph, there are only N states, and at most $N - 1$ decisions per state, so the DP solution has polynomial complexity $O(N^2)$. We previously gave examples of problems where the size hence complexity was factorial and exponential. Such problems are said to be *intractable*. The fact that in many cases problem size is not polynomial is known as the *curse of dimensionality* which afflicts dynamic programming. For some problems, such as the traveling salesman problem, this intractability is associated with the problem itself, and any algorithm for solving such problems is likewise intractable.

Regardless of whether a problem is tractable or not, it is also of interest to reduce the complexity of any algorithm for solving the problem. We noted from the start that a given problem, even simple ones like linear search, can be solved by different DPFEs. Thus, in general, we should always consider alternative formulations, with the objective of reducing the dimensionality of the state space as a major focus.

1.1.14 Greedy Algorithms

For a given state space, an approach to reducing the dimensionality of a DP solution is to find some way to reduce the number of states for which $f(S)$ must actually be evaluated. One possibility is to use some means to determine the optimal decision $d \in D(S)$ without evaluating each member of the set $\{R(S,d) \circ f(T(S,d))\}$; if the value $R(S,d) \circ f(T(S,d))$ need not be evaluated, then $f(S')$ for next-state $S' = T(S,d)$ may not need to be evaluated. For example, for some problems, it turns out that

$$\text{opt}_{d \in D(S)}\{R(S,d)\} = \text{opt}_{d \in D(S)}\{R(S,d) \circ f(T(S,d))\}, \qquad (1.47)$$

If this is the case, and we solve $\text{opt}_{d \in D(S)}\{R(S,d)\}$ instead of $\text{opt}_{d \in D(S)}$ $\{R(S,d) \circ f(T(S,d))\}$, then we only need to evaluate $f(T(S,d))$ for N states, where N is the number of decisions. We call this the *canonical* greedy algorithm associated with a given DPFE. A *noncanonical* greedy algorithm would be one in which there exists a function Φ for which

$$\text{opt}_{d \in D(S)}\{\Phi(S,d)\} = \text{opt}_{d \in D(S)}\{R(S,d) \circ f(T(S,d))\}. \qquad (1.48)$$

Algorithms based on optimizing an auxiliary function Φ (instead of $R \circ f$) are also called "heuristic" ones.

For one large class of greedy algorithms, known as "priority" algorithms [7], in essence the decision set D is ordered, and decisions are made in that order.

Regrettably, there is no simple test for whether optimal greedy policies exist for an arbitrary DP problem. See [38] for a further discussion of greedy algorithms and dynamic programming.

1.1.15 Probabilistic DP

Probabilistic elements can be added to a DP problem in several ways. For example, rewards (costs or profits) can be made random, depending for example on some random variable. In this event, it is common to simply define the reward function $R(S,d)$ as an expected value. In addition, next-states can be random. For example, given the current state is S and the decision is d, if $T_1(S,d))$ and $T_2(S,d))$ are two possible next-states having probabilities p_1 and p_2, respectively, then the probabilistic DPFE would typically have the form

$$f(S) = \min_{d \in D(S)} \{R(S,d) + p_1.f(T_1(S,d)) + p_2.f(T_2(S,d))\}. \qquad (1.49)$$

It is common for probabililistic DP problems to be staged. For finite horizon problems, where the number of stages N is a given finite number, such DPFEs can be solved just as any nonserial DPFE of order 2. In Chap. 2, we give several examples. However, for infinite horizon problems, where the state space is not finite, iterative methods (see [13]) are generally necessary. We will not discuss these in this book.

For some problems, rather than minimizing or maximizing some total reward, we may be interested instead in minimizing or maximizing the probability of some event. Certain problems of this type can be handled by defining base conditions appropriately. An example illustrating this will also be given in Chap. 2.

1.1.16 Nonoptimization Problems

Dynamic programming can also be used to solve nonoptimization probems, where the objective is not to determine a sequence of decisions that *optimizes* some numerical function. For example, we may wish to determine *any* sequence of decisions that leads from a given goal state to one or more given target states. The Tower of Hanoi problem (see [57, p.332–337] and [58]) is one such example. The objective of this problem is to move a tower (or stack) of N discs, of increasing size from top to bottom, from one peg to another peg using a third peg as an intermediary, subject to the constraints that on any peg the discs must remain of increasing size and that only "basic" moves of one disc at a time are allowed. We will denote the basic move of a disc from peg x to peg y by $< x, y >$. For this problem, rather than defining $f(S)$ as the minimum or maximum value of an objective function, we define $F(S)$ as a *sequence* of basic moves. Then $F(S)$ is the concatenation of the sequence of moves for certain subproblems, and we have

$$F(N, x, y) = F(N-1, x, z) F(1, x, y) F(N-1, z, y). \tag{1.50}$$

Here, the state $S = (N, x, y)$ is the number N of discs to be moved from peg x to peg y using peg z as an intermediary. This DPFE has no *min* or *max* operation. The value of $F(S)$ is a sequence (or string), not a number. The idea of solving problems in terms of subproblems characterizes DP formulations.

The DPFE (1.50) is based on the observation that, to move m discs from peg i to peg j with peg k as an intermediary, we may move $m-1$ discs from i to k with j as an intermediary, then move the last disc from i to j, and finally move the $m-1$ discs on k to j with i as an intermediary. The goal is $F(N, i, j)$, and the base condition is $F(m, i, j) =< i, j >$ when $m = 1$. These base conditions correspond to basic moves. For example, for $N = 3$ and pegs A, B,and C,

$$\begin{aligned} F(3, A, B) &= F(2, A, C) F(1, A, B) F(2, C, B) \\ &= <A,B><A,C><B,C><A,B><C,A><C,B><A,B> . \end{aligned}$$

In this book, our focus is on numerical optimization problems, so we will consider a variation of the Tower of Hanoi problem, where we wish to determine the *number* $f(N)$ of required moves, as a function of the number N of discs to be moved. Then, we have

$$f(N) = 2f(N-1) + 1, \tag{1.51}$$

The base condition for this DPFE is $f(1) = 1$. This recurrence relation and its analytical solution appear in many books, e.g., [53].

It should be emphasized that the foregoing DPFE has no explicit optimization operation, but we can add one as follows:

$$f(N) = \mathrm{opt}_{d \in D}\{2f(N-1) + 1\}, \tag{1.52}$$

where the decision set D has, say, a singleton "dummy" member that is not referenced within the optimand. As another example, consider

$$f(N) = \mathrm{opt}_{d \in D}\{f(N-1) + f(N-2)\}, \tag{1.53}$$

with base conditions $f(1) = f(2) = 1$. Its solution is the N-th Fibonacci number.

In principle, the artifice used above, of having a dummy decision, allows general recurrence relations to be regarded as special cases of DPFEs, and hence to be solvable by DP software. This illustrates the generality of DP and DP tools, although we are not recommending that recurrence relations be solved in this fashion.

1.1.17 Concluding Remarks

The introduction to dynamic programming given here only touches the surface of the subject. There is much research on various other aspects of DP, including formalizations of the class of problems for which DP is applicable, the theoretical conditions under which the Principle of Optimality holds, relationships between DP and other optimization methods, methods for reducing the dimensionality, including approximation methods, especially successive approximation methods in which it is hoped that convergence to the correct answer will result after a reasonable number of iterations, etc.

This book assumes that we can properly formulate a DPFE that solves a given discrete optimization problem. We say a DPFE (with specified base conditions) is *proper*, or properly formulated, if a solution exists and can be found by a finite computational algorithm. Chap. 2 provides many examples that we hope will help readers develop new formulations for their problems of interest. Assuming the DPFE is proper, we then address the problem of numerically solving this DPFE (by describing the design of a software tool for DP. This DP tool has been used for all of our Chap. 2 examples. Furthermore, many of our formulations can be adapted to suit other needs.

1.2 Computational Solution of DPFEs

In this section, we elaborate on how to solve a DPFE. One way in which a DPFE can be solved is by using a "conventional" procedural programming language such as Java. In Sect. 1.2.1, a Java program to solve Example 1.1 is given as an illustration.

1.2.1 Solution by Conventional Programming

A simple Java program to solve Example 1.1 is given here. This program was intentionally written as quickly as possible rather than with great care to reflect what a nonprofessional programmer might produce. A central theme of this book is to show how DP problems can be solved with a minimum amount of programming knowledge or effort. The program as written first solves the DPFE (1.23) [Method S] recursively. This is followed by an iterative procedure to reconstruct the optimal policy. It should be emphasized that this program does not generalize easily to other DPFEs, especially when states are sets rather than integers.

```
class dpfe {

  public static double[][] b= {
      { 999.,   .2,   .5,   .3, 999., 999., 999., 999.},
      { 999., 999., 999., 999.,   1.,   .6, 999., 999.},
      { 999., 999., 999., 999.,   .4, 999.,   .6, 999.},
      { 999., 999., 999., 999., 999.,   .4,   1., 999.},
      { 999., 999., 999., 999., 999., 999., 999.,   .9},
      { 999., 999., 999., 999., 999., 999., 999.,  1.5},
      { 999., 999., 999., 999., 999., 999., 999.,   .6},
      { 999., 999., 999., 999., 999., 999., 999., 999.}
    } ; //branch distance array
  public static int N = b.length;         //number of nodes
  public static int[] ptr = new int[N];   //optimal decisions

  public static double fct(int s) {
    double value=999.; ptr[s]=-1;
    if (s==N-1) {value=0.0; }               // target state
    else
      for (int d=s+1; d<N; d++)               // for s>d
        if (b[s][d]<999.)                       // if d=succ(s)
          if (value>b[s][d]+fct(d))               // if new min
            { value=b[s][d]+fct(d); ptr[s]=d; }  //reset
    return value;
  } //end fct

  public static void main(String[] args) {
    System.out.println("min="+fct(0));       //compute goal
    int i=0; System.out.print("path:"+i);
    while (ptr[i]>0) {                       //reconstruction
      System.out.print("->"+ptr[i]);
      i=ptr[i];
    }
  } // end main
} // end dpfe
```

A *recursive* solution of the DPFE was chosen because the DPFE itself is a recursive equation, and transforming it to obtain an iterative solution is not a natural process. Such a transformation would generally take a significant amount of effort, especially for nonserial problems. On the other hand, a major disadvantage of recursion is the inefficiency associated with recalculating $f(S)$ for states S that are next-states of many other states. This is analogous to the reason it is preferable to solve the Fibonacci recurrence relation iteratively rather than recursively. Although finding an iterative solution for the Fibonacci problem is easy, and it also happens to be easy for the linear search problem, in general we cannot expect this to be the case for DP problems.

1.2.2 The State-Decision-Reward-Transformation Table

This book will describe an alternative to conventional programming, as illustrated above, based on the ability to automatically generate the state space for a given DPFE. Recall that, for Example 1.1, the state space is the set

$$\{\{a,b,c\},\{b,c\},\{a,c\},\{a,b\},\{c\},\{b\},\{a\},\emptyset\}$$

or

$$\{\{0,1,2\},\{1,2\},\{0,2\},\{0,1\},\{2\},\{1\},\{0\},\emptyset\}$$

if we give the decisions a, b, and c the numerical labels 0,1,2 instead.

The state space for a given DPFE can be generated in the process of producing the State-Decision-Reward-Transformation (SDRT) table. The SDRT table gives, for each state S and for each decision d in the decision space $D(S)$, the reward function $R(S,d)$ and the transformation function(s) $T(S,d)$ for each pair (S,d), starting from the goal state S^*. $T(S,d)$ allows us to generate next-states. For Example 1.1, the SDRT table is given in Table 1.1.

As each next-state S' is generated, if it is not already in the table, it is added to the table and additional rows are added for each of the decisions in $D(S')$. If a base-state is generated, which has no associated decision, no additional rows are added to the table.

Given the SDRT table, for a serial DPFE, we can easily construct a state transition system model whose nodes are the states. For Example 1.1, the (Boolean) adjacency matrix for this state transition model is as follows:

Table 1.1. SDRT Table for Linear Search Example

state	decision	reward	next-states
$\{0,1,2\}$	$d=0$	0.2	$(\{1,2\})$
$\{0,1,2\}$	$d=1$	0.5	$(\{0,2\})$
$\{0,1,2\}$	$d=2$	0.3	$(\{0,1\})$
$\{1,2\}$	$d=1$	1.0	$(\{2\})$
$\{1,2\}$	$d=2$	0.6	$(\{1\})$
$\{0,2\}$	$d=0$	0.4	$(\{2\})$
$\{0,2\}$	$d=2$	0.6	$(\{0\})$
$\{0,1\}$	$d=0$	0.4	$(\{1\})$
$\{0,1\}$	$d=1$	1.0	$(\{0\})$
$\{2\}$	$d=2$	0.9	$(\emptyset)$
$\{1\}$	$d=1$	1.5	$(\emptyset)$
$\{0\}$	$d=0$	0.6	$(\emptyset)$

	1	2	3	4	5	6	7	8
1	0	1	1	1	0	0	0	0
2	0	0	0	0	1	1	0	0
3	0	0	0	0	1	0	1	0
4	0	0	0	0	0	1	1	0
5	0	0	0	0	0	0	0	1
6	0	0	0	0	0	0	0	1
7	0	0	0	0	0	0	0	1
8	0	0	0	0	0	0	0	0

The weighted adjacency matrix whose nonzero elements are branch labels is

	1	2	3	4	5	6	7	8
1	0	0.2	0.5	0.3	0	0	0	0
2	0	0	0	0	1.0	0.6	0	0
3	0	0	0	0	0.4	0	0.6	0
4	0	0	0	0	0	0.4	1.0	0
5	0	0	0	0	0	0	0	0.9
6	0	0	0	0	0	0	0	1.5
7	0	0	0	0	0	0	0	0.6
8	0	0	0	0	0	0	0	0

The row and column numbers or indices shown $(1,\ldots,8)$ are not part of the matrix itself; in programming languages, such as Java, it is common to start indexing from zero $(0,\ldots,7)$ instead of one.

Later in this book we show that nonserial DPFEs can be modeled in a similar fashion using a generalization of state transition systems called Petri nets.

1.2.3 Code Generation

The adjacency matrix obtained from the SDRT table associated with a DPFE, as described in Sect. 1.2.2, provides the basis for a DP program generator, i.e., a software tool that automatically generates "solver code", specifically, a sequence of assignment statements for solving a DPFE using a conventional programming language such as Java. We illustrate this solver code generation process in this section.

Given a weighted adjacency matrix, for example, the one given above, we can obtain the numerical solution of the DPFE by defining an assignment statement for each row of the matrix which sets a variable a_i for row i equal to the minimum of terms of the form $c_{i,j} + a_j$, where j is a successor of i.

```
a1=min{.2+a2,.5+a3,.3+a4}
a2=min{1.+a5,.6+a6}
a3=min{.4+a5,.6+a7}
a4=min{.4+a6,1.+a7}
a5=min{.9+a8}
a6=min{1.5+a8}
a7=min{.6+a8}
a8=0
```

These assignment statements can be used in a conventional nonrecursive computer program (in any procedural programming language) to calculate the values a_i. The statements should be compared with the equations of Example 1.1 [Method S]. As in that earlier example, evaluating the values a_i yields the following results: $a_8 = 0, a_7 = 0.6, a_6 = 1.5, a_5 = 0.9, a_4 = \min(1.9, 1.6) = 1.6, a_3 = \min(1.3, 1.2) = 1.2, a_2 = \min(1.9, 2.1) = 1.9, a_1 = \min(2.1, 1.7, 1.9) = 1.7$; note that $a_1 = 1.7$ is the goal. These assignment statements must of course be "topologically" reordered, from last to first, before they are executed.

1.2.4 Spreadsheet Solutions

Above, we showed the basis for a DP program generator that automatically generates a sequence of assignment statements for solving a DPFE using a conventional programming language. We show in this section how a spreadsheet that solves a DPFE can be automatically generated.

The assignment statements given Sect. 1.2.3 for the linear search problem can also be rewritten in the form

```
=min(.2+A2,.5+A3,.3+A4)
=min(1.+A5,.6+A6)
=min(.4+A5,.6+A7)
=min(.4+A6,1.+A7)
=min(.9+A8)
=min(1.5+A8)
```

```
=min(.6+A8)
0
```

which when imported into the first column of a spreadsheet will yield the same results as before; cell A1 of the spreadsheet will have 1.7 as its computed answer. One advantage of this spreadsheet solution is that "topological" sorting is unnecessary.

In this spreadsheet program, only the lengths of the shortest paths are calculated. To reconstruct the optimal policies, i.e. the sequence of decisions that yield the shortest paths, more work must be done. We will not address this reconstruction task further in this Chapter.

The foregoing spreadsheet has formulas that involve both the minimization and addition operations. A simpler "basic" spreadsheet would permit formulas to have only one operation. Suppose we define an intermediary variable a_k for each of the terms $c_{i,j} + a_j$. Then we may rewrite the original sequence of statements as follows:

```
a1=min(a9,a10,a11)
a2=min(a12,a13)
a3=min(a14,a15)
a4=min(a16,a17)
a5=min(a18)
a6=min(a19)
a7=min(a20)
a8=0
a9=.2+a2
a10=.5+a3
a11=.3+a4
a12=1.+a5
a13=.6+a6
a14=.4+a5
a15=.6+a7
a16=.4+a6
a17=1.+a7
a18=.9+a8
a19=1.5+a8
a20=.6+a8
```

As above, we may also rewrite this in spreadsheet form:

```
=min(A9,A10,A11)
=min(A12,A13)
=min(A14,A15)
=min(A16,A17)
=min(A18)
=min(A19)
=min(A20)
```

```
0
=.2+A2
=.5+A3
=.3+A4
=1.+A5
=.6+A6
=.4+A5
=.6+A7
=.4+A6
=1.+A7
=.9+A8
=1.5+A8
=.6+A8
```

This basic spreadsheet is a tabular representation of the original DPFE, and is at the heart of the software system we describe in this book. This software automatically generates the following equivalent spreadsheet from the given DPFE:

```
0
=B1+0.9
=MIN(B2)
=B3+1
=B3+0.4
=B1+1.5
=MIN(B6)
=B7+0.6
=B7+0.4
=MIN(B4,B8)
=B10+0.2
=B1+0.6
=MIN(B12)
=B13+0.6
=B13+1
=MIN(B5,B14)
=B16+0.5
=MIN(B9,B15)
=B18+0.3
=MIN(B11,B17,B19)
```

(Only Column B is shown here.) The different ordering is a consequence of our implementation decisions, but does not affect the results.

1.2.5 Example: SPA

As another illustration, that we will use later in this book since it is a smaller example that can be more easily examined in detail, we consider the shortest path in an acyclic graph (SPA) problem, introduced as Example 1.6 in Sect. 1.1.10. The SDRT table is as follows:

```
StateDecisionRewardTransformationTable
(0) [d=1] 3.0 ((1)) ()
(0) [d=2] 5.0 ((2)) ()
(1) [d=2] 1.0 ((2)) ()
(1) [d=3] 8.0 ((3)) ()
(2) [d=3] 5.0 ((3)) ()
```

From this table, we can generate solver code as a sequence of assignment statements as follows:

```
A1=min(A2+3.0,A3+5.0)
A2=min(A3+1.0,A4+8.0)
A3=min(A4+5.0)
A4=0.0
```

Simplifying the formulas, so that each has only a single (minimization or addition) operation, we may rewrite the foregoing as follows:

```
A1=min(A5,A6)
A2=min(A7,A8)
A3=min(A9)
A4=0.0
A5=A2+3.0
A6=A3+5.0
A7=A3+1.0
A8=A4+8.0
A9=A4+5.0
```

As in the case of the preceding linear search example, these assignment statements must be topologically sorted if they are to be executed as a conventional sequential program. (This sorting is unnecessary if they are imported into a Column A of a spreadsheet.) Rearranging the variables (letting B9=A1, B7=A2, B4=A3, etc.), we have:

```
B1=0.0
B2=B1+8.0
B3=B1+5.0
B4=min(B3)
B5=B4+5.0
B6=B4+1.0
B7=min(B6,B2)
B8=B7+3.0
B9=min(B8,B5)
```

These assignment statements can be executed as a conventional sequential program. Alternatively, importing them into Column B, we arrive at the following spreadsheet solver code:

```
=0.0
=B1+8.0
=B1+5.0
=min(B3)
=B4+5.0
=B4+1.0
=min(B6,B2)
=B7+3.0
=min(B8,B5)
```

1.2.6 Concluding Remarks

It is not easy to modify the above Java or spreadsheet "solver code" to solve DP problems that are dissimilar to linear search or shortest paths. Conventional programming and hand-coding spreadsheets, especially for problems of larger dimension, are error-prone tasks. The desirability of a software tool that automatically generates solver code from a DPFE is clear. That is the focus of this book.

1.3 Overview of Book

In Chap. 2, we discuss numerous applications of DP. Specifically, we formulate a DPFE for each of these applications. For many applications, we provide alternate formulations as well. This compendium of examples shows the generality and flexibility of dynamic programming as an optimization method and of the DP2PN2Solver software tool described in this book for solving dynamic programming problems.

In Chap. 3, we describe gDPS, a text-based specification language for dynamic programming problems. gDPS serves as the input language for the DP2PN2Solver tool. Its syntax is given in BNF form. In effect, a gDPS source program is a transliteration of a DPFE.

In Chap. 4, we show how each of the DPFEs given in Chap. 2 can be expressed in the gDPS language of Chap. 3. The result is a set of computer programs for solving the DP problems given in Chap. 2.

In Chap. 5, we define Bellman nets, a class of Petri nets, which serve as a useful model of DPFEs. Petri nets, hence also Bellman nets, may be regarded as a class of directed graphs.

In Chap. 6, we show how the DPFEs in Chap. 2 or Chap. 4 can be represented as a Bellman net.

In Chap. 7, we describe the overall structure of DP2PN2Solver, a "compiler" tool whose (source code) input is a DPFE and whose (object code) output is "solver code", i.e., a program which when executed solves the DPFE. The first phase of this tool translates a DPFE into its Bellman net representation, and the second phase translates the Bellman net into solver code. In Sect. 7.2, we show the internal Bellman net representations for the DPFEs in Chap. 2. Unlike the graphical representation in Chap. 6, the internal representation is tabular, well suited as an intermediary between Phases 1 and 2.

In Chap. 8, we describe Phase 1 of our DP tool, which parses DPFE source code and outputs its Bellman net representation. This Bellman net is produced indirectly: the parser generates intermediate "builder code" which when executed outputs the Bellman net.

In Chap. 9, we describe Phase 2 of our DP tool, which inputs a Bellman net and produces solver code which when executed outputs a numerical solution of the associated DPFE. This solver code can be Java code, a spreadsheet, or an XML-coded Petri net (using a Petri net simulation language). In the case of Java code, the solver code produced does not directly calculate results, but consists instead of calls to systems routines that perform the calculations.

In Chap. 10, we show the numerical outputs obtained by executing the Java solver codes produced by Phase 2 from the Bellman nets for each of the problems of Chap. 2 and Chap. 4.

In Chap. 11, we show the numerical outputs obtained by executing the spreadsheet and XML solver codes produced by Phase 2 for some sample Bellman nets.

Chapter 12 concludes the book with a brief summary and a discussion of current research into ways to improve and extend our DP software tool.

Appendix A provides supplementary program listings that we include for completeness, detailing key portions of our software tool, for example.

Appendix B is a User's Guide for our tool, including downloading and installation instructions.

2

Applications of Dynamic Programming

Dynamic programming has been applied to numerous areas in mathematics, science, engineering, business, medicine, information systems, bioinformatics, among others. There are few disciplines in which there are no optimization problems to which DP has been applied. In this book, we focus on applications to operations research and computer science. Most of these applications can be classified into a few major groups: applications involving graph routing, sequencing, selection, partitioning, production and distribution, divide-and-conquer, probabilistic or weighted transitions, string processing, and nonoptimization problems. These groups are neither disjoint nor complete; some applications fit in more than one group, and others fit in none of them. (In the following, the parenthesized labels identify specific problems we discuss in this Chapter.)

1. Graph *routing* problems are associated with finding shortest (or longest) *paths* in graphs. The problem is especially simple when the graph is acyclic (SPA, Sect. 2.43). Special cases include graphs where the nodes are grouped in stages (SCP, Sect. 2.38) or in separate lines (ASMBAL, Sect. 2.4). A more complicated problem allows the graph to be cyclic (SPC, Sect. 2.44), and even have negative branch labels. For cyclic graphs, we may wish to find longest simple paths (LSP, Sect. 2.26) or shortest Hamiltonian paths (TSP, Sect. 2.47). For some applications, neither the source nor the target are specified; instead, we are to find the "all-pairs shortest-paths" (APSP, Sect. 2.2) from each node p to each other node q, for all pairs (p, q). The problem of finding a path whose maximal branch is minimal (MINMAX, Sect. 2.28) is of special interest since it involves non-additive costs. Many other classes of problems can be formulated as graph routing problems whose branch distances must generally be derived rather than being explicitly given in a distance array.
2. *Sequencing* problems are associated with finding the optimal ordering or *permutation* of a set of objects. The set of objects may be data to be placed in an array for linear searching (LINSRC, Sect. 2.24), a set of program

A. Lew and H. Mauch: *Applications of Dynamic Programming*, Studies in Computational Intelligence (SCI) **38**, 45–100 (2007)
www.springerlink.com

files to be placed on a tape (PERM, Sect. 2.32), or a set of processes to be scheduled. In *scheduling* problems, processes may be jobs to be scheduled for execution on a CPU to minimize average turnaround time (SPT, Sect. 2.45), or multitask jobs to be scheduled in flowshop fashion (FLOWSHOP, Sect. 2.12), or disk file access requests to be scheduled to minimize total seek times (SEEK, Sect. 2.39). Sequencing problems can also viewed as a class of assignment problems where each object is assigned to some position in the sequence (ASSIGN, Sect. 2.5).

3. *Selection* problems are associated with finding optimal (usually proper) *subsets* of a given set of objects. One example is the problem of finding a spanning tree of a graph that has minimum weight (MWST, Sect. 2.29); this tree consists of a subset of branches with minimum total weight that spans the graph while satisfying the eligibility constraint of acyclicness. In *knapsack* problems, objects have values and weights, and the objective is to select a subset with optimal total value such that the total weight does not exceed the size of the knapsack, where there may be at most one object of a given type (KS01, Sect. 2.20) or multiple objects of a given type (KSINT, Sect. 2.22 and RDP, Sect. 2.36). In *scheduling* problems, a subset of jobs to be executed is selected taking into account eligibility constraints based upon, for example, deadlines (DEADLINE, Sect. 2.8) or start and stop times (INTVL, Sect. 2.16). Integer linear programming problems can be solved using the same DP approach adopted for knapsack problems (ILP, Sect. 2.14), and in turn knapsack problems can be solved by formulating them as integer linear programming problems (ILPKNAP, Sect. 2.15). Some covering problems can also be formulated as a variation of a knapsack problem (COV, Sect. 2.7; KSCOV, Sect. 2.21).
4. *Partitioning* or *clustering* problems are characterized by having a set S of objects of size $|S| = N$ that are to be grouped into disjoint subsets, called a *partition*, in an optimal fashion. The *rank* K of the partition, i.e., the number of disjoint subsets or clusters ($1 \leq K \leq N$), either may be given in advance or may be left to be optimally determined. A *sequential* partitioning problem, also known as a *segmentation* problem, is one in which S is considered an ordered sequence of objects, and nonadjacent objects can only be in a cluster or *segment* if all intervening objects are in the same segment. Examples of such segmentation problems include curve fitting (SEGLINE, Sect. 2.41) and program pagination (SEGPAGE, Sect. 2.41). A partitioning problem may also be viewed as that of assigning each object to one of a given number of clusters. In one example of such an assignment problem (ASSIGN, Sect. 2.5), we assume the special case where $K = N$. It is more common, however, for K to be well less than N. Some covering problems, as mentioned above, can also be formulated as a variation of a partitioning problem.
5. *Distribution* problems are characterized by having a set of objects that are to be distributed from a set of suppliers to a set of receivers in an optimal fashion, where there are specified costs associated with destination y

receiving objects from source x. In transportation problems, this is modeled by having a *flow* from x to y (TRANSPO, Sect. 2.46). In some simpler special cases, there is one source and multiple destinations, and distribution costs depend on the allotments, i.e. the number of objects allocated to each destination (ALLOT, Sect. 2.1; ODP, Sect. 2.31). Another special case is that of assignment problems where sources must be matched (paired) with destinations (ASSIGN, Sect. 2.5). We also note that many distribution problems can be modeled as integer linear programming problems, and handled as above. Furthermore, it is common for distribution problems to have probabilistic elements as well, so that they must be handled as below.

6. *Production* or *inventory* or *replacement* problems are characterized by having a resource, units of which can be produced (by a supplier) or demanded and consumed (by a receiver) at each of a series of stages, where there are specified costs of producing or not producing additional units of resource to replace consumed ones, or to add to existing inventory to meet future demand (PROD, Sect. 2.34; INVENT, Sect. 2.17; REPLACE, Sect. 2.37; LOT, Sect. 2.25). Investment problems (INVEST, Sect. 2.18; INVESTWLV, Sect. 2.19) can be modeled in like fashion, where the amount that can be invested corresponds to the inventory, and gains and losses from investments correspond to demands.
7. Optimal *binary tree* problems are associated with situations where a decision divides the original problem into two subproblems whose solutions can be solved separately and combined to obtain the solution of the original problem. This is similar to what are known as *divide-and-conquer* algorithms. The main characteristic is that each decision leads to multiple next-states rather than a single one. Examples include constructing an optimal binary search tree (BST, Sect. 2.6), constructing an optimal alphabetic radix or prefix-code tree (ARC, Sect. 2.3), determining the best way to multiply a chain of matrices (MCM, Sect. 2.27), and determining the best way to obtain a set of order statistics (SELECT, Sect. 2.42). The DPFEs for these problems are nonserial (second-order).
8. *Probabilistic* problems have probabilities associated with next-states. Unlike the aforementioned divide-and-conquer problems, in probabilistic problems each decision results in single next-state determined by chance from a set of alternatives. The DPFEs for these problems (INVEST, Sect. 2.18; INVESTWLV, Sect. 2.19; PROD, Sect. 2.34; PRODRAP, Sect. 2.35) are nonserial. Many probabilistic problems are production or investment problems, where the state is the current inventory or balance, the decision is how much to produce or invest, and the next-states may increase or decrease, sometimes based on random external factors such as demand. We note that there are also serial DP problems where probabilities are not associated with next-states, but instead are associated with the reward function (RDP, Sect. 2.36); in the simplest case, the reward is an expected value computed from given probabilities.

9. Probabilistic problems are only one class of problems in which next-states are weighted. Boolean weights are used, for example, for a scheduling problem (INTVL3, Sect. 2.16) and a routing problem (APSPFW, Sect. 2.2). Such weights can also be based on "discounting" (DPP, Sect. 2.9).
10. Problems involving *string processing* are also of interest, especially in the fields of text processing and computational biology. We discuss the problems of "editing" or transforming one string to another (EDP, Sect. 2.10) and of finding "matching" subsequences (LCS, Sect. 2.23).
11. *Nonoptimization* problems, in the context of sequential decision processes, are those where decisions are not made for the purpose of optimizing some objective function. Instead, decisions are made only to satisfy certain constraints, especially so as to ensure that a specific base state is reached. For example, DP can be used to solve recurrence relations, such as for the Fibonacci sequence (FIB, Sect. 2.11) and for the Tower of Hanoi problem (HANOI, Sect. 2.13). In addition, DP can also be used to determine sequences of moves that solve some problem, such as for certain puzzles (HANOI, Sect. 2.13; NIM, Sect. 2.30; POUR, Sect. 2.33).

There are of course other ways to categorize the various applications. One other way is by discipline. For example, applications covered in a typical operations research textbook are (SCP, ALLOT, ALLOTm, ILP, PRODRAP, INVESTWLV) in [21, Chap. 11] or (NIM, POUR, SCP, INVENT, ALLOTf, KSINT, REPLACE, DDP, TSP, MINMAX, ALLOTm, LOT) in [63, Chap. 18]. Applications covered in a typical computer science textbook [10, Chap. 15] include (ASMBAL, MCM, LSP, LCS, BST). In this book, we treat all of these applications, among many others, to demonstrate the generality of our formalism and the utility of our DP software tool for solving DP problems.

In this chapter, we show how each of the problems referred to here can be solved using dynamic programming. Specifically, we formulate a DPFE for each problem. In later chapters, these DPFEs are expressed in the text-based language gDPS, which serves as the input source programs for our DP software tool that translates them into object programs whose execution solves the problems. We may view gDPS both as a mathematical specification language and as a programming language.

It should be noted that a number of the problems included here can be solved much more efficiently by greedy algorithms. In these cases, we present less efficient DP solutions in part to demonstrate the generality of DP as a method to solve optimization problems, and in part to provide a large sample of DP formulations. Having such a large sample for reference, it should be easier to formulate DP solutions to new problems. Furthermore, the set of examples of the use of our software tool is therefore also large, which should make it easier to learn to use this tool. Finally, it should be emphasized that we often can modify DP to solve variations of problems for which greedy algorithms are inapplicable.

The following sections present the problems in alphabetical order of their names.

2.1 Optimal Allotment Problem (ALLOT)

The optimal allotment problem is that of deciding how to distribute a limited amount of resources to a set of users of these resources, where there are specified costs or profits associated with allotting units of the resource to users. The optimal distribution problem ODP (Sect. 2.31) may be regarded as an allotment problem. The problem may also be regarded as a variation of the knapsack problem KSINT (Sect. 2.22).

Assume there are M total units of the resource, and let $C(k,d)$ be the cost or profit associated with allotting d units to user k, where $d = 0, \ldots, M$ and $k = 1, \ldots, N$. Suppose we make the allotment decisions in stages, initially allotting d_1 units to user 1, then d_2 units to user 2, etc. This arbitrary sequencing $1, 2, \ldots, N$ can be assumed since only the quantities allotted matter, not the sequence in which they are made. We define the state (k,m) as remaining number m of units of resource at stage k. The cost of deciding to allot d units at stage k to user k is $C(k,d)$. The next-state is $(k+1, m-d)$. The DPFE is

$$f(k,m) = \min_{d \in \{0,\ldots,m\}} \{C(k,d) + f(k+1, m-d)\}. \tag{2.1}$$

The goal is to find $f(1,M)$ with base-conditions $f(N+1,m) = 0$ when $m \geq 0$. If we allow $d > m$ (bounding d by M instead), we may use the additional base-condition $f(N+1,m) = \infty$ when $m < 0$ to prevent allotting more resources than is available.

For instance, let $M = 4$, $N = 3$ and

$$(C_{k,d})_{k \in \{1,2,3\}; d \in \{0,\ldots,4\}} = \begin{pmatrix} \infty & 1.0 & 0.8 & 0.4 & 0.0 \\ \infty & 1.0 & 0.5 & 0.0 & 0.0 \\ \infty & 1.0 & 0.6 & 0.3 & 0.0 \end{pmatrix}.$$

Then $f(1,M) = 1.0 + 0.5 + 1.0 = 2.5$ for the optimal sequence of allotments $d_1 = 1$, $d_2 = 2$, $d_3 = 1$.

This simple allotment problem can be generalized in several ways.

- In the ALLOTt problem, in [21, Example 3, pp.549–552], allotment decisions and their costs are defined in separate *tables*. (ODP and ALLOTm also have tabular costs.)
- In the ALLOTf problem, in [63, Example 5, pp.975–977], the costs are defined nontabularly, i.e., by general *functions*.
- In the ALLOTm problem, in [63, Example 14, pp.998–999], costs are *multiplicative* rather than additive. (ALLOTt also has multiplicative costs.)

There are also probabilistic DP problems where allotment costs are random variables; for many such problems, we may simply use the expected values of these costs in nonprobabilistic DPFEs.

2.2 All-Pairs Shortest Paths Problem (APSP)

In Sect. 1.1.11, we discussed the *All-Pairs Shortest Paths* (APSP) problem of finding the shortest path from any node p to any other node q, where p and q are arbitrary nodes in a set S. We may of course use any general designated-source or target-state shortest path algorithm, such as (SPC, Sect. 2.44), varying p and q repetitively. APSP algorithms can be more efficient since it is possible to reuse calculations when shortest paths are computed in batches rather than individually. Such reuse requires that calculations be suitably ordered. (We do not address this efficiency issue here.) In our implementation, specifically in our generation of the solution tree, subtrees are recalculated.

In the DPFEs given in Sect. 1.1.11, the goal is to compute $F(k,p,q)$ which is regarded as a matrix equation where p and q are row and column subscripts for a matrix $F^{(k)}$. The DPFE gives $F^{(k)}$ in terms of the matrix $F^{(k-1)}$. The relaxation DPFE (1.43) is:

$$F(k,p,q) = \min_{r \in S}\{b(p,r) + F(k-1,r,q)\}, \tag{2.2}$$

for $k > 0$, with $F(0,p,q) = 0$ if $p = q$ and $F(0,p,q) = \infty$ if $p \neq q$. Recall that we assume here that $b(p,p) = 0$ for all p.

An alternative DPFE is the Floyd-Warshall DPFE (1.44):

$$F(k,p,q) = \min\{F(k-1,p,q), F(k-1,p,k) + F(k-1,k,q)\}, \tag{2.3}$$

for $k > 0$, where k is the highest index of the set of nodes. The base cases are $F(0,p,q) = 0$ if $p = q$ and $F(0,p,q) = b(p,q)$ if $p \neq q$. Both DPFEs define a matrix of values $F^{(k)}$, which is to be determined for one pair of subscripts (p,q) at a time (rather than in a batch).

These two formulations differ is one significant way. In the former case (APSP), the decision space is the set of nodes S. The problem can be solved in much the same fashion as for SPC, except that the target is a parameter rather than fixed. In the latter case (APSPFW), the decision space is Boolean, reflecting whether or not a path shorter than the prior one has been found. The Floyd-Warshall DPFE can be reformulated using transition weights as follows:

$$F(k,p,q) = \min_{d \in \{0,1\}}\{(1-d).F(k-1,p,q) + d.F(k-1,p,k) + d.F(k-1,k,q)\}. \tag{2.4}$$

For example, consider the graph whose adjacency matrix C, whose entries are the branch distances b, is

$$C = \begin{pmatrix} \infty & 3 & 5 & \infty \\ \infty & \infty & 1 & 8 \\ \infty & 2 & \infty & 5 \\ \infty & \infty & \infty & \infty \end{pmatrix}$$

where $C[p][q] = b(p, q) = \infty$ if there is no branch from node p to node q. This is the same example given in Sect. 1.1.10, and is also used for SPC, as displayed in Fig. 2.8.

For APSP, which uses (2.2), assuming $p = 0$ and $q = 3$, we find that $F(3, 0, 3) = 9$. From the solution obtained during the course of this calculation, we also find that $F(2, 1, 3) = 6$ and $F(1, 2, 3) = 5$. However, some values such as for $F(3, 0, 2) = 4$ must be found separately, such as by assuming $p = 0$ and $q = 2$.

For APSPFW, which uses (2.4), assuming $p = 0$ and $q = 3$, we find that $F(3, 0, 3) = 9$. From the solution obtained during the course of this calculation, we also find that $F(2, 0, 3) = 9$ and $F(1, 0, 3) = 11$, among many other values. However, some values such as for $F(3, 1, 3) = 6$ must be found separately, such as by assuming $p = 1$ and $q = 3$.

2.3 Optimal Alphabetic Radix-Code Tree Problem (ARC)

The optimal *alphabetic* radix code tree problem [23] is a variation of the Huffman code tree problem. The Huffman code tree can be obtained by the "greedy" heuristic of choosing to combine in a subtree the two nodes having smallest weight, and replacing these two nodes a and b by a new node c having a weight equal to the sum of the weights of nodes a and b. The alphabetic variation constrains each choice to that of combining only adjacent nodes, with each new node being placed in the same position as the nodes it replaces. Given this constraint, the greedy heuristic is no longer necessarily optimal, but the problem can be solved using DP. Since each tree can be represented as a parenthesized list of its leaves in inorder, the problem is equivalent to finding an optimal parenthesization, and can be solved in a fashion similar to BST and MCM. How costs can be attributed separably to the individual decisions is the key to solving this problem. We note that a Huffman tree may be regarded as constructed from the bottom up, where the first decision is to choose the leaves that should be initially combined. On the other hand, the alphabetic radix code tree will be constructed from the top down, where the first decision is to choose the root of the overall tree, which partitions the leaves into left and right subsets. The main problem is to define separable partitioning costs.

A radix code tree has one useful property that provides the key to solving the optimization problem. The total cost of the tree is the sum of the costs of the internal nodes, and the cost of an internal node is the sum of the leaves in the subtree rooted at that node. Internal nodes, of course, correspond to decisions. Thus, we let the cost of an internal node of an alphabetic radix code tree equal the sum of the leaves in the subtree rooted at that node. This cost is the same regardless of how the leaves are to be partitioned. We may then solve the constrained problem by using a DPFE of the same top-down form as for MCM. Given $S = (w_0, w_1, \ldots, w_{n-1})$ is a *list* (i.e., an ordered sequence) of weights associated with leaves, we define the state to be a pair (i, j), which represents the list $(w_i, w_2, \ldots, w_j)$. Then

$$f(i,j) = \min_{i \leq d < j} \{c(i,j,d) + f(i,d) + f(d+1,j)\} \text{ if } i < j, \tag{2.5}$$

where $c(i,j,d) = \sum_{k=i}^{j} w_k$. The goal is to find $f(0, n-1)$ with base condition $f(i,j) = 0$ when $i = j$.

For example, if $S = (1,2,3,4)$ initially, the optimal tree is $((1,2),3),4)$ and $f(S) = 3+6+10 = 19$. From initial state $(0,3)$ representing the list $(1,2,3,4)$, the initial decision is $d = 2$ which results in two next-states, state $(0,2)$ representing the list $(1,2,3)$ and state $(3,3)$ representing the list (4); the cost of this decision is the sum $1+2+3+4 = 10$. As a second example, if $S = (2,3,3,4)$ initially, the optimal tree is $((2,3),(3,4))$ and $f(S) = 5+7+12 = 24$.

2.4 Assembly Line Balancing (ASMBAL)

In what are known collectively as *assembly line balancing* problems, a product can be assembled by going through a series of processing stations. There are usually costs associated with processing step, and other costs associated with going from one station to another. In general, it is desirable to "balance" the workload among the processing stations in a fashion that depends upon how the various costs are defined. We discuss here one simple example of such assembly line balancing problems.

This "scheduling" example, described in [10, pp.324–331], may be regarded as a variation of the shortest path problem for acyclic graphs (SPA, see Sect. 2.43) or the staged version (SCP, see Sect. 2.38). In this problem, the nodes of the graph are grouped into stages and also into lines. Transitions can only be made from a node at stage k and line i to a node at stage $k+1$ and line j, at a cost $c(k,i,j)$. Usually, $c(k,i,i) = 0$, i.e. at any stage k there is a cost associated with switching lines, but not for staying in the same line. In addition to a transition cost c (associated with branches), there may be another cost v (associated with nodes). An initial state s and terminal state t are also defined; for simplicity, we adopt the convention that s and t are in line 0, as opposed to a special third line. Then the initial decision is made from s to enter line j at a cost $c(0,0,j)$, and a final decision is made from any line j to enter t at a cost $c(N,j,0)$.

The graph associated with the assembly line balancing problem is acyclic, so it can be topologically sorted by stages. We number the nodes in such a topological order from 0 to 13, with 0 as the initial state s, 13 as the terminal state t. Assume the node costs for the 14 nodes are

$$v = (0, 7, 8, 9, 5, 3, 6, 4, 4, 8, 5, 4, 7, 0),$$

and the branch costs b are given by the following weighted adjacency matrix:

$$\begin{pmatrix}
\infty & 2 & 4 & \infty & \infty & \infty & \infty & \infty & \infty & \infty & \infty & \infty & \infty & \infty \\
\infty & \infty & \infty & 0 & 2 & \infty & \infty & \infty & \infty & \infty & \infty & \infty & \infty & \infty \\
\infty & \infty & \infty & 2 & 0 & \infty & \infty & \infty & \infty & \infty & \infty & \infty & \infty & \infty \\
\infty & \infty & \infty & \infty & \infty & 0 & 3 & \infty & \infty & \infty & \infty & \infty & \infty & \infty \\
\infty & \infty & \infty & \infty & \infty & 1 & 0 & \infty & \infty & \infty & \infty & \infty & \infty & \infty \\
\infty & \infty & \infty & \infty & \infty & \infty & \infty & 0 & 1 & \infty & \infty & \infty & \infty & \infty \\
\infty & \infty & \infty & \infty & \infty & \infty & \infty & 2 & 0 & \infty & \infty & \infty & \infty & \infty \\
\infty & \infty & \infty & \infty & \infty & \infty & \infty & \infty & \infty & 0 & 3 & \infty & \infty & \infty \\
\infty & \infty & \infty & \infty & \infty & \infty & \infty & \infty & \infty & 2 & 0 & \infty & \infty & \infty \\
\infty & \infty & \infty & \infty & \infty & \infty & \infty & \infty & \infty & \infty & \infty & 0 & 4 & \infty \\
\infty & \infty & \infty & \infty & \infty & \infty & \infty & \infty & \infty & \infty & \infty & 1 & 0 & \infty \\
\infty & \infty & \infty & \infty & \infty & \infty & \infty & \infty & \infty & \infty & \infty & \infty & \infty & 3 \\
\infty & \infty & \infty & \infty & \infty & \infty & \infty & \infty & \infty & \infty & \infty & \infty & \infty & 2 \\
\infty & \infty & \infty & \infty & \infty & \infty & \infty & \infty & \infty & \infty & \infty & \infty & \infty & \infty
\end{pmatrix}$$

where $b(i, j) = \infty$ if there is no branch from node i to node j.

For $1 \leq i \leq 12$, node i is the $((i + 1)/2)$-th node in line 0 if i is odd, whereas node i is the $(i/2)$-th node in line 1 if i is even. Note that there are $N = 6$ stages (not counting s and t). Fig. 2.1 shows the assembly line instance under consideration.

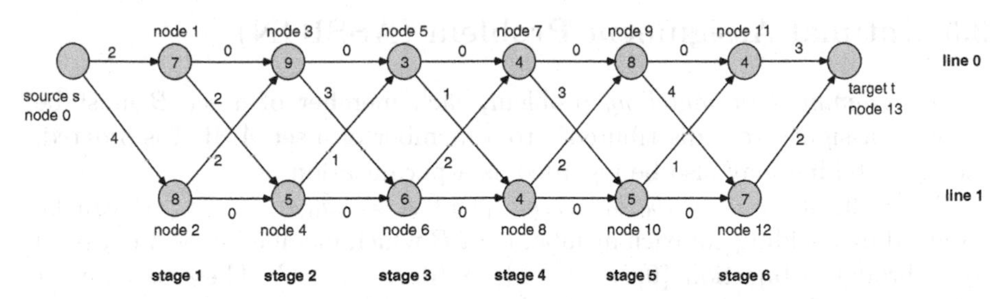

Fig. 2.1. Instance of an Assembly Line Balancing Problem

Such assembly line balancing problems can be solved by a slight modification to the solution to the stagecoach problem (SCP, Sect. 2.38). The main change to SCP is that of adding each node-cost v to all of its outgoing (or alternatively all of its incoming) branches. That is, we would define the net cost or reward function by $R(k, i, j) = v(k, i) + c(k, i, j)$. Furthermore, the next-state function is $T(k, i, j) = (k + 1, j)$. The goal is $(0, 0)$. The base-condition is $f(k, i) = 0$ when $k > N$. We note that in this staged formulation, only the non-infinite members of b need to be stored in an array. The successors of each node are constrained to be nodes in the next stage, which can be determined as a function of their subscripts as opposed to being specified in a table.

In the common case where there is a cost of zero for staying in the same line, $c(k,i,i) = 0$, then only the nonzero costs have to be provided as part of the problem data. A modified reward function $R'(k,i,j)$ could test for whether $i = j$ and if so yield a value of 0, instead of looking up its value in a table or array. Thus, this problem can be solved using the staged DPFE

$$f(k,i) = \min_j \{R'(k,i,j) + f(k+1,j)\}, \tag{2.6}$$

for $f(0,0)$ given the base condition $f(N+1,i) = 0$. For the above example, $f(0,0) = (0+2)+(7+2)+(5+1)+(3+1)+(4+0)+(5+1)+(4+3) = 38$, for the optimal sequence of line changes 0,1,0,1,1,0, not counting the convention that we start and end in line 0.

The assembly line balancing problem can also be solved as an ordinary shortest path in an acyclic graph (SPA, Sect. 2.43) problem, where the reward function adds node-costs to branch-costs as above. In effect, we can simply ignore stage numbers. For the above example, we would solve the DPFE

$$f(i) = \min_j \{v(i) + b(i,j) + f(j)\}, \tag{2.7}$$

for $f(s)$ given the base condition $f(t) = 0$. For the above example, we obtain the goal $f(0) = 38$. The shortest path from $s = 0$ to $t = 13$ goes through the sequence of nodes 0,1,4,5,8,10,11,13, and has length 38.

2.5 Optimal Assignment Problem (ASSIGN)

In an *assignment* or *matching* problem, each member of a set B must be uniquely assigned (or "distributed") to a member of a set A. If A is ordered, then a matching may also be regarded as a permutation of A.

A permutation $B = (b_0, b_1, \ldots, b_{n-1})$ of $A = (a_0, a_1, \ldots, a_{n-1})$ can be obtained by deciding, for each member b_i of B which member a_j of A to assign to it through a bijection $\{0, \ldots, n-1\} \to \{0, \ldots, n-1\}$. These assignment decisions can be made in any order, so we let i be a stage number and let $c(i,j)$ be the cost of assigning a_j to b_i at stage i. Since we require unique assignments, at each stage we keep track of the members of A that have not yet been assigned. This set S is incorporated into our definition of state. Specifically, the state is (k,S) where k is a stage number and S is a set. A decision in state (k,S) chooses a member $d \in S$, at a cost $C(k,S,d)$ that in general may also be a function of S. The next-state is $(k+1, S - \{d\})$. The DPFE is

$$f(k,S) = \min_{d \in S} \{C(k,S,d) + f(k+1, S - \{d\})\}.$$

Within this framework, we can solve a variety of different assignment or distribution problems by using different definitions of the cost function. For example, let $C(k,S,d) = \sum_{i \notin S}(w_i) + w_d$. Then the optimal assignment corresponds to the solution of the SPT problem (see Sect. 2.45). (We note that $\sum_{i \notin S}(w_i) = \text{TTLWGT} - \sum_{i \in S}(w_i)$, where TTLWGT is the sum of all the weights.) Thus, the optimization problem can be solved using the DPFE

$$f(k,S) = \min_{d \in S}\{\sum_{i \notin S}(w_i) + w_d + f(k+1, S - \{d\})\}. \tag{2.8}$$

The base-condition is $f(k,S) = 0$ when $k = n+1$ or $S = \emptyset$. The goal is find $f(1,S^*)$, where S^* is the originally given set of N processes.

For instance, if $S^* = \{0,1,2\}$ with weights $(w_0, w_1, w_2) = (3,5,2)$, then $f(1,S^*) = 2 + 5 + 10 = 17$ for the optimal sequence of assignment decisions $d_1 = 2, d_2 = 0, d_3 = 1$. Thus, the optimal permutation of $A^* = (3,5,2)$ is $B^* = (2,3,5)$.

2.6 Optimal Binary Search Tree Problem (BST)

This problem is described in [10, pp.356–362]. Assume a set of n data items $X = \{x_0, \ldots, x_{n-1}\}$ and a total order defined on these items is given. The access probability of a data item x_i is $p(x_i)$ or p_i for short. (Note that $\sum_{i=0}^{n-1} p_i = 1$.)

The task is to build a *binary search tree* that has minimal cost, where the cost of the tree is defined as

$$\sum_{i=0}^{n-1}(p_i \text{level}(x_i))$$

and level(x_i) denotes the level (depth) of the node corresponding to data item x_i in the tree. Note that items can be stored in internal nodes of the tree, not only in leaves.

We give two alternative approaches to solve this problem using DP. In the first formulation we define the state to be the set S of items to be arranged in the tree. The DP functional equation can be expressed as

$$f(S) = \begin{cases} \min_{\alpha \in S}\{f(S_l) + f(S_r) + r(\alpha, S)\} & \text{if } S \neq \emptyset \\ 0 & \text{if } S = \emptyset, \end{cases} \tag{2.9}$$

where $S_l = \{x \in S : x < \alpha\}$ is the remaining set of items that are smaller than the decision α (and thus appear to the left of α) and $S_r = \{x \in S : x > \alpha\}$ is the remaining set of items that are larger than α (and thus appear to the right of α) and the cost of the decision is defined as

$$r(\alpha, S) = \sum_{x \in S} p(x).$$

Using an alternative base case the DP functional equation can be expressed as

$$f(S) = \begin{cases} \min_{\alpha \in S}\{f(S_l) + f(S_r) + r(\alpha, S)\} & \text{if } |S| > 1 \\ p(x) & \text{if } S = \{x\}, \end{cases}$$

The goal is to compute $f(X)$.

A second DP functional equation formulates the problem by defining a state to be a pair of integers providing the start index and the end index of the data items to be arranged (an approach similar to the DP model for the MCM problem in Sect. 2.27.) For this formulation we require, without loss of generality, that the data items $X = (x_0, \ldots, x_{n-1})$ are already ordered. Then the DP functional equation can be stated as

$$f(i,j) = \begin{cases} \min_{k \in \{i,\ldots,j\}} \{f(i,k-1) + f(k+1,j) + \sum_{l=i}^{j} p_l\} & \text{if } i \leq j \\ 0 & \text{if } i > j. \end{cases} \tag{2.10}$$

Using an alternative base case the DP functional equation can be expressed as

$$f(i,j) = \begin{cases} \min_{k \in \{i,\ldots,j\}} \{f(i,k-1) + f(k+1,j) + \sum_{l=i}^{j} p_l\} & \text{if } i < j \\ p_i & \text{if } i = j. \end{cases}$$

In this second model the goal is to compute $f(0, n-1)$.

We consider the following instance of this problem. We have the following 5 data items, listed in their lexicographical order: (A, B, C, D, E). Their respective search probabilities are $(0.25, 0.05, 0.2, 0.4, 0.1)$. The optimal value of this instance is $f(X) = 1.9$ (in terms of the second DP model: $f(0,4) = 1.9$), which corresponds to the optimal binary search tree depicted in Fig. 2.2.

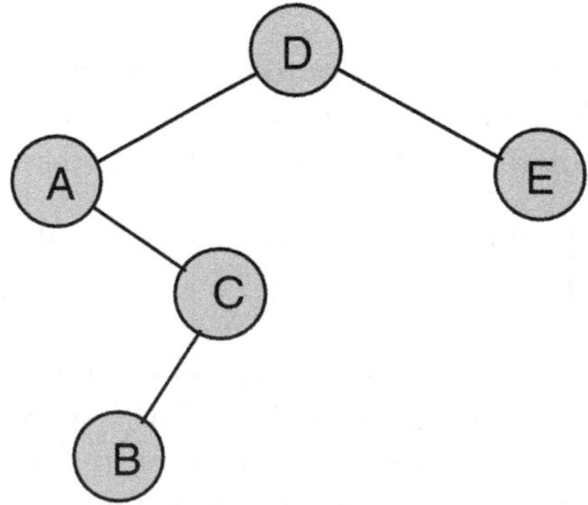

Fig. 2.2. Optimal Binary Search Tree for the Example Instance

This problem sometimes appears in the literature in slightly modified versions. One straightforward generalization introduces "dummy keys" representing values not in X, which takes care of unsuccessful searches [10, p.357]. A second generalization considers arbitrary weights (e.g. real-valued weights) instead of probabilities. Another variation is to require that data items only be stored in the leaves of the tree, not in the internal nodes.

2.7 Optimal Covering Problem (COV)

This optimal *covering* problem is taken from [54, p.17]. Given are k different sized shrubs that need to be protected from frost. Assume the shrubs are sorted by size such that shrub 0 is the smallest, and shrub $k-1$ is the largest. The cost to manufacture a cover for shrub size i is denoted c_i. However, due to manufacturing constraints, covers will be manufactured in no more than n different sizes, where $n \leq k$. Larger covers can protect smaller bushes. The objective is to select the n sizes which enable one to cover all shrubs at least cost.

Let j denote the number of cover sizes that have not been chosen yet and let l denote the largest shrub of the ones which are still under consideration. The DP functional equation for this problem can be expressed as

$$f(j,l) = \begin{cases} \min\limits_{d \in \{j-2,\ldots,l-1\}} \{(l-d)c_l + f(j-1,d)\} & \text{if } j > 1 \\ (l+1)c_l & \text{if } j = 1. \end{cases}$$

The goal is to compute $f(n, k-1)$.

Consider an instance of this problem with $k = 10$ shrubs, cover size costs $(c_0, \ldots, c_9) = (1, 4, 5, 7, 8, 12, 13, 18, 19, 21)$ and a manufacturing constraint of $n = 3$ cover sizes. Then the optimal policy is to order the manufacturing of cover sizes of 9, 6, and 4 (thus covering bushes 9, 8, and 7 with cover size 9, covering bushes 6 and 5 with cover size 6, and covering bushes 4 through 0 with cover size 4) at a total cost of $f(3,9) = 129$.

2.8 Deadline Scheduling Problem (DEADLINE)

The deadline scheduling problem (see [22, pp.206–212] and [10, pp.399–401]) is that of choosing the optimal subset of a set of unit-time processes to be executed on a single processor, each process having a specified deadline and profit, where its profit is earned if the process completes execution before its deadline. The optimal subset is the one whose total earned profit is maximal. The unit-time assumption means that each process completes execution one time unit after it starts. This scheduling problem can be solved using a greedy algorithm, but here we show that it can also be solved using DP in a fashion similar to many other scheduling problems. The state (k, S) is a stage number

k and a set S of processes that have not yet been considered. A decision d is a member of S. The next-state is $(k+1, S-\{d\})$. The cost of choosing process d is either its profit or 0 depending upon whether its inclusion in the set of scheduled processes yields a "feasible" set, i.e. a set in which each process meets its deadline. This feasibility test is easier to implement if the set of processes is given in increasing order of deadlines. Assuming this ordering by deadlines, a chosen subset of processes would be executed in this sequential order; since we assumed unit-time processes, the j-th scheduled process terminates at time j. Thus, a sequence of processes $S = \{1, \ldots, k\}$ of size k is *feasible* if $j \leq t_j$ for $1 \leq j \leq k$, in which case each process j in S can terminate before its deadline time t_j.

The optimization problem can be solved using the DPFE

$$f(k, S) = \max_{d \in S}\{c(d|S) + f(k+1, S-\{d\})\}, \tag{2.11}$$

where $c(d|S) = w_d$ if choosing to include d in the set of scheduled processes is "feasible", else $c(d|S) = 0$. Our goal is to solve for $f(1, S^*)$ given the base case $f(k, S) = 0$ for $k = N+1$ or $S = \emptyset$.

Assume a set of jobs $S^* = \{0,1,2,3,4\}$ having profits $p = \{10, 15, 20, 1, 5\}$ and deadline times $t = \{1,2,2,3,3\}$. Then $f(1, \{0,1,2,3,4\}) = 15 + 20 + 5 + 0 + 0 = 40$ for the optimal sequence of decisions $d_1 = 1, d_2 = 2, d_3 = 4, d_4 = 0, d_5 = 3$.

2.9 Discounted Profits Problem (DPP)

This Discounted Profits Problem (DPP) is described in [64, pp.779–780]. It is an intertemporal optimization problem that can be solved with DP. By incorporating the time value of money into the model, we get what is often referred to as a "discounted" DP problem.

Assume we are given a lake with an initial population of b_1 fish at the beginning of year 1. The population at the beginning of year t is denoted b_t. By selling x_t fish during year t a revenue $r(x_t)$ is earned. The cost of catching these fish is $c(x_t, b_t)$ and depends also on the number of fish in the lake. Fish reproduce, and this is modeled by a constant reproduction rate s — in [64, pp.779–780] it is assumed $s = 1.2$. That is, at the beginning of a year there are 20% more fish in the lake than at the end of the previous year. The finite planning horizon extends through the years $1, \ldots, T$ during which we assume a constant interest rate y. The decision variable x_t denotes the number of fish to be caught and sold in year t. The goal is to maximize the net profit (in year 1 dollars) that can be earned during the years $1, \ldots, T$ within the planning horizon. Typically for this type of decision problem there is a tradeoff of current benefits against future benefits.

A state in this DP model is a pair (t, b) representing the current year t and the fish population b at the beginning of the year. The DP functional equation becomes

$$f(t,b) = \begin{cases} \max\limits_{x_t \in \{0,\ldots,b\}} \{\, r(x_t) - c(x_t, b) \\ \qquad\qquad + \frac{1}{1+y} f(t+1, \lfloor s(b - x_t) \rfloor)\} & \text{if } t \le T \\ 0 & \text{if } t = T+1. \end{cases}$$

and the goal is to compute $f(1, b_1)$.

For instance, let $T = 2, y = 0.05, s = 2$ (for sake of simplicity we use $s = 2$ instead of $s = 1.2$) and set the initial fish population to $b_1 = 10$ (to be interpreted in thousands). For simplicity, let the revenue function be defined linearly as $r(x_t) = 3x_t$ and let the cost function be defined linearly as $c(x_t, b_t) = 2x_t$. Then the maximal net profit (in year 1 dollars) is $f(1, 10) \approx 19.05$ resulting from the decisions $x_1 = 0$ and $x_2 = 20$ (i.e. no fish is harvested in the first year, all the fish is harvested in the second year).

2.10 Edit Distance Problem (EDP)

The edit distance problem (EDP) is called "the most classic inexact matching problem solved by dynamic programming" in [16, p.215]. It is also known as the "string editing problem" [22, pp.284–286]. A variant of this problem is described in [10, pp.364–367].

Let Σ be a finite alphabet. Given two strings $x \in \Sigma^m$ and $y \in \Sigma^n$, say $x = x_1 \cdots x_m$ and $y = y_1 \cdots y_n$. The task is to transform string x into string y by using three elementary types of edit operations.

- A *delete* operation D deletes a single character from a string at a cost of $c(D)$.
- An *insert* operation I inserts a single character from Σ into a string at a cost of $c(I)$.
- A *replacement* (also called *substitution*) operation R replaces a single character of a string with a character from Σ at a cost of $c(R)$. (Usually, replacing with the same character gets assigned a cost of 0.)

The goal is to find an *edit sequence*, a sequence of edit operations that performs the transformation from x to y at a minimal cost. The cost of an edit sequence is defined as the sum of the cost of its edit operations.

The following recursive equation computes the minimal cost $f(i,j)$ for X_i, the length-i prefix substring of x, and Y_j, the length-j prefix substring of y.

$$f(i,j) = \begin{cases} jD & \text{if } i = 0 \\ iI & \text{if } j = 0 \\ \min\{\, f(i-1,j) + c(D), \\ \qquad f(i,j-1) + c(I), \\ \qquad f(i-1,j-1) + c(R)\} & \text{if } i > 0 \text{ and } j > 0 \end{cases}$$

where the cost function c might be defined as

$$\begin{aligned} c(D) &= c_D && \text{for deleting any character, at any position} \\ c(I) &= c_I && \text{for inserting any character, at any position} \\ c(R) &= \begin{cases} 0 & \text{if } x_i = y_j \text{ (matching characters)} \\ c_R & \text{if } x_i \neq y_j \text{ (a true replacement)} \end{cases} \end{aligned}$$

taking care of the fact that the replacement cost for matching characters should be 0. A generalization of the EDP called "alphabet-weight edit distance" [16, 42] makes the cost of an edit operation dependent upon which character from the alphabet gets deleted, inserted or replaced.

The recursive relationship can be restated as a DP functional equation solving the EDP as follows.

$$f(X_i, Y_j) = \begin{cases} jD & \text{if } i = 0 \\ iI & \text{if } j = 0 \\ \min_{d \in \{D,I,R\}} \{f(t(X_i, Y_j, d)) + c(d)\} & \text{if } i > 0 \text{ and } j > 0. \end{cases} \tag{2.12}$$

where the transformation function is defined by

$$\begin{aligned} t(X_i, Y_j, D) &= (X_{i-1}, Y_j) \\ t(X_i, Y_j, I) &= (X_i, Y_{j-1}) \\ t(X_i, Y_j, R) &= (X_{i-1}, Y_{j-1}). \end{aligned}$$

The goal is to compute $f(x, y)$, the cost of a minimal cost edit sequence.

Consider the problem instance $x =$ "CAN" and $y =$ "ANN" with insertion cost $c_I = 1$, deletion cost $c_D = 1$ and replacement cost $c_R = 1$ from [16, p.223]. There are several minimal cost edit sequences with cost $f(x, y) = 2$, as can be seen by considering the derivations

- CAN $\vdash_R$ CNN $\vdash_R$ ANN,
- CAN $\vdash_D$ AN $\vdash_I$ ANN,
- CAN $\vdash_I$ CANN $\vdash_D$ ANN.

This problem is closely related to the longest common subsequence (LCS) problem of section 2.23.

2.11 Fibonacci Recurrence Relation (FIB)

A *recurrence relation* of the form $f(i) = F(f(1), \ldots, f(i-1))$ can be viewed as a DPFE

$$f(i) = \text{opt}_d\{F(f(1), \ldots, f(i-1))\},$$

where F is independent of a decision d. For example, consider the recurrence relation for the Fibonacci sequence, $f(i) = f(i-1) + f(i-2)$, for $i \geq 3$. Its solution, the N-th Fibonacci number $f(N)$ can be found by solving the DPFE

$$f(i) = \text{opt}_{d \in S}\{f(i-1) + f(i-2)\}, \tag{2.13}$$

where S is a set having a singleton "dummy" value. The goal is $f(N)$, and the base cases are $f(1) = f(2) = 1$. For $N = 7$, $f(7) = 13$.

2.12 Flowshop Problem (FLOWSHOP)

The flowshop problem (see [22, pp.301–306] and [4, pp.142–145]) is a process scheduling problem where each process i has two tasks, A and B, that must be performed in order, A before B. The tasks are performed on separate processors, where the sequence chosen for the processes must also be the sequence in which the individual tasks are performed. For example, assume the execution times for the A tasks $p_i = \{3, 4, 8, 10\}$, and the execution times for the B tasks $q_i = \{6, 2, 9, 15\}$, where $i = \{0, 1, 2, 3\}$. If the processes are executed in the order given, then the start and completion times for the tasks would be as shown in the time chart provided in Table 2.1.

Table 2.1. Time Chart for FLOWSHOP Instance

processor 1	A_1: 0–3	A_2: 3–7	A_3: 7–15	A_4: 15–25	
processor 2		B_1: 3–9	B_2: 9–11	B_3: 15–24	B_4: 25–40

The two rows in Table 2.1 indicate, as a function of time, the tasks that execute and their start and completion times. We emphasize that B_3 was delayed from time 11 when B_2 completed because A_3 completes at time 15. The overall cost for a schedule is the time at which the last task completes, which for the above example is 40. The flowshop problem is to find the schedule that has minimum overall cost. We observe that the overall cost is the sum of three times:

1. the execution times of all of the A tasks, denoted c_1;
2. the execution time of the final B task that is executed, denoted c_2;
3. any delay incurred for this final B task, denoted c_3.

In the foregoing example, $c_1 = 25$, $c_2 = 15$, and $c_3 = 0$, and their sum is 40.

To solve this problem by DP, we must define separable costs for each decision. We do so by adopting a virtual stage approach. If task d is chosen next, we define the cost of this decision as the execution time of its A task p_d; the execution time of its B task q_d will be deferred by adding it to a virtual-stage variable k. Specifically, we define the state as (k, S), where the virtual-stage k is the time that would elapse between completion of the last A and B tasks that have already been scheduled, and S is the set of processes that have not yet been chosen. This elapsed time equals q_d if there is no delay, i.e., if $k < p_d$. A decision d is a member of S. The cost of choosing process d in state (k, S) is p_d. The delay associated with decision d is $\max(k - p_d, 0)$. Thus, the next-state is $(\max(k - p_d, 0) + q_d, S - \{d\})$. We conclude that the problem can be solved using the DPFE

$$f(k, S) = \min_{d \in S}\{p_d + f(\max(k - p_d, 0) + q_d, S - \{d\})\}. \tag{2.14}$$

The goal is to find $f(0, S^*)$, where S^* is the originally given set of N processes. The base-condition is $f(k, S) = k$ when $S = \emptyset$. For the above example, $f(0, S^*) = 38$ for the optimal sequence of decisions $d_1 = 0, d_2 = 2, d_3 = 3, d_4 = 1$.

2.13 Tower of Hanoi Problem (HANOI)

The *Tower of Hanoi* problem (discussed in Sect. 1.1.16) is that of moving a tower of N discs, of increasing size from top to bottom, from one peg to another peg using a third peg as an intermediary, where on any peg the discs must remain of increasing size. The number of required basic moves (of a single disc), $f(N)$, is known to satisfy the recurrence relation $f(i) = 2f(i-1)+1$, for $i \geq 2$, where $f(i)$ is the number of basic moves required to move i discs. Thus, in a fashion analogous to the Fibonacci problem (FIB, Sect. 2.11), the solution to the Tower of Hanoi problem can be found by solving the DPFE

$$f(i) = opt_{d \in S}\{2f(i-1)+1\}, \tag{2.15}$$

where S is a set having a singleton dummy value. The goal is $f(N)$, and the base case is $f(1) = 1$. For $N = 3$, $f(3) = 7$.

The recurrence relation (2.15) gives the number of moves required to move the given N discs according to the Tower of Hanoi rules. The actual sequence of moves can also be found using a nonoptimization DP formulation, as discussed in Sect. 1.1.16. The DPFE (1.50) given there, whose solution is a sequence of moves, uses concatenation rather than addition. Therefore, it cannot be solved directly by a *numerical* DP solver. However, the DPFE can be modified as follows:

$$\begin{aligned} f(m,i,j,k) = \mathrm{opt}_{d \in S}\{&f(m-1,i,k,j) + f(1,i,j,k) \\ &+ f(m-1,k,j,i)\}, \end{aligned} \tag{2.16}$$

where S is a set having a singleton dummy value. In this additive DPFE, $f(m,i,j,k)$ is the total number of basic moves required to move a tower of m discs from i to j. We note that the preceding DPFE (2.15) can be derived from this latter DPFE (2.16).

The DPFE (2.16) is based upon the observation that, to move m discs from peg i to peg j with peg k as an intermediary, we may move $m-1$ discs from i to k with j as an intermediary, then move the last disc from i to j, and finally move the $m-1$ discs on k to j with i as an intermediary. The goal is $f(N,1,2,3)$, and the base condition is $f(m,i,j,k) = 1$ when $m = 1$. The cost of each dummy decision is zero, but each of the basic moves, associated with the base cases, contributes a cost of one to the overall total. For $N = 3$, $f(3,1,2,3) = 7$.

Since the basic moves (of a single disc) correspond to the base-states $f(1,i,j,k)$, it is possible to deduce the optimal sequence of these moves by examining the base-states that are reached. (This reconstruction process is not simple, and will be omitted here.)

2.14 Integer Linear Programming (ILP)

A general problem statement of Integer Linear Programming (ILP) can be found e.g. in [21, 49]. Here, we assume in addition that all entries of vectors c and b and the matrix A are nonnegative integers, and we consider a maximization problem with '$\leq$' constraints:

$$\begin{aligned} & \max c^T x \\ & \text{s.t.} \quad Ax \leq b \\ & x_1, \ldots, x_n \in \mathbf{N} \cup \{0\} \end{aligned}$$

Such an ILP problem can be solved with DP in various ways. Two approaches are given here. For the first formulation, let a state be a set S of index-value pairs. Each such pair represents the assignment (already made in previous decisions) of an x_j to a particular value [37, 40]. A decision is made at stage j by assigning a value from a discrete set of feasible decisions D to x_{j+1}. The DP functional equation is

$$f(j, S) = \begin{cases} \max\limits_{x_{j+1} \in D} \{c_{j+1} x_{j+1} + f(j+1, S \cup \{(j+1, x_{j+1})\})\} & \text{if } j < n \\ 0 & \text{if } j = n. \end{cases} \tag{2.17}$$

and the goal becomes to compute $f(0, \emptyset)$.

The following second formulation of the ILP as a DP problem has the advantage that there is a chance of overlapping subproblems to occur. Let a state be a pair $(j, (y_1, \ldots, y_m))$ of the index and an m-tuple representing the slack y_i for each of the constraints at the current state. As before, a decision is made at stage j by assigning a value from a discrete set of feasible decisions D to x_{j+1}. The DP functional equation is

$$\begin{aligned} & f(j, (y_1, \ldots, y_m)) \\ & = \begin{cases} \max\limits_{x_{j+1} \in D} \{c_{j+1} x_{j+1} \\ \quad + f(j+1, (y_1 - a_{1,j+1} x_{j+1}, \ldots, y_m - a_{m,j+1} x_{j+1}))\} & \text{if } j < n \\ 0 & \text{if } j = n. \end{cases} \end{aligned} \tag{2.18}$$

and the goal becomes to compute $f(0, (b_1, \ldots, b_m))$. The advantage of this formulation is that different sequences of previous decisions might lead to a common state with identical slack across all constraints. In this fortunate case of an overlapping subproblem, we save the recomputation of that state.

For both DP formulations the set of feasible decisions D depends on the current state and it is calculated by

$$D = D(j,(y_1,\ldots,y_m)) = \{0,\ldots,\min\{\lfloor\frac{y_1}{a_{1,j}}\rfloor,\ldots,\lfloor\frac{y_m}{a_{m,j}}\rfloor\}\}$$

where a term $\frac{y_i}{0}$ (with a zero in the denominator) should be interpreted as ∞, since in that case there is no upper bound for that particular variable-constraint combination. (Note that for the first DP formulation the current slack values y_i can be computed from the assignment of values to the decision variables made so far.) The assumption that the entries of A be nonnegative is crucial and allows us to bound the set D from above. Otherwise, it does not work; e.g. a constraint like $5x_1 - 3x_2 \leq 35$ does *not* imply a decision set of $\{0,\ldots,7\}$ for the x_1 variable. Note that after assuming $A \geq 0$ we must also assume $b \geq 0$ for the ILP problem to have feasible solutions.

Consider the following problem instance from [37, 40]. Let $c = (3,5)$, $b = (4,12,18)$, and

$$A = \begin{pmatrix} 1 & 0 \\ 0 & 2 \\ 3 & 2 \end{pmatrix}.$$

Then the optimal solution is $(x_1,x_2) = (2,6)$ with maximal function value $f(0,4,12,18) = 36$.

2.15 Integer Knapsack as ILP Problem (ILPKNAP)

The integer knapsack problem can also be formulated as an ILP (Sect. 2.14). For instance, let the capacity be 22, and let there be $n = 3$ classes of objects, (A,B,C), with values $(v_0,v_1,v_2) = (15,25,24)$ and weights $(w_0,w_1,w_2) = (10,18,15)$. This problem instance can be modeled using the objective function coefficients $c = (15,25,24)$, the right hand side constraint vector $b = (22,1,1,1)$, and the constraint matrix

$$A = \begin{pmatrix} 10 & 18 & 15 \\ 1 & 0 & 0 \\ 0 & 1 & 0 \\ 0 & 0 & 1 \end{pmatrix}$$

So we may use the ILP formulation to solve the knapsack problem. It is optimal to pick object B once, and the optimal value of the knapsack is $f(0,22,1,1,1) = 25$.

2.16 Interval Scheduling Problem (INTVL)

The interval scheduling problem [30, pp.116–121] (also known as the activity selection problem [10, pp.371–378]) is that of choosing the optimal subset P^* of a set $P = \{0,\ldots,N-1\}$ of N processes (or activities) to be executed on a single processor, each process i having a specified interval (s_i, t_i), consisting of a start time s_i and a termination time t_i, during which process i is eligible to execute, and also having a weight w_i that is the profit gained if this eligible process is scheduled (selected) for processing. Since there is a single processor, the intervals of the selected processes cannot overlap. This problem can be solved using the nonserial DPFE

$$f(p,q) = \max_{d \in P}\{f(p, s_d) + c(d|p,q) + f(t_d, q)\}, \tag{2.19}$$

where $c(d|p,q) = w_d$ if d is "eligible", i.e. $p \leq s_d$ and $t_d \leq q$; else $c(d|p,q) = 0$. The goal is to solve for $f(0,T)$ given the base cases $f(p,q) = 0$ for $p \geq q$, where $T \geq \max_i\{t_i\}$.

Assume we are given a set of processes $P = \{0,1,2,3,4,5\}$ with start times $(s_0,\ldots,s_5) = (9,8,3,5,2,1)$, termination times $(t_0,\ldots,t_5) = (12,11,10,7,6,4)$, and weights $(w_0,\ldots,w_5) = (1,2,7,4,4,2)$. For $T = 20$, the goal is $f(0,20) = 8$, where the optimal decisions are to select processes 1, 3, and 5 to be included in P^*. We note that the initial decision can be to choose any one of these three processes; the other two processes would be chosen in subsequent decisions.

For a given state (p,q), each process d in P is eligible for inclusion in P^* if its start and termination times are within the interval (p,q). As an alternative to defining the state as an interval and test d in P for eligibility, we may instead incorporate in the definition of the state the set $S \subset P$ of eligible processes in the interval. This yields the nonserial ("INTVL2") DPFE

$$f(S,p,q) = \max_{d \in S}\{f(S_L, p, s_d) + c(d|p,q) + f(S_R, t_d, q)\}, \tag{2.20}$$

where S_L and S_R are the subsets of S consisting of the eligible processes in the respective left and right subintervals. The goal is to solve for $f(P,0,T)$ for $T \geq \max_i\{t_i\}$, given the base cases $f(S,p,q) = 0$ when $p \geq q$ or when $S = \emptyset$.

For the preceding problem, the goal is $f(\{0,1,2,3,4,5\},0,20) = 8$, where the optimal decisions again are to select processes 1, 3, and 5 to be included in P^*. For the initial decision d to choose process 1 with start time 8 and termination time 11, the left next-state is $(\{3,4,5\},0,8)$, since processes 3, 4, and 5 terminate before time $s_d = 8$, and the right next-state is $(\emptyset,11,20)$, since no processes start after time $t_d = 11$. If in state $(\{3,4,5\},0,8)$, the next decision is to choose process 3 with start time 5 and termination time 7, and the next-states are $(\{5\},0,5)$ and $(\emptyset,7,8)$.

As previously noted, processes 1, 3, and 5 are all included in the final set P^* of scheduled processes, but they can be selected in any order. Therefore, if j and k are both in P^*, with $j < k$, we may arbitrarily choose to always select process k first. If we do so, then although the foregoing DPFE is nonserial, the processes that are selected for inclusion in P^* can be made serially in decreasing order of their index positions in the set P.

Suppose we consider the processes in the increasing order 0,1,2,3,4,5. For each process i, we make a boolean decision $d_i = 1$ if we include i in P^*, or

$d_i = 0$ if not. Then the optimal sequence of decisions is $d_0 = 0, d_1 = 1, d_2 = 0, d_3 = 1, d_4 = 0, d_5 = 1$, i.e., processes 1, 3, and 5 are selected, resulting in a total cost of $w_1 + w_3 + w_5 = 2 + 4 + 2 = 8$.

One alternate formulation assumes that the set P is sorted in increasing order of termination time and decisions are made in reverse of this ordering. If the decision is made to include process i in P^*, then processes that terminate after s_i are ineligible and may be disregarded. It is thus useful to precalculate the set $E_i = \{j \mid t_j \leq s_i\}$ of eligible processes that terminate before process i and let $\pi(i) = \max(E_i)$, which is equal to zero if E_i is empty.

Assume we are given a set of processes $P = \{0, 1, 2, 3, 4, 5\}$ with start times $(s_0, \ldots, s_5) = (1, 2, 5, 3, 8, 9)$, termination times $(t_0, \ldots, t_5) = (4, 6, 7, 10, 11, 12)$, and weights $(w_0, \ldots, w_5) = (2, 4, 4, 7, 2, 1)$. (The processes are the same as given in the prior example.) Since P is sorted by termination times, we may use $\pi(i)$ to eliminate many ineligible decisions. For this example, $\pi = \{0, 0, 1, 0, 3, 3\}$.

The interval scheduling problem can then be solved using the serial ("INTVL1") DPFE

$$f(k) = \max\{w_{k-1} + f(\pi(k-1)), f(k-1)\}, \tag{2.21}$$

where $\pi(i)$ is as defined above. In this DPFE, at each stage k, process k is considered, and the decision whether to schedule (select) the process or not is Boolean; the cost of a decision equals the process weight if a process is selected, or equals zero if not. (Note the subscripts used for w and π are $k-1$ since we assume these arrays are indexed from zero.) The goal is to solve for $f(N)$ given the base case $f(0) = 0$.

Solving the DPFE, we have $f(6) = f(5) = 8, f(5) = 2 + f(3) = 8, f(3) = 4 + f(1) = 6$, and $f(1) = 2 + f(0) = 2$. The sequence of optimal decisions are $d_1 = 0, d_2 = 1, d_3 = 1, d_4 = 1$, so we conclude that process 5 is not included in P^*, but processes 4, 2, and 0 are.

We note in conclusion that, using transition weights, the foregoing is equivalent to the ("INTVL3") DPFE

$$f(k) = \max_{d \in \{0,1\}} \{d \cdot (w_{k-1} + f(\pi(k-1))) + (1-d) \cdot f(k-1)\}, \tag{2.22}$$

The numerical solution is the same as before.

2.17 Inventory Problem (INVENT)

In an *inventory* problem, there is a product that can be acquired (either produced or purchased) at some specified cost per unit, and that is consumed based upon demands at specified times. There is also an inventory "holding" cost for storing products that are not consumed, and a penalty cost in case the demand cannot be satisfied. We may formulate such an inventory problem as an N-stage sequential decision process, where at each stage k a decision must be made to acquire x units at an acquisition cost $C(k,x)$, that may depend on the stage k and on the number of units acquired x. The state is (k,s), where k is the stage number and s is the size of the inventory, i.e. how many units of the product are available at the start of the stage. The demand $D(k)$ generally depends on the stage. If the decision in state (k,s) is to acquire x units, the next-state (k',s') is $(k+1, s+x-D(k))$. The cost associated with this state transition is the acquisition cost $C(k,x)$ plus the inventory holding cost $I(k,s,x)$ for $s>0$. $C(k,x)$ often includes an overhead or setup cost if $x>0$, in addition to a per-unit cost. $I(k,s,x)$ for $s<0$ represents a penalty for not satisfying a demand of size s. Restrictions on capacity CAP (how many units may be acquired or produced in each stage) and an inventory limit LIM (how large the inventory may be in each stage) may also be imposed.

For some inventory problems, such as probabilistic production problems as discussed in Sect. 2.34, demand is assumed to be random. In this section, we consider inventory problems where it is assumed that the future sequence of demands $(D_0,\ldots,D_{N-1})$ is fixed and known in advance. The DPFE for such a problem is

$$f(k,s) = \min_{x}\{C(k,x) + I(k,s,x) + f(k+1, s+x-D(k))\}.$$

with base condition $f(k,s)=0$ when $k=N$. We may combine C and I to obtain a reward function $R(k,s,x) = C(k,x) + I(k,s,x)$. The goal is to find $f(0,s_0)$, where s_0 is the initial inventory.

If we also assume that enough must be acquired to always meet demand, or equivalently that the penalty for not satisfying the demand (resulting in a negative inventory) is infinite, then in state (k,s) the amount chosen to be acquired x must be such that the inventory to be held $s' = s+x-D(k)$ is nonnegative. Thus, at the k-th stage, we require that $x \geq D(k)-s$. This constraint may be incorporated into how the decision space is defined. The DPFE for such an inventory problem is

$$f(k,s) = \min_{x\geq D(k)-s}\{C(k,x) + I(k,s,x) + f(k+1, s+x-D(k))\}.$$

An instance of such an inventory problem is described in [64, pp.758–763], where it is assumed that $N=4, C(k,x)=3+x$ for $x>0$, $C(k,0)=0, CAP=5, I(k,s,x)=(s+x-D(k))/2$ for $s\geq 0, LIM=4, D(0)=1, D(1)=3, D(2)=2, D(3)=4$, and $s_0=0$. The goal is to find $f(0,s_0)$ with base condition $f(N,s)=0$. The solution is $f(0,0)=20$. The optimal decisions are $x_1=1, x_2=5, x_3=0$, and $x_4=4$.

2.18 Optimal Investment Problem (INVEST)

There are many examples of investment or "gambling" problems (such as INVESTWLV, Sect. 2.19), where at each state a decision is made to invest a certain amount from one's "bankroll", and whether the next-state is a gain or loss is probabilistically determined. Such investment problems may be treated as production problems (such as PROD, Sect. 2.34), where the inventory s is one's bankroll, and the next-state — say, $s' = s + d$ or $s' = s - d$ for a gain or loss, respectively — is probabilistically determined. Of course, the gains can reflect favorable odds, e.g., $s' = s + 2d$, and the losses can be modified, e.g., $s' = s - d + 1$. Furthermore, the gain and loss probabilities may be functions of the size of the investment i. Formulated as a production problem, the production and inventory costs would be zero.

For one such investment problem [8, problem 19.29], the DPFE is

$$f(k, s) = \max_i \{p_i f(k+1, s+i) + (1-p_i) f(k+1, s-i+1)\}. \quad (2.23)$$

Given an initial amount of s_0, the goal is to compute $f(1, s_0)$ with base-condition $f(k, s) = s$ when $k = N$.

For instance, assume gain probabilities p_i of 1, 0.2, 0.4 for a decision $i = 0$, 1, and 2, respectively. Then for $s_0 = 2$ and $N = 4$, we have $f(1, 2) = 2.6$ where the optimal *initial* decision is to invest 1 unit.

2.19 Investment: Winning in Las Vegas Problem (INVESTWLV)

The winning in Las Vegas problem as described in [20, p.423] is a probabilistic DP problem (see Sect. 1.1.15). Each round of a game involves betting x_n chips and then either, with a probability of p, winning an additional x_n chips, or, with a probability of $1 - p$, losing the bet. Given the time to play R rounds, starting with s_1 chips, and a target amount of t chips the objective is to find a policy that maximizes the probability of holding at least t chips after R rounds of betting. The decision variable x_n represents the number of chips to bet in round n where $n \in \{1, \ldots, R\}$.

The DP model for this problem defines a state as a pair (n, s_n) consisting of the round number (i.e. the stage) and the number of chips in hand s_n to begin round n of play. The DP functional equation is

$$f(n, s_n) = \begin{cases} \max\limits_{x_n \in \{0,\ldots,s_n\}} \{ (1-p) f(n+1, s_n - x_n) \\ \qquad\qquad + p f(n+1, s_n + x_n)\} & \text{if } n \le R \\ 0 & \text{if } n > R \text{ and } s_n < t \\ 1 & \text{if } n > R \text{ and } s_n \ge t \end{cases}$$

and the goal of the computation is $f(1, s_1)$.

Consider the following problem instance from [20, p.423]. Let $p = 2/3, R = 3, s_1 = 3$ and $t = 5$. Then the optimal policy has a probability of $f(1, 3) = 20/27 \approx 0.7407$ for reaching the target of 5 chips within 3 rounds when starting with 3 chips. The optimal policy is to bet $x_1 = 1$ chip in round 1. Then

- if round 1 is won, bet $x_2 = 1$ in round 2, then
  - if round 2 is won, bet $x_3 = 0$ in round 3
  - if round 2 is lost, bet $x_3 = 2$ or $x_3 = 3$ in round 3
- if round 1 is lost, bet $x_2 = 1$ or $x_2 = 2$ in round 2, then
  - if round 2 is won,
    - bet $x_3 = 2$ or 3 (for $x_2 = 1$)
    - bet $x_3 = 1, 2, 3$ or 4 (for $x_2 = 2$)
  - if round 2 is lost, the target cannot be reached any more.

2.20 0/1 Knapsack Problem (KS01)

Given a knapsack capacity $c \in \mathbf{N}$ and n objects $A = \{a_0, \ldots, a_{n-1}\}$, each having a nonnegative value $v(a_i) \in \{x \in \mathbf{R} : x \geq 0\}$ and a weight $w(a_i) \in \mathbf{N}$, the 0/1 knapsack problem (KS01) asks to pack the knapsack with a subset of objects from A, such that the sum of the weight of the chosen objects does not exceed the capacity c, and such that the sum of the values of the chosen objects is maximal. Let x_i denote the binary decision variable of whether to include (i.e. $x_i = 1$) object a_i, or not (i.e. $x_i = 0$). For brevity, we write v_i for $v(a_i)$ and w_i for $w(a_i)$. KS01 can also be formally stated as an integer LP problem:

$$\begin{array}{ll} \max & z = \sum_{i=0}^{n-1} v_i x_i \\ \text{s.t.} & \sum_{i=0}^{n-1} w_i x_i \leq c \\ & x_i \in \{0,1\} \quad \forall i \in \{0, \ldots, n-1\}. \end{array}$$

The KS01 problem can be interpreted as a multistage decision problem in a straightforward manner. For each object a_i, a decision x_i has to be made whether to include a_i, or not. Let the stage number i denote the index of the object currently under consideration and let w denote the space currently available in the knapsack. The DP functional equation is

$$f(i, w) = \begin{cases} 0 & \text{if } i = -1 \text{ and } 0 \leq w \leq c \\ -\infty & \text{if } i = -1 \text{ and } w < 0 \\ \max\limits_{x_i \in \{0,1\}} \{x_i v_i + f(i-1, w - x_i w_i)\} & \text{if } i \geq 0. \end{cases} \tag{2.24}$$

The goal is to compute $f(n-1, c)$. The above DP functional equation can be slightly improved by introducing a decision set

$$D = D(i, w) = \begin{cases} \{0\} & \text{if } w_i > w \\ \{0, 1\} & \text{if } w_i \leq w. \end{cases}$$

that disallows infeasible decisions instead of punishing illegal states with a $-\infty$ assignment. Replacing the static decision set $\{0, 1\}$ used in (2.24) we get

$$f(i,w) = \begin{cases} 0 & \text{if } i = -1 \text{ and } 0 \le w \le c \\ \max_{x_i \in D}\{x_i v_i + f(i-1, w - x_i w_i)\} & \text{if } i \ge 0. \end{cases} \quad (2.25)$$

For instance, let $c = 22$, $n = 3$, $A = (a_0, a_1, a_2)$, $(v_0, v_1, v_2) = (25, 24, 15)$, and $(w_0, w_1, w_2) = (18, 15, 10)$. Then it is optimal to pick only object a_0, in the final ($i = 0$) stage, and the optimal value of the knapsack is $f(2, 22) = 25$.

The integer knapsack problem (KSINT, see Sect. 2.22), a variation of KS01 that allows integral decision variables x_i, can be reduced to KS01, and vice versa.

2.21 COV as KSINT Problem (KSCOV)

Covering problems, such as COV (see Sect. 2.7), can also be formulated as an integer knapsack problem, and solved by a variation of KSINT (see Sect. 2.22) where the weights are all equal to 1. The objective is to include in the knapsack the cover size chosen for each of a set of n different object (shrub) sizes, numbered consecutively from 1 to n, where there may be only M different cover sizes, each size having a different weight (or cost). For example, given $n = 10$, we may choose 3 covers of size 10, which covers objects of sizes 8,9, and 10, then choose 2 covers of size 7, which covers objects of sizes 6 and 7, and, assuming only $M = 3$ cover sizes are allowed, finally choose 5 covers necessarily of size 5, which covers the remaining objects of sizes 1 to 5. If the weights of the $n = 10$ cover sizes are $(1, 4, 5, 7, 8, 12, 13, 18, 19, 21)$, in increasing order of size, then the decisions to include 3 covers of size 10, 2 covers of size 7, and 5 covers of size 5 in the knapsack have a total weight of $63 + 26 + 40 = 129$.

Let the state be defined as (k, s), where k is a stage number and s is the largest remaining object to cover. The decision d must be a number between 1 and s. We have $d = 1$ if only one object size is to be covered by the choice d, and $d = s$ if all of the remaining objects are to be covered by the choice d. The cost of this decision is $d \cdot \text{weight}_s$. The next-state is $(k + 1, s - d)$. The DPFE for this problem is

$$f(k,s) = \min_{d \in \{1,\ldots,s\}} \{d \cdot \text{weight}_s + f(k+1, s-d)\}. \quad (2.26)$$

The goal is $f(1, n)$. The base-conditions are $f(k, s) = 0$ when $s = 0$, and $f(k, s) = \infty$ when $k > M$ and $s > 0$. For the above example, $f(1, 10) = 129$ for the optimal sequence of decisions $d_1 = 3$, $d_2 = 2$, and $d_3 = 5$.

2.22 Integer Knapsack Problem (KSINT)

The integer knapsack problem (KSINT) is a generalization of the 0-1 knapsack problem (KS01, see Sect. 2.20) where multiple objects of the same "class" (weight and cost) can be included in the knapsack. The problem can be solved by a slight variation of KS01. Specifically, instead of defining the decision space $D(S)$ as the set $\{0,1\}$, we would define $D(S)$ as the set $\{0,1,\ldots,m\}$ where m can be a fixed constant, or m can be a function of the state S, or of the capacity c, or both. For example, $m_k = \lfloor c/w_k \rfloor$ constitutes an upper bound on the number of objects of class k that can be included.

Assume there are n classes numbered 0, 1, ..., $n-1$. If the objects are considered in a fixed order, in stages according to the class number of the objects, and the decision at each stage k is that of choosing how many of each object of class k to include in the knapsack, then the associated DPFE is

$$f(k,w) = \max_{x_k \in \{0,\ldots,m_k\}} \{x_k v_k + f(k-1, w - x_k w_k)\}, \tag{2.27}$$

where k is the stage number (starting at $k = n-1$ and ending at $k = 0$), w is remaining capacity of the knapsack at stage k, x_k is the number of objects of class k to include in the knapsack, $x_k v_k$ is the value (or profit) for including these objects, and $w - x_k w_k$ is the next remaining capacity; in general, nonnegative integer x_k may be bounded by m_k. The base cases are $f(k,w) = 0$ for $k = -1$ or $w < 0$. Our goal is to solve for $f(n-1,c)$, where c is the capacity of the knapsack.

For example, let capacity $c = 22$, $n = 3$, values $(v_0, v_1, v_2) = (24, 15, 25)$ and weights $(w_0, w_1, w_2) = (15, 10, 18)$. Then $f(2,22) = 30$ for decisions $x_2 = 0, x_1 = 2, x_0 = 0$.

2.23 Longest Common Subsequence (LCS)

The longest common subsequence (LCS) problem is discussed in [10, pp.350–356] and in [16, p.227], where it is treated as a special case of a weighted optimal alignment or string similarity problem.

Definition 2.1 (subsequence). *Given a sequence $x = (x_1, x_2, \ldots)$ and a strictly monotonically increasing indexing sequence $i = (i_1, i_2, \ldots)$ of natural numbers. Then the sequence $(x_{i_1}, x_{i_2}, \ldots)$ is called a* subsequence *of x.*

Note that a subsequence is not the same as a substring, where it is required that the characters must be contiguous.

Given two finite sequences $x = (x_1, \ldots, x_m)$ and $y = (y_1, \ldots, y_n)$ the goal is to find a subsequence $z = (z_1, \ldots, z_l)$ of both x and y (called *common subsequence*) that is of maximum length.

Let X_i denote the length-i prefix substring of x and let Y_j denote the length-j prefix substring of y. In [10, p.352] the following "recursive formula" is given to solve the problem, where $f(i,j)$ is the length of an LCS of X_i and Y_j.

$$f(i,j) = \begin{cases} 0 & \text{if } i = 0 \text{ or } j = 0 \\ f(i-1, j-1) + 1 & \text{if } i \text{ and } j > 0 \text{ and } x_i = y_j \\ \max\{f(i, j-1), f(i-1, j)\} & \text{if } i \text{ and } j > 0 \text{ and } x_i \neq y_j. \end{cases} \quad (2.28)$$

When interpreting equation (2.28) as a DP functional equation two shortcomings can be seen. First, the way it is formulated, it is not clear at all what the decision space is, and how decisions are made at each stage. Second, it is somehow hidden implicitly how X_i and Y_j are part of the current state.

A more precise formulation requires a clear definition of a state and the introduction of a decision space. A state in the LCS problem is actually the pair (X_i, Y_j). The short notation of (i, j) declaring a pair of indices as the state as used in equation (2.28) implicitly assumes the global presence of the referenced input strings x and y. This becomes obvious when we define the current decision set D as a function of the current state and when we define the transformation function and the reward function as a function of the current state and the decision d taken. Observe that the decision space $\mathcal{D} = \{d_1, d_2, d_{12}\}$ consists of the following 3 decisions:

1. d_1 = chop last character from X_i only
2. d_2 = chop last character from Y_j only
3. d_{12} = chop last character from both X_i and Y_j

Depending on the current state (X_i, Y_j) the *current* decision set D is defined as

$$D(X_i, Y_j) = \begin{cases} \{d_1, d_2\} & \text{if last char. of } (X_i, Y_j) \text{ don't match } (x_i \neq y_j) \\ \{d_{12}\} & \text{if last characters of } (X_i, Y_j) \text{ match } (x_i = y_j). \end{cases}$$

The transformation function is defined by

$$\begin{aligned} t(X_i, Y_j, d_1) &= (X_{i-1}, Y_j) \\ t(X_i, Y_j, d_2) &= (X_i, Y_{j-1}) \\ t(X_i, Y_j, d_{12}) &= (X_{i-1}, Y_{j-1}). \end{aligned}$$

The reward function is defined by

$$\begin{aligned} r(X_i, Y_j, d_1) &= 0 \\ r(X_i, Y_j, d_2) &= 0 \\ r(X_i, Y_j, d_{12}) &= 1. \end{aligned}$$

Now the DP functional equation can be expressed as

$$f(X_i, Y_j) = \begin{cases} 0 & \text{if } X_i = \epsilon \text{ or } Y_j = \epsilon \\ \max_{d \in D(X_i, Y_j)} \{f(t(X_i, Y_j, d)) + r(X_i, Y_j, d)\} & \text{otherwise,} \end{cases} \quad (2.29)$$

where ϵ denotes the empty string.

An alternative DP functional equation is given in [16, pp.227–228]. Using our notation it can be expressed as

$$f(X_i, Y_j) = \begin{cases} 0 & \text{if } X_i = \epsilon \text{ or } Y_j = \epsilon \\ \max\{ f(X_{i-1}, Y_{j-1}) + \delta_{x_i, y_j}, & \\ \qquad f(X_{i-1}, Y_j), f(X_i, Y_{j-1})\} & \text{otherwise,} \end{cases} \tag{2.30}$$

where

$$\delta_{x,y} = \begin{cases} 1 \text{ if } x = y \\ 0 \text{ if } x \neq y \end{cases}$$

denotes Kronecker's delta.

The DP functional equation (2.29) is slightly more efficient than the DP functional equation (2.30) because not all entries of the matrix

$$[f(X_i, Y_j)]_{0 \le i \le m, 0 \le j \le n}$$

are computed. The time complexity is still $O(mn)$ however. In any case, the goal is to compute $f(x, y) = f(X_m, Y_n)$.

Consider the following instance of this problem. Let $x = X_7 = (a, b, c, b, d, a, b)$ and $y = Y_6 = (b, d, c, a, b, a)$. Then the maximum length of a common subsequence is $f(X_7, Y_6) = 4$ and the LCS's are (b, d, a, b), (b, c, a, b) and (b, c, b, a).

2.24 Optimal Linear Search Problem (LINSRC)

The optimal linear search problem is that of determining the optimal permutation of a set of data having specified probabilities so as to minimize the expected linear search time. It was discussed extensively in Sects. 1.1.4 and 1.1.5 of Chap. 1 to illustrate the basic concepts of dynamic programming.

One of many DPFEs to solve this problem is (1.23).

2.25 Lot Size Problem (LOT)

In a variation of inventory or production problems (Sects. 2.17 and 2.34), called the "lot size" problem in [61], rather than allow the amount acquired x to be equal to any integer from $D(k) - s$ up to, say, the sum of all of the remaining demands, $\sum_{j=k}^{N-1} D(j)$, instead x can be restricted to be equal to a partial sum $\sum_{j=k}^{m} D(j)$ for some m. This restriction significantly reduces the size of the decision space, hence also the state space, while still guaranteeing optimality.

An instance of this problem is described in Winston [63, pp.1001–1005], where it is assumed that $N = 5, C(k, x) = 250 + 2x$ for $x > 0$, $C(k, 0) = 0, I(k, s) = s$ for $s \ge 0, D(0) = 220, D(1) = 280, D(2) = 360, D(3) = 140$, and $D(3) = 270$. The goal is to find $f(1)$ with base condition $f(k) = 0$ when $k > N$. The solution is $f(1) = 3680$. The optimal decisions are $x_1 = 0, x_2 = 0, x_3 = 1$, and $x_4 = 0$.

2.26 Longest Simple Path Problem (LSP)

Given a directed weighted graph $G = (V, E)$, a start vertex $s \in V$ and a target vertex $t \in V$. The task is to find a *simple path* (i.e. all vertices in the path are distinct) from s to t having maximal length. The reason the problem asks for a simple path is that there does not exist a maximal length nonsimple path from s to t if a positive length cycle is reachable from s. Let the edges be given in the form of an adjacency matrix $C = (c_{i,j})$; an entry of $-\infty$ indicates that there is no edge. The DP functional equation can be expressed as

$$f(S, v) = \begin{cases} \max\limits_{d \notin S}\{f(S \cup \{d\}, d) + c_{v,d}\} & \text{if } v \neq t \\ 0 & \text{if } v = t \end{cases}$$

where the length of a maximal length path is computed as $f(\{s\}, s)$. In the above DP functional equation, a state is a pair (S, v) where S can be interpreted as the set of vertices already visited and v as the current vertex. A minor improvement is to disallow decisions that lead to a $-\infty$-weighted edge by pruning the decision set accordingly (instead of punishing f with a $-\infty$ value.)

Consider the problem instance from [10, p.343] of finding the longest path from node 0 to node 3 in the graph of figure 2.3 defined by the weighted adjacency matrix

$$C = \begin{pmatrix} -\infty & 1 & -\infty & 1 \\ 1 & -\infty & 1 & -\infty \\ -\infty & 1 & -\infty & 1 \\ 1 & -\infty & 1 & -\infty \end{pmatrix}$$

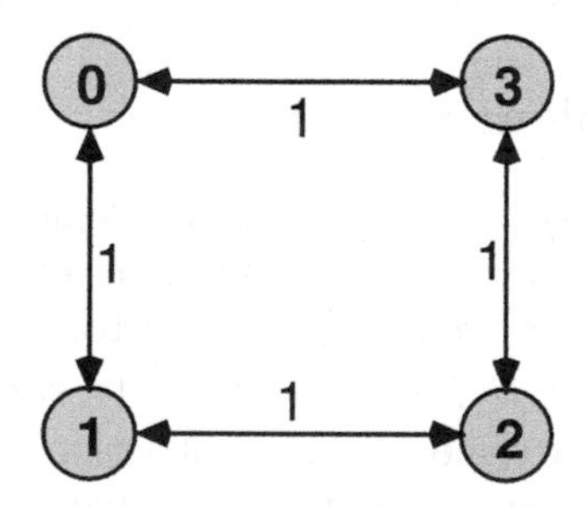

Fig. 2.3. A Longest Simple Path Problem Instance

The longest simple path has length $f(\{0\}, 0) = 3$ and the path is $(0, 1, 2, 3)$.

In [10, p.343], the LSP problem was cited as an example of one that lacked the *optimal substructure* property, hence could not be solved by DP, at least using the same formulation as for SPC. However, we showed above that the

LSP problem can be solved by DP using the "path-state" approach. One of the virtues of DP, as has been shown in numerous other cases (such as for INTVL), is that there may be many different ways to formulate the solution of a problem, and if one formulation is problemmatical there may be other formulations that are not.

2.27 Matrix Chain Multiplication Problem (MCM)

This problem is described in [10, pp.331–338].

Given a product $A_1 A_2 \cdots A_n$ of matrices of various (but still compatible) dimensions, the goal is to find a parenthesization, which minimizes the number of componentwise multiplications. The dimensions of A_i are denoted by d_{i-1} and d_i. The DP functional equation can be expressed as

$$f(i,j) = \begin{cases} \min\limits_{k \in \{i,\ldots,j-1\}} \{f(i,k) + f(k+1,j) + d_{i-1} d_k d_j\} & \text{if } i < j \\ 0 & \text{if } i = j. \end{cases}$$

The total number of componentwise multiplications is computed as $f(1,n)$.

For instance, let $n = 4$ and $(d_0, d_1, d_2, d_3, d_4) = (3, 4, 5, 2, 2)$. Apply the DP functional equation in a bottom up fashion to compute $f(1,4) = 76$. By keeping track which arguments minimize the min-expressions in each case, one arrives at the optimal parenthesization $((A_1(A_2 A_3))A_4)$.

2.28 Minimum Maximum Problem (MINMAX)

Given an acyclic graph with weighted branches, the shortest path problem is to find the path from source s to target t whose (weighted) length is minimal. Each path is a sequence of branches, and its length is the sum of the weights of the branches in the path. The *maximal link* in any path is the maximum of the weights of the branches in the path. The MINMAX problem is that of finding the path from s to t whose maximal link is minimal. To use DP to solve this problem, ordinarily we would construct the MINMAX path by making a sequence of decisions for the branches in the path, and attribute to each of these decisions a separable cost. However, for this problem these cost are not additive (or multiplicative, as for RDP), as is usually the case. Despite this, the problem can still be solved using DP.

One way to solve this MINMAX problem using DP is by adopting a path-state formulation, letting the cost C of each decision be zero, and deferring to the base case for each path the determination of the maximal link in the path. That is, we would use the DPFE

$$f(p_1, \ldots, p_i) = \min_{q \notin \{p_1, \ldots, p_i\}} \{f(p_1, \ldots, p_i, q)\},$$

with base conditions $f(p_1,\ldots,p_i) = \text{maxlink}(p_1,\ldots,p_i)$ when $p_i = t$, where maxlink(P) is the maximum weight among the branches connecting the nodes in the set $P = \{p_1,\ldots,p_i\}$. Rather than letting the cost C of each decision equal 0 and including these costs in the base-condition, we may add a stage number k to the state and define the cost of a decision in path-state S to be zero except in the last stage when the cost is set equal to the maximal link. This, of course, is an inefficient enumerative approach. However, it illustrates the generality and flexibility of DP as an optimization methodology.

Since we are only interested in the maximal link in a path rather than the sequencing of the branches in the path, in the above DPFE, the path-state can be represented as an unordered set of nodes, together with the last node p_i in the path; the latter is needed to restrict the decision set to successors q of p_i. By so doing, we arrive at the DPFE

$$f(k,\{p_1,\ldots,p_i\},p_i) = \min_{q\notin\{p_1,\ldots,p_i\}} \{C(k,\{p_1,\ldots,p_i\},p_i) + f(k+1,\{p_1,\ldots,p_i,q\},q)\}, \qquad (2.31)$$

where $C(k,\{p_1,\ldots,p_i\},p_i) = 0$, as defined above. The base condition is $f(k,\{p_1,\ldots,p_i\},p_i) = \text{maxlink}(p_1,\ldots,p_i)$ when $p_i = t$. The maximal link can be calculated given an unordered set of nodes in a path since, if the graph is topologically sorted, there can be only one sequencing of the nodes in the path.

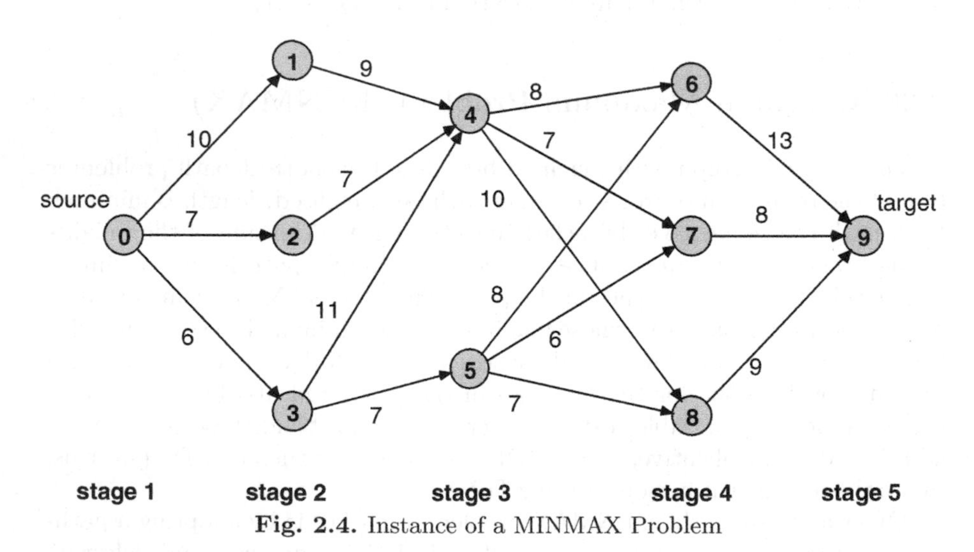

Fig. 2.4. Instance of a MINMAX Problem

For instance, consider the acyclic graph shown in Fig. 2.4, which is from [64, p.786]. Here, the nodes have been renumbered to start from 0. This graph can be represented by the following weighted adjacency matrix.

$$B = \begin{pmatrix} \infty & 10 & 7 & 6 & \infty & \infty & \infty & \infty & \infty & \infty \\ \infty & \infty & \infty & \infty & 9 & \infty & \infty & \infty & \infty & \infty \\ \infty & \infty & \infty & \infty & 7 & \infty & \infty & \infty & \infty & \infty \\ \infty & \infty & \infty & \infty & 11 & 7 & \infty & \infty & \infty & \infty \\ \infty & \infty & \infty & \infty & \infty & \infty & 8 & 7 & 10 & \infty \\ \infty & \infty & \infty & \infty & \infty & \infty & 8 & 6 & 7 & \infty \\ \infty & \infty & \infty & \infty & \infty & \infty & \infty & \infty & \infty & 13 \\ \infty & \infty & \infty & \infty & \infty & \infty & \infty & \infty & \infty & 8 \\ \infty & \infty & \infty & \infty & \infty & \infty & \infty & \infty & \infty & 9 \\ \infty & \infty & \infty & \infty & \infty & \infty & \infty & \infty & \infty & \infty \end{pmatrix}$$

The solution is $f(1, 0, \{0\}) = 8$. One sequence of nodes in an optimal path is $(0, 2, 4, 7, 9)$. Another one is $(0, 3, 5, 7, 9)$.

In conclusion, we note that the MINMAX problem, in principle, can also be solved in a nonenumerative fashion, using a staged DPFE of the form

$$f(k, i) = \min_j \{\max(b(i, j), f(k + 1, j))\},$$

for $f(1, 0)$ with base-condition $f(N - 1, i) = b(i, N)$. However, this uses a nonadditive cost function.

2.29 Minimum Weight Spanning Tree Problem (MWST)

The minimum weighted spanning tree (MWST) problem is to find, among all spanning trees of an undirected graph having numerically weighted branches, the one having minimum weight, i.e. whose sum of branch weights in the tree is minimal. By definition, a subgraph of G is a *spanning tree* T if T spans G (connects all the nodes in G) and is acyclic. A graph G having N nodes must necessarily have at least $N - 1$ branches in order for G to be connected, hence for it to have a spanning tree at all. Spanning trees can always be constructed by choosing one of its branches at a time from G to include in the tree-so-far, starting from the empty state, until exactly $N - 1$ branches have been chosen; a subgraph of G having fewer branches would not span the graph, and one having more branches would be cyclic. Thus, this problem may be regarded as a multistage decision process where exactly $N - 1$ decisions must be made.

It should be noted that although a tree, as a special case of a graph, is most commonly represented by an adjacency matrix, for the MWST application it is more convenient to represent a tree by a vector. If G consists of an ordered set of N branches B, then a spanning tree or a partially constructed "tree-so-far" T of G consists of a subset of B. The *characteristic vector* representation of T has the form $(t_1, t_2, \ldots, t_N)$, where $t_i = 1$ if the i-th branch of B is in T, else $t_i = 0$.

To solve the minimum weighted spanning tree problem by DP, we define the state as (S, k), where S is the set of branches of G that have not been chosen for inclusion in T and k is a stage number that counts how many branches have been chosen for inclusion in T. The complement of S is the tree-so-far, and k is its size. A decision d is a member of S, subject to the "eligibility" constraint that adding d to the tree-so-far does not create a cycle. The cost $C(S,d)$ of choosing branch d either equals the weight of the branch w_d, or equals ∞ if the branch is ineligible. The next-state is $(S - \{d\}, k+1)$. Thus, the DPFE is

$$f(S,k) = \min_{d \in S}\{C(S,d) + f(S - \{d\}, k+1)\}. \tag{2.32}$$

The base-condition is $f(S, N-1) = 0$ for any S. The goal is to find $f(S^*, 0)$ for S^* equal to the set of all of the branches of G.

For example, consider the graph of Fig. 2.5 with branches 0,1,2,3,4 having weights 5,4,3,2,1, respectively, whose weighted adjacency matrix is

$$\begin{pmatrix} \infty & 1 & 3 & \infty \\ 1 & \infty & 2 & 4 \\ 3 & 2 & \infty & 5 \\ \infty & 4 & 5 & \infty \end{pmatrix}$$

Then $f(\{0,1,2,3,4\},0) = 4+2+1 = 7$ for the optimal sequence of decisions $d_1 = 1$, $d_2 = 3$, and $d_3 = 4$.

In conclusion, we note that the eligibility test can be implemented by determining "connected components" (cf. [29]) after each decision. Alternatively, eligibility can be tested by predetermining all cycles. Furthermore, other eligibility constraints, such as on the degree of each node in the spanning tree, as might arise for a networking application, can be accommodated by this DP formulation. (If such generalizations were not a consideration, the optimal decision can be made using a greedy policy, such as that of Kruskal or Prim, as described for example in [29].)

2.30 The Game of NIM (NIM)

In most applications of dynamic programming, our objective is to make an optimal sequence of decisions. In any given state, our immediate objective is to make the optimal current decision, after which some next-state(s) is entered. In many "games", some adversary may make the next decision, which may affect the state in which we make our subsequent decision. Numerous board games, such as checkers and chess, are examples of such games, have been tackled using DP, as mentioned in [3]. A far simpler example is NIM, well known in artificial intelligence literature; for a survey, see [5]. The game of NIM is also described in [63, Chap. 18].

In NIM, the state s is the number of objects, say, matchsticks, that are present. A decision, when it is our turn to play, is that of removing a certain

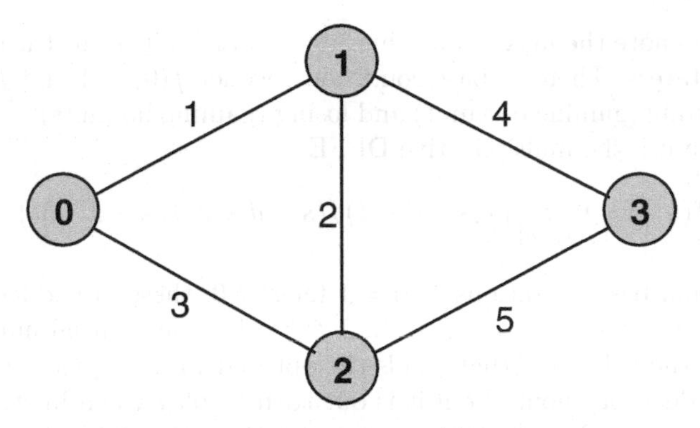

Fig. 2.5. Instance of a MWST Problem

number of matchsticks, from 1 to some limit (say) 3. After our turn, our adversary also has the option of removing 1, 2, or 3 matchsticks, after which we make another decision. The player who removes the last matchstick is considered either the winner or the loser. We assume the latter. Thus, if the state is 2, 3, or 4 when it is our turn to play, the optimal (winning) decision is to remove 1, 2, or 3, resp., leaving our adversary with a state of 1, a losing position. When it is our turn, if $s = 1$, we necessarily must remove the one remaining matchstick, hence lose, gaining no reward; if $s = 0$, we win and gain a reward of one "unit". The optimization problem is that of deciding what our decision should be in an arbitrary state $s > 0$ to maximize our reward.

For $s > 1$, we proceed to formulate a DPFE by first observing that in state s, if our decision is to remove d, then the next-state for our adversary will be $s - d$. The next-state when it is our turn to make a decision again will be $s-d-1$, $s-d-2$, or $s-d-3$, depending upon the decision of our adversary. The decision of our adversary is unknown, but must also be 1, 2 or 3.

One way to proceed is to assume our adversary's decision is made randomly, in which case we may use a probabilistic DP formulation of the form

$$f(s) = \max_{d \in \{1,2,3\}} \{p_1.f(s-d-1) + p_2.f(s-d-2) + p_3.f(s-d-3)\}.$$

(If these decisions are equally likely, the probabilities may be factored out and cancelled, leaving a nonprobabilistic DPFE.) If our adversary plays optimally, we make the worst-case assumption that, whatever our decision, our adversary will act to minimize our reward, leading us to the DPFE

$$f(s) = \max_{d \in \{1,2,3\}} \{\min(f(s-d-1), f(s-d-2), f(s-d-3))\}. \quad (2.33)$$

This formulation is analogous to that for MINMAX (Sect. 2.28), but since the values of f are Boolean (so a MIN operation is equivalent to an AND operation), we use a multiplicative DPFE instead, as we show below.

Let $f(s)$ denote the maximum achievable reward if it is our turn to make a decision in state s. Then, as base conditions, we let $f(0) = 1$ and $f(1) = 0$ to indicate winning (gaining one unit) and losing (gaining no units), respectively. For $s \geq 2$, we use the multiplicative DPFE

$$f(s) = \max_{d \in \{1,2,3\}} \{f(s-d-1)f(s-d-2)f(s-d-3)\} \qquad (2.34)$$

with additional base conditions $f(s) = 1$ for $s < 0$; these are added in lieu of the restriction $d < s$. The goal is to find $f(M)$ for some initial number M of matchsticks; the value of d that yields the optimal value of $f(M)$ corresponds to what our decision should be if it is our turn to play when in state M.

For instance, for $M = 9$, $f(9) = 0$, which means that 9 is a losing position in that if our adversary plays optimally, we cannot win regardless of our next decision. However, for $M = 10$, $f(10) = 1$, which means that 10 is a winning position in that we are guaranteed to win regardless of how our adversary plays provided we play optimally ourselves; the optimal initial decision is $d = 1$, so that our adversary is placed in the losing position 9. Our next decision would depend on that of our adversary, i.e., on whether M is 6, 7, or 8 in our next turn. Similarly, for larger M, say, $M = 29$ and $M = 30$, we have $f(29) = 0$ and $f(30) = 1$.

2.31 Optimal Distribution Problem (ODP)

Distribution problems are related to the "allotment" problem (see Sect. 2.1). For illustrative purposes we use the following simple problem instance from [43]. (This instance is tabular, like ALLOTt.) Suppose we need to borrow \$ 6 million in $t = 0$. Each of our creditors gives us several financing alternatives, as shown in Table 2.2. It is not possible to negotiate other options, e.g. to request intermediate amounts not mentioned in the table. It is not possible to choose two alternatives from the same creditor. Alternative 0 is to not borrow from that creditor. Creditor 0 considers larger credit amounts as more risky and charges more interest. Creditor 1 has excessive amounts of cash to invest and gives us incentives to borrow larger amounts. Creditor 2 only deals with large amounts. The goal is to minimize the net present value (NPV) of our total future payments.

We present the DP approach to this problem for illustrative purposes. It is a 3-stage decision process, since we have to choose exactly one alternative a_i where $i = 0, 1, 2$ from each of the three creditors. Define the state x at stage i as the amount of money secured so far from creditors 0 through $i-1$. Let $y_i(a_i)$ denote the NPV of the future payments to creditor i under alternative a_i. Let $c_i(a_i)$ denote the principal amount of the loan from creditor i under alternative a_i. For $i = 0, 1, 2$ the DP functional equation is

$$f(i, x) = \min_{a_i} \{y_i(a_i) + f(i+1, x + c_i(a_i))\}.$$

Table 2.2. A financing problem

	cash flow in $t = 0$ $c_i(a_i)$	NPV of future payments $y_i(a_i)$
creditor 0		
alternative 0	0	0
alternative 1	1	4
alternative 2	2	12
alternative 3	3	21
creditor 1		
alternative 0	0	0
alternative 1	1	6
alternative 2	2	11
alternative 3	3	16
alternative 4	4	20
creditor 2		
alternative 0	0	0
alternative 1	3	16
alternative 2	4	22

The base case is

$$f(3, x) = \begin{cases} \infty & \text{if } x < 6 \\ 0 & \text{if } x \geq 6. \end{cases}$$

which assigns a penalty of ∞ if our decisions lead to a total principal amount that is less than the required \$ 6 million. An alternative base case

$$f(2, x) = \begin{cases} \infty & \text{if } 0 \leq x \leq 1 \\ 22 & \text{if } x = 2 \\ 16 & \text{if } 3 \leq x \leq 5 \\ 0 & \text{if } 6 \leq x \leq 7. \end{cases}$$

terminates the recursion one stage earlier. As an alternative to having infinite values as penalty terms we could prune the current decision set by applying constraints that prevent total principal amounts of less than \$ 6 million.

The optimal total NPV associated with our problem instance can be computed as $f(0, 0) = 31$, which is associated with the decisions to choose alternative 1 from creditor 0, alternative 2 from creditor 1, and alternative 1 from creditor 2.

2.32 Optimal Permutation Problem (PERM)

This permutation or ordering problem is described in [22, pp.229–232]. It is equivalent to the optimal linear search problem. For the simple problem instance discussed here, we assume that there are 3 programs $\{p_0, p_1, p_2\}$ of length $\{5, 3, 2\}$, respectively, which are to be stored sequentially as files on a single tape. Assume all programs are retrieved equally often and the tape head is positioned at the front of the tape before each retrieval operation. The goal is to minimize the mean retrieval time.

There is a greedy algorithm that easily leads to the optimal storage order. We present the DP approach to this problem. Define a state S as the subset of all programs that still need to be stored, and let $l(x)$ denote the length of program x. One approach leads to the DP functional equation

$$f(S) = \min_{x \in S}\{l(x) \cdot |S| + f(S - \{x\})\}$$

with the base case $f(\emptyset) = 0$.

The optimal total cost associated with our problem instance can be computed as $f(\{p_0, p_1, p_2\}) = 17$, which corresponds to the program order (p_2, p_1, p_0) on the tape.

2.33 Jug-Pouring Problem (POUR)

A "jug pouring" problem [3] can be solved using DP. In the simplest version of this problem, there are two jugs A and B of different capacities P and Q, respectively, and the problem is to obtain in one of the jugs T units (say, ounces). Assume we have a third jug C of capacity $R = P + Q$, which is initially full, whereas the original two jugs are empty. There are six possible decisions: (1) fill A from C, (2) fill B from C, (3) empty A into C, (4) empty B into C, (5) transfer A to B, and (6) transfer B to A. In the latter two cases, the transferring operation terminates when one of the jugs is empty or full, whichever occurs first.

This problem can be formulated as a sequential decision process; in each state, one of the six decisions is made. We define the state as (s, i, j, k), where s is a stage number and i, j, and k are the contents of jugs A, B, and C, respectively. The initial goal state is $(1, 0, 0, R)$ and the base states are (n, i, j, k) when $i = T$ or $j = T$. While we are interested in any sequence of decisions that transforms the system from the goal state to a base state, it is just as convenient to find the shortest such sequence, i.e., the sequence having the fewest number of decisions. Letting the cost of a decision equal 1, the problem can be solved using the DPFE

$$f(s, i, j, k) = \min_{d}\{1 + f(s + 1, i', j', k')\}, \tag{2.35}$$

where d is one of the six possible decisions, and i', j', and k' are the contents of the jugs if decision d is made in state (s, i, j, k).

For example, let $P = 9$, $Q = 4$, and $T = 6$. Then the goal is $f(1, 0, 0, 13)$ and the base condition is $f(n, i, j, k)$ when i or j equals 6. In this event,

$f(1,0,0,13) = 8$ and the terminating state is $(9,6,4,3)$ following a sequence of 8 decisions.

We note in conclusion that this formulation permits cycling, e.g., we can pour back and forth between jugs. Rather than deriving a more complicated formulation to avoid this cycling, either we can impose constraints on possible decisions, or we can set an upper limit on the stage number to terminate the recursion.

2.34 Optimal Production Problem (PROD)

In a *production* problem, there is a product which can be produced at some specified cost and which can be consumed based upon user demand. There may also be an inventory cost for storing products that are not consumed, and a penalty cost in case the demand cannot be satisfied. The optimal production problem is an N-stage sequential decision process, where at each stage k a decision must be made to produce i units at a production cost $C(k,i)$, that may depend on the stage k and on the number of units produced i. The state is (k,s), where k is the stage number and s is the size of the inventory, i.e. how many units of the product are available at the start of the stage. The demand $D(k)$ generally depends on the stage. If the decision in state (k,s) is to produce i units, the next-state is $(k+1, s+i-D(k))$. The cost associated with this state transition is the production cost $C(k,i)$ plus the inventory cost $I(k,s)$ for $s > 0$. $I(k,s)$ for $s < 0$ represents a penalty for not satisfying a demand of size s.

The DPFE for this production problem is

$$f(k,s) = \min_i \{C(k,i) + I(k,s) + f(k+1, s+i-D(k))\},$$

with base condition $f(k,s) = I(k,s)$ if $k > N$. We may combine C and I to obtain a reward function $R(k,s,i) = C(k,i) + I(k,s)$. The goal is to find $f(1, s_0)$, where s_0 is the initial inventory.

It is common for the demand to be random. For example [8, problem 19.14], for all k, let $D(k) = 1$ with probability p and $D(k) = 2$ with probability $1-p$. The DPFE is then probabilistic,

$$\begin{aligned} f(k,s) = \min_i \{&C(k,i) + I(k,s) + p \cdot f(k+1, s+i-1) \\ &+ (1-p) \cdot f(k+1, s+i-2)\}. \end{aligned} \tag{2.36}$$

For the above probabilistic assumption, also assume that the production cost, for i, limited to 0, 1, or 2, is given by $C(k,0) = 0$, $C(k,1) = 10$, and $C(k,2) = 19$, for all k, and that the inventory/penalty function is $I(i) = 1.1i$ for $i > 0$ and $I(i) = -1.5i$ for $i < 0$. If $p = 0.6$, then $f(1,0) = 42.244$ where the optimal initial decision is to produce 2 units, leading to next-states $(2,1)$ and $(2,0)$ for demands 1 and 2, respectively. (Continuing, $f(2,1) = 19.9$ and $f(2,0) = 28.26$.)

2.35 Production: Reject Allowances Problem (PRODRAP)

The reject allowances problem (PRODRAP) as described in [20, p.421] is a probabilistic DP problem (see Sect. 1.1.15). It is an extension of production problems (see Sect. 2.34), where here it is also assumed that in order to produce an item that meets stringent quality standards a manufacturer may have to produce more than one item to obtain an acceptable item. The number of extra items produced in a production run is called the *reject allowance*. Any item produced will be defective with probability p (and acceptable with probability $1-p$). In a production run, up to lot size L items can be produced. There is only time for R production runs. Let c_m be the marginal production cost per item, let c_s be the fixed setup cost for a production run, and let c_p denote the penalty costs if the manufacturer cannot deliver an acceptable item. The decision variable x_n denotes the number of items produced during production run n. The goal becomes to minimize the total expected cost.

A DP model of this problem consists of a state (n) representing the production run number. Contrary to [20, p.421] we do not include the number of acceptable items still needed into the state, since it is unnecessary as we are only looking for a single acceptable item.

The DP functional equation becomes

$$f(n) = \begin{cases} \min\limits_{x_n \in \{0,\dots,L\}} \{K(x_n) + c_m x_n + p^{x_n} f(n+1)\} & \text{if } n \le R \\ c_p & \text{if } n = R+1. \end{cases}$$

where the function

$$K(x_n) = \begin{cases} 0 & \text{if } x_n = 0 \\ c_s & \text{if } x_n > 0 \end{cases}$$

describes the fact that there is no setup cost if no items are to be produced in a production run.

The problem instance from [20, p.421] lets $p = 0.5, L = 5, R = 3, c_m = 1, c_s = 3, c_p = 16$. The minimal total expected cost is $f(1) = 6.75$ and stems from the following policy for the decisions x_n. Produce $x_1 = 2$ items on the first production run; if none is acceptable, then produce either $x_2 = 2$ or $x_2 = 3$ items on the second production run; if none is acceptable, then produce either $x_3 = 3$ or $x_3 = 4$ items on the third production run.

2.36 Reliability Design Problem (RDP)

The reliability design problem (RDP) problem appears in [1, p.113] as an exercise; the example instance used here is from [22, pp.295–298]. Given a set $D = \{d_0, \dots, d_{n-1}\}$ of n device types for n corresponding stages, the task is to design a system that consists of n stages connected to each other in a serial fashion. If one or more stages of the system fail, then the system as a whole fails. Thus the reliability of the system as a whole is the product of the reliabilities of the individual stages. A reliability r and a discretized cost c is assigned to each device type according to

$$\begin{aligned} r : D &\rightarrow [0;1] \\ d_i &\mapsto r(d_i) \end{aligned}$$

and

$$\begin{aligned} c : D &\rightarrow \mathbf{N} \\ d_i &\mapsto c(d_i) \end{aligned}$$

and for shorter notation we write r_i instead of $r(d_i)$ and c_i instead of $c(d_i)$. To improve the reliability of a particular stage, components can be duplicated and arranged in a parallel fashion. Now, as long as at least one component of that stage is working, the stage is considered working. The reliability of a stage becomes $1-(1-r_i)^{m_i}$, where the integer decision variables m_i denotes the number of components d_i allocated to stage i. How many devices should be allocated to each stage in order to maximize the reliability of the system, but still stay within a given budget $b \in \mathbf{N}$?

Let $x \in \mathbf{N}$ be the amount of money left to spend. The DP functional equation for this problem can be expressed as

$$f(i,x) = \max_{m_i \in M(i,x)} \{(1-(1-r_i)^{m_i}) \cdot f(i-1, x - c_i m_i)\}$$

with the base cases $f(-1,x) = 1.0$ where $x \leq b$. The goal is to compute the maximum reliability of the system as $f(n-1,b)$. Note that the DPFE of this problem is not additive, but multiplicative.

The decision space $M(i,x)$ for a decision variable m_i has the lower bound of 1, since the system needs at least 1 device at each stage; it also has the upper bound of

$$u(i,x) = \lfloor \frac{x - \sum_{j=0}^{i-1} c_j}{c_i} \rfloor$$

which can be justified by the fact that we have to reserve $\sum_{j=0}^{i-1} c_j$ to cover the cost of at least 1 device for the stages we have yet to decide on. Thus, $M(i,x) = \{1,\ldots,u(i,x)\}$. The upper bound $u(i,x)$ is an improvement to the one given in [22, p.297], which is static and does not take into account previously made decisions.

Given a problem instance with $n = 3$ stages, costs $(c_0, c_1, c_2) = (30, 15, 20)$, reliabilities $(r_0, r_1, r_2) = (0.9, 0.8, 0.5)$ and a budget $b = 105$, the optimal sequence of decisions is $m_2 = 2$, $m_1 = 2$, $m_0 = 1$, resulting in a system with reliability $f(2, 105) = 0.648$.

2.37 Replacement Problem (REPLACE)

A *replacement* problem (REPLACE) is similar to production or inventory problems, where objects that are demanded must be replaced whether by producing them or purchasing them, at a given replacement or acquisition cost C. We also associate with objects that are not replaced a total maintenance cost $tm(d)$, that depends on its "usage" time (i.e. the number of stages d between its acquisition and its replacement), and we associate with objects that are replaced a salvage value $v(d)$, that also depends on its usage time d. Note that the total maintenance cost $tm(d)$ for an object that is maintained for d stages is $\sum_{i=1}^{d} m_i$, where m_i is the incremental cost for the i-th stage.

To formulate this replacement problem using DP, we define the state to be the time or stage number k at which a replacement is made, where the initial state is 0. A fixed number of stages N is assumed. If at stage k it is decided to make the *next* replacement at stage k', then the usage time is $d = k' - k$. The DPFE for such a replacement problem is

$$f(k) = \min_{d}\{C + tm(d) + v(d) + f(k + d)\},$$

where the next-state is $k + d$ resulting from a decision $d > 0$. The goal is to find $f(0)$. The base conditions are $f(k) = 0$ for all $k \geq N$.

An instance of such a replacement problem is described in [64, p.774–777], where it is assumed that $N = 5, C = 1000, m(1) = 60, m(2) = 80, m(3) = 120, v(1) = 800, v(2) = 600, v(3) = 500$. From the incremental maintenance costs m, the total maintenance costs are $tm(1) = 60$, $tm(2) = 140$, $tm(3) = 260$. It is also assumed that the lifetime of an object is $L = 3$, after which its maintenance cost is infinite and its salvage value is zero; i.e. $tm(4) = \infty$ and $v(4) = 0$. This is equivalent to restricting the decision space for d to range from 1 to L, subject to the constraint $k + d \leq N$. The solution is $f(0) = 1280$. One sequence of optimal decisions is $d_1 = 1, d_2 = 1$, and $d_3 = 3$ with next-states 1, 2, and 5.

2.38 Stagecoach Problem (SCP)

The stagecoach problem (SCP) is a special case of the problem of finding the shortest path in an acyclic graph (SPA) (see Sect. 2.43). It is also known as the "multistage graph problem" [22, p.257].

Definition 2.2 (Multistage Graph). *A multistage graph $G = (V, E)$ is a directed graph, in which the vertices are partitioned into $k \geq 2$ disjoint sets $V_0, \ldots, V_{k-1}$ and the following two properties hold.*

- *If (u, v) is an edge in E and $u \in V_i$ then $v \in V_{i+1}$.*
- $|V_0| = |V_{k-1}| = 1$.

According to [20, p.398]

> The stagecoach problem is a literal prototype of dynamic programming problems. In fact, this example was purposely designed to provide a literal physical interpretation of the rather abstract structure of such problems.

The stagecoach problem is to find a shortest path in a weighted multistage graph from the single-source node $s \in V_0$ (i.e. located at stage 0), to the single-target node $t \in V_{k-1}$ (i.e. located at stage $k-1$). By definition, there cannot be cycles in a multistage graph. Without loss of generality, assume that $V = \{0, \ldots, n-1\}$, $s = 0$, and $t = n-1$. The edges are given in the form of an adjacency matrix $C = (c_{i,j})$.

For illustrative purposes, this problem can also be solved with a DP approach slightly different from that for the SPA problem. The difference is that the stage number g together with the node number x serves as the state description (g, x), whereas in the SPA problem the node number alone serves as the state description (because there are no explicit stage numbers). A decision d is the next node to go to, and the cost (reward) of the decision is $c_{x,d}$. The DP functional equation reads

$$f(g,x) = \begin{cases} \min\limits_{d \in V_{g+1}} \{f(g+1,d) + c_{x,d}\} & \text{if } x < n-1 \\ 0 & \text{if } x = n-1. \end{cases} \tag{2.37}$$

Here $f(g, x)$ is the length of the shortest path from a node $x \in V_g$, located at stage g, to the target node $n-1$. The goal is to compute $f(0, 0)$.

Consider the following problem instance, taken from [64, p.753], of finding the shortest path from node 0 to node 9 in the graph of Fig. 2.6 defined by the weighted adjacency matrix

$$C = \begin{pmatrix} \infty & 550 & 900 & 770 & \infty & \infty & \infty & \infty & \infty & \infty \\ \infty & \infty & \infty & \infty & 680 & 790 & 1050 & \infty & \infty & \infty \\ \infty & \infty & \infty & \infty & 580 & 760 & 660 & \infty & \infty & \infty \\ \infty & \infty & \infty & \infty & 510 & 700 & 830 & \infty & \infty & \infty \\ \infty & \infty & \infty & \infty & \infty & \infty & \infty & 610 & 790 & \infty \\ \infty & \infty & \infty & \infty & \infty & \infty & \infty & 540 & 940 & \infty \\ \infty & \infty & \infty & \infty & \infty & \infty & \infty & 790 & 270 & \infty \\ \infty & \infty & \infty & \infty & \infty & \infty & \infty & \infty & \infty & 1030 \\ \infty & \infty & \infty & \infty & \infty & \infty & \infty & \infty & \infty & 1390 \\ \infty & \infty & \infty & \infty & \infty & \infty & \infty & \infty & \infty & \infty \end{pmatrix}$$

The shortest path length is computed as $f(0, 0) = 2870$ and the shortest path is $(0, 1, 4, 7, 9)$.

2.39 Seek Disk Scheduling Problem (SEEK)

Assume that a disk scheduler is to handle file data requests A, B, and C to tracks (or cylinders) 100, 50, and 190, respectively, and that the disk read/write head is initially at track 140. Assume it takes 1 time-unit for the disk to *seek* (move) a distance of one track. If the file data requests are handled in first-come-first-serve order (A, B, C), then the total seek time

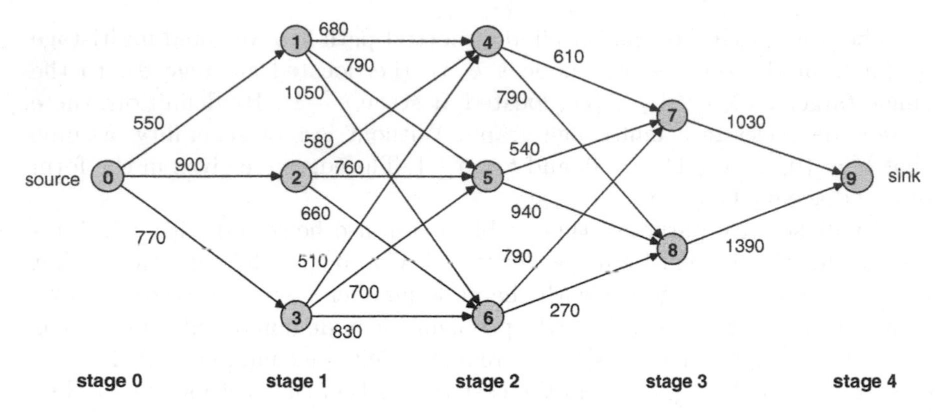

Fig. 2.6. Instance of a Stagecoach Problem

is 40+50+140=230 time-units. Suppose the requests S are to be serviced in the order which minimizes total seek time instead. This problem can be solved using DP as follows. Associate with each request $s \in S$ for file data the position or track number t_s where the data resides. For instance, $(t_A, t_B, t_C) = (100, 50, 190)$. Let the state (S, x) be the set of requests S that remain to be scheduled, together with the present position x of the disk head. A decision d is a member of S. The cost $R(S, x, d)$ is the seek distance from x to t_d, which equals $|x - t_d|$. The next-state $T(S, x, d)$ is $(S - \{x\}, t_d)$. Then the DPFE for this problem is

$$f(S,x) = \begin{cases} \min_{d \in S}\{f(S - \{x\}, t_d) + |x - t_d|\} & \text{if } S \neq \emptyset \\ 0 & \text{if } S = \emptyset. \end{cases}$$

In the example instance, the goal is to compute the minimal seek time $f(\{A, B, C\}, 140) = 190$, which can be achieved by the order (C, A, B).

2.40 Segmented Curve Fitting Problem (SEGLINE)

In (SEGPAGE, Sect. 2.41), given an ordered set $S = \{s_1, s_2, \ldots, s_N\}$ of size N, we define a *segment* of S to be a nonempty subset of S containing only adjacent members of S, $\{s_i, \ldots, s_j\}$ for $1 \leq i \leq j \leq N$. Letting $S = \{A, B, C, D\}$ be denoted by (ABCD), its segments are (A), (B), (C), (D), (AB), (BC), (CD), (ABC), (BCD), and $(ABCD)$. The possible partitions or clusterings or *segmentations* of S are: $\{ABCD\}$, $\{A, BCD\}$, $\{ABC, D\}$, $\{AB, CD\}$, $\{AB, C, D\}$, $\{A, BC, D\}$, $\{A, B, CD\}$, and $\{A, B, C, D\}$. These segmentations can be found by scanning the string "$ABCD$" from left to right, and deciding whether or not to insert a *breakpoint* (denoted by a comma in between characters) at each stage of the scan.

In a variation of this comma-inserting process, suppose during a scan of a string that we decide whether or not each character (rather than location between characters) is to be a breakpoint. Each such breakpoint marks the end of one segment and the beginning of the next segment, with the character placed in both segments. For the string "$ABCD$", segment breakpoints can occur for character 'B' or 'C' or both or neither, and the respective segmentations of S are: $\{AB, BCD\}$, $\{ABC, CD\}$, $\{AB, BC, CD\}$, and $\{ABCD\}$. We have then a sequential decision process where the state s is the set of locations or "breakpoints" already chosen, and where a decision d chooses where the next breakpoint is to be. The corresponding DPFE for this optimal segmentation problem has the general form

$$f(s) = \min_d \{c(d, s) + f(s + d)\},$$

where $c(d, s)$ is the cost of placing the next breakpoint at d when in state s.

As an example, consider the problem of fitting a curve $y = f(x)$ of a specified type (such as polynomial) through a given data-set S of N data-points $\{(x_i, y_i)\}$. In general, we would like $f(x_i)$ to be as close to y_i as possible, for each value of x_i; each difference $e_i = f(x_i) - y_i$ is an approximation error. It is common to wish to minimize the maximum magnitudes of the approximation errors, or to minimize the mean of the sum of the squared errors. The latter are known as the *least-squares errors* (lse). Polynomial approximation to obtain least-squares errors is also known as the *regression* problem, discussions of which can be found in many books, such as [35]. For simplicity, we assume here that the curve is a straight line $y = mx + b$, where m is the slope of the line and b is the y-intercept. The least-squares or *regression* line has slope and y-intercept given by the formulas:

$$m = \frac{\frac{1}{N}\sum x_i y_i - \frac{1}{N}\sum x_i \frac{1}{N}\sum y_i}{\frac{1}{N}\sum x_i^2 - \left(\frac{1}{N}\sum x_i\right)^2} \tag{2.38}$$

$$b = \frac{1}{N}\sum y_i - m\frac{1}{N}\sum x_i \tag{2.39}$$

For instance, let $S = \{(0, 0), (1, 1), (2, 3), (3, 0)\}$. Then the regression line is $y = 0.2x + 0.7$, for which the error is $lse(0, 3) = 1.45$. Furthermore, for $S = \{(0, 0), (1, 1), (2, 3)\}$, the regression line is $y = 1.5x - 0.1667$, for which the error is $lse(0, 2) = 0.0556$, and for $S = \{(1, 1), (2, 3), (3, 0)\}$, the regression line is $y = -0.5x + 2.3333$, for which the error is $lse(1, 3) = 1.3889$. For $S = \{(0, 0), (1, 1)\}$, $S = \{(1, 1), (2, 3)\}$, and $S = \{(2, 3), (3, 0)\}$, the regression line is $y = x$, $y = 2x - 1$, and $y = -3x + 9$, respectively, for each of which the errors are $lse(0, 1) = lse(1, 2) = lse(2, 3) = 0$.

Since $N = 4$, we know that a least-squares cubic polynomial (of degree 3) is "interpolating", i.e., it passes exactly through the four points, so that the error is 0. However, cubic polynomials are more complex than quadratic (degree 2) or linear (degree 1) ones. So which is preferable depends upon what cost we attach to the degree.

Alternatively, there may well be circumstances where we may prefer to fit a polygonal rather than polynomial curve through a given set of data. A "polygonal" curve of "degree" i is a sequence of i straight lines. One such polygonal curve is the sequence of three lines that connects (0,0) to (1,1), (1,1) to (2,3), and (2,3) to (3,0). It also is interpolating. However, just as in the case of polynomials, polygonal curves, as defined here, need not pass through any of the data points, in general; furthermore, we do not require continuity, i.e., the lines need not meet.

In addition to the polygonal curve of degree 1 with error 1.45, and the polygonal curve of degree 3 with error 0, there are also polygonal curves of degree 2 that may or may not be preferable. Which is preferable again depends upon what cost we attach to the degree. Here, we will assume that this degree cost is a constant K times the degree; equivalently, we assume each break has a cost of K.

Determining the optimal polygonal curve is a segmentation problem of the type described above, and can be solved using DP [2]. Given an ordered set of N data points, say, $S = (A, B, C, D)$, we may scan the sequence S from left to right, and at each stage decide whether to end a segment and start a new segment. For example, at stage B, we must decide whether to end the first segment, by fitting a line between A and B, and then considering the remainder (B, C, D), or alternatively to not end the first segment at B; in the latter event, at the next stage C, we must decide whether or not to end the segment, by fitting a line through A, B, and C, and then considering the remainder (C, D). The cost of deciding to end a segment is the least-squares error associated fitting a line through the points in that segment, found using regression formulas, *plus* the cost of a break K.

The DPFE is

$$f(s) = \min_{d \in \{s+1,\ldots,N-1\}} \{lse(s,d) + K + f(d)\}, \tag{2.40}$$

The goal is to find $f(0)$. The base-condition is $f(N-1) = 0$.

Consider the above problem instance, with error-cost matrix lse as given.

$$lse = \begin{pmatrix} \infty & 0 & 0.0556 & 1.45 \\ \infty & \infty & 0 & 1.3889 \\ \infty & \infty & \infty & 0 \\ \infty & \infty & \infty & \infty \end{pmatrix}$$

For $K = 10$, $f(0) = K + 1.45 = 11.45$; there is only one segment, the single regression line having error 1.45. For $K = 1$, $f(0) = 2K + 0.556 = 2.0556$; there are two segments, one the regression line passing through $(0,0)$, $(1,1)$, and $(2,3)$ having error 0.0556, and the other the line connecting $(2,3)$ and $(3,0)$ having error 0.0. For $K = 0.01$, $f(0) = 3K + 0 = 0.03$; there are three segments, the interpolating polygonal curve having error 0.0.

In the above formulation, the number of segments is not specified in advance. For a given breakcost K, the optimal number of segments is determined as part of the problem solution. In some situations, the number of segments may be constrained by some limit regardless of K; the problem is to find the best polygonal curve having exactly (or at most) LIM segments, where K may or may not be also specified (as some nonzero constant). Let $f(i, s)$ denote the optimal cost of a polygonal curve having i segments (or less). The DPFE is

$$f(i, s) = \min_{d \in \{s+1,\ldots,N-1\}} \{lse(s, d) + K + f(i-1, d)\}, \qquad (2.41)$$

The goal is to find $f(LIM, 0)$. The base-conditions are $f(i, s) = 0.0$ when $i = 0$ and $s = N-1$, and $f(i, s) = \infty$ when $i = 0$ and $s < N-1$. In addition, $f(i, s) = \infty$ when $i > 0$ and $s = N-1$ if LIM is an exact limit; or $f(i, s) = 0.0$ when $i > 0$ and $s = N-1$ if LIM is an upper limit. (This former uses the "fixed-time" approach, whereas the latter uses "relaxation".)

For the above example, for an exact limit LIM=2 and $K = 10$, the solution is $f(2, 0) = 20.0556$; there are two segments (with cost $2K = 20.0$), one the regression line passing through $(0, 0)$, $(1, 1)$, and $(2, 3)$ having error 0.0556, and the other the line connecting $(2, 3)$ and $(3, 0)$ having error 0.0. If the limit LIM=2 is just an upper bound instead, the solution is $f(2, 0) = 11.45$ since the single-segment curve has lower cost. The three-segment interpolating curve is precluded regardless of its cost since it violates the limit.

There are numerous other variations of this segmentation problem. For example, other error criteria (besides least-squares) would simply change how the error cost matrix is calculated. Furthermore, rather than segmented lines, we may use segmented parabolas. For some applications, it is desirable for the segmented curve to be continuous and differentiable. In fact, we may require the lines to not only meet at the breakpoints, but to meet "smoothly" (i.e., so that the slopes and perhaps higher derivatives at the breakpoints match). This is related to what is known as *spline* approximation. However, we will not discuss these variations any further here.

2.41 Program Segmentation Problem (SEGPAGE)

Given an ordered set $S = \{s_1, s_2, \ldots, s_N\}$ of size N, we define a *segment* of S to be a nonempty subset of S satisfying the "linearity" constraint that it may contain only adjacent members of S, $\{s_i, \ldots, s_j\}$ for $1 \leq i \leq j \leq N$. Letting $S = \{A, B, C, D\}$ be denoted more simply by $(ABCD)$, its segments are (A), (B), (C), (D), (AB), (BC), (CD), (ABC), (BCD), and $(ABCD)$. As a counterexample, (BD) and (ABD) are not segments, because they both lack an intervening member of S, namely, C.

There are numerous ways to partition an ordered set $S = (ABCD)$ into disjoint segments depending on the size (or "rank") K of the partition P.

1. If $K = 1$, then all the members of S must be in the same and only segment, so that there is a single partition P which has this one segment: $P = \{ABCD\}$.
2. If $K = 2$, then either one segment has one member of S and the other segment has the remaining three members, or each of the two segments have two members of S. In the former case, there are two possible partitions $P_1 = \{A, BCD\}$, $P_2 = \{ABC, D\}$, and in the latter case there is only possible partition: $P_3 = \{AB, CD\}$.
3. If $K = 3$, then two members of S must be in one segment and the remaining two members of S must be in their own segments, so that there are a total of three segments. There are three ways to choose two adjacent members of S to be in the same segment, so there are three possible partitions for $K = 3$: $P_1 = \{AB, C, D\}$, $P_2 = \{A, BC, D\}$, $P_3 = \{A, B, CD\}$.
4. If $K = 4$, then each member of S must be in its own segment, so that there is a single partition P which has a total of four segments: $P = \{A, B, C, D\}$.

(For a general N, the number of possible partitions in each of these cases can be determined by combinatorial analysis; see [53], for example.) The optimal segmentation problem is that determining which of all the possible partitions minimizes or maximizes some objective.

We observe that each of the partitions or segmentations given above can be represented as an ordered list of the members of S separated by zero or more commas. Thus, a partition of S into segments can be done by making a sequence of decisions on where these commas should be placed. Since only the locations of the commas are significant, rather than the timings of their placements, we arbitrarily assume the commas are placed from left to right. (We also assume "dummy" commas or breakpoints at the beginning and end of S.) We have then a sequential decision process where the state s is the set of locations or "breakpoints" at which there are already commas, and where a decision d chooses where the next breakpoint is to be. The corresponding DPFE for the optimal segmentation problem has the general form

$$f(s) = \min_d \{c(d, s) + f(s + d)\},$$

where $c(d, s)$ is the cost of adding the next breakpoint at d when in state s.

As an example, consider the *pagination* problem that arises in virtual memory systems. A program consists of a sequence of program blocks, which must be partitioned into pages so as to minimize interpage references (i.e., references between program blocks in one page to program blocks in another page). There must be a page size limit m, which restricts how many program blocks can fit into a single page; otherwise, if all blocks were placed in the same page, there would be no interpage references at all. If it is assumed, for simplicity, as in [28], that the partitioning is sequential, then this pagination problem is equivalent to the segmentation problem, and it is solvable using a DPFE of the foregoing form where the cost function c must be suitably defined. If the program is scanned sequentially, say, from left to right, we may define the state simply as the location of the prior break; at each stage of the scan, we decide whether or not to add another break. If the program consists of N blocks, there are $N+1$ stages or locations (in between and before and after these blocks) where breaks may be located. The cost $C(p,q)$ of adding a break at stage q given that the prior break was at stage p can be calculated from knowledge of reference patterns between program blocks. Specifically, the entries in this cost matrix, which are associated with branch execution frequencies, can be derived from a Markov chain model of the program, as mentioned in [28], from which the following example is taken.

$$C = \begin{pmatrix}
\infty & 0 & 2 & 82 & 82 & 82 & 82 & 2 & 202 & 202 & 2 & 42 & 42 & 2 & 2 & 0 & 0 \\
\infty & \infty & 2 & 82 & 82 & 82 & 82 & 2 & 202 & 202 & 2 & 42 & 42 & 2 & 2 & 0 & 0 \\
\infty & \infty & \infty & 82 & 82 & 82 & 82 & 2 & 202 & 202 & 2 & 42 & 42 & 2 & 2 & 0 & 0 \\
\infty & \infty & \infty & \infty & 42 & 82 & 82 & 2 & 202 & 202 & 2 & 42 & 42 & 2 & 2 & 0 & 0 \\
\infty & \infty & \infty & \infty & \infty & 42 & 42 & 2 & 202 & 202 & 2 & 42 & 42 & 2 & 2 & 0 & 0 \\
\infty & \infty & \infty & \infty & \infty & \infty & 41 & 1 & 201 & 201 & 2 & 42 & 42 & 2 & 2 & 0 & 0 \\
\infty & \infty & \infty & \infty & \infty & \infty & \infty & 1 & 201 & 201 & 2 & 42 & 42 & 2 & 2 & 0 & 0 \\
\infty & \infty & \infty & \infty & \infty & \infty & \infty & \infty & 201 & 201 & 2 & 42 & 42 & 2 & 2 & 0 & 0 \\
\infty & \infty & \infty & \infty & \infty & \infty & \infty & \infty & \infty & 101 & 2 & 42 & 42 & 2 & 2 & 0 & 0 \\
\infty & \infty & \infty & \infty & \infty & \infty & \infty & \infty & \infty & \infty & 2 & 42 & 42 & 2 & 2 & 0 & 0 \\
\infty & \infty & \infty & \infty & \infty & \infty & \infty & \infty & \infty & \infty & \infty & 42 & 42 & 2 & 2 & 0 & 0 \\
\infty & \infty & \infty & \infty & \infty & \infty & \infty & \infty & \infty & \infty & \infty & \infty & 22 & 2 & 2 & 0 & 0 \\
\infty & \infty & \infty & \infty & \infty & \infty & \infty & \infty & \infty & \infty & \infty & \infty & \infty & 2 & 2 & 0 & 0 \\
\infty & \infty & \infty & \infty & \infty & \infty & \infty & \infty & \infty & \infty & \infty & \infty & \infty & \infty & 2 & 0 & 0 \\
\infty & \infty & \infty & \infty & \infty & \infty & \infty & \infty & \infty & \infty & \infty & \infty & \infty & \infty & \infty & 0 & 0 \\
\infty & \infty & \infty & \infty & \infty & \infty & \infty & \infty & \infty & \infty & \infty & \infty & \infty & \infty & \infty & \infty & 0 \\
\infty & \infty & \infty & \infty & \infty & \infty & \infty & \infty & \infty & \infty & \infty & \infty & \infty & \infty & \infty & \infty & \infty
\end{pmatrix}$$

Given this cost matrix, the optimal segmentation problem can be solved using the DPFE

$$f(p) = \min_{p+1 \le q \le p+m} \{C(p,q) + f(q)\}. \tag{2.42}$$

The goal is to find $f(0)$. The base-condition is $f(N) = 0$ where $N = 16$. The solution is $f(0) = 87$. An optimal sequence of decisions for the breakpoints is $d_1 = 1$, $d_2 = 5$, $d_3 = 7$, $d_4 = 10$, $d_5 = 13$, and $d_6 = 15$. (There are also dummy breakpoints at $d_0 = 0$ and $d_7 = 16$.) The optimal segmentation of $(s_1 s_2 s_3 s_4 s_5 s_6 s_7 s_8 s_9 s_{10} s_{11} s_{12} s_{13} s_{14} s_{15})$ is $(s_1 s_2 s_3 s_4, s_5 s_6, s_7 s_8 s_9, s_{10} s_{11} s_{12}, s_{13} s_{14}, s_{15})$.

In conclusion, we note that since the cost matrix is triangular, the pagination problem is equivalent to that of finding the shortest path in an acyclic graph (SPA). This is a consequence of our linearity constraint. Without this linearity assumption, the general partitioning or clustering problem would have exponential complexity since there are $(O(2^N))$ subsets of a set of size N, each subset being possibly in the optimal partition.

2.42 Optimal Selection Problem (SELECT)

The (single) "selection" problem is that of finding the k-th smallest member (also known as the k-th "order statistic") of an unordered set A of size N. (The choice $k = N$ selects the largest member and $k = N/2$ selects the median.) Assuming "random access" memory, if A is an ordered array, selection takes constant $O(1)$ time, since it requires an amount of time independent of N to perform what would in essence be a subscripting operation. However, to achieve constant-time selection may require additional time to sort A, usually $O(N \log N)$ or $O(N^2)$. On the other hand, it has been shown that a selection algorithm for unsorted data can be designed (using a divide-and-conquer strategy [10, Chap. 9]) that takes only linear $O(N)$ time.

Consider the *set* selection problem where we wish to find the set $S \subset A$ of the k_1-th, k_2-th, ..., k_m-th smallest members of a given unordered set A. We may of course use the linear-time selection algorithm m different times, which requires $O(mN)$ time. We observe that once we find the k_d-th member x of A, we can partition $A - \{x\}$ into two subsets, A_1 and A_2, consisting of those members of A less than x and greater than x, respectively. This selection and partitioning can be done in $O(N)$ time. Thus we can recursively perform set selection for left and right subsets A_1 and A_2, both of which have sizes less than N; A_1 has size $k_d - 1$, and A_2 has size $N - k_d$. The k_i-th smallest of A (of size N) for $i < d$ is the k_i-th smallest of A_1 (of size $k_d - 1$), whereas the k_i-th smallest of A for $i > d$ is the $(k_i - k_d)$-th smallest of A_2 (of size $N - k_d$). Thus, making a decision d, to find the k_d-th member x of A, also partitions S into subsets $S_1 = \{k_1, \ldots, k_{d-1}\}$ and $S_2 = \{k_{d+1}, \ldots, k_m\}$ associated with A_1 and A_2, respectively.

The *optimal* set selection problem is that of deciding the optimal way to find the m different values of A for a given S. (Note that finding the actual values in A is not part of this problem.) The optimal solution can be represented as a binary tree whose root is the optimal initial choice of d, and whose subtrees are the optimal trees for the left and right subproblems. This resembles other optimal binary tree problems (e.g., BST and MCM), as noted by Brown [9]. The DPFE is

$$f(\{k_i, \ldots, k_j\}|p, q) = \min_{k_d \in \{k_i, \ldots, k_j\}} \{(q - p + 1) + f(\{k_i, \ldots, k_{d-1}\}|p, k_d - 1) + f(\{k_{d+1}, \ldots, k_j\}|k_d + 1, q)\}.$$

The goal is to find $f(S|1, N)$ with base condition $f(S|p, q) = 0$ if $|S| = 0$. If the set $\{k_i, \ldots, k_j\}$ is represented by the pair (i, j), the DPFE can be rewritten in the form

$$f((i, j)|p, q) = \min_{i \leq d \leq j} \{(q - p + 1) + f((i, d - 1)|p, k_d - 1) + f((d + 1, j)|k_d + 1, q)\}. \quad (2.43)$$

The goal is to find $f((1, m)|1, N)$ with base condition $f((i, j)|p, q) = 0$ if $i > j$.

Let $S = \{3, 6, 8, 10\}$ and $N = 10$. The optimal tree has 6 at the root, $\{3\}$ in its left subtree, and $\{8, 10\}$ is its right subtree. The cost of the decision to choose 6 as the overall root is equal to the cost of selecting the 6-th smallest of a set $\{1, \ldots, 10\}$ of size 10, which equals 10. The cost of the decision to choose 3 as the root of the left subtree is equal to the cost of selecting the 3-rd smallest of a set $\{1, \ldots, 5\}$ of size 5, which equals 5. The cost of the decision to choose 8 as the root of the right subtree is equal to the cost of selecting the 2-nd smallest of a set $\{7, \ldots, 10\}$ of size 4, which equals 4, plus the cost of the decision to choose 10 as the root of the right subtree of the right subtree; this final quantity is equal to the cost of selecting the 2-nd smallest of a set $\{9, 10\}$ of size 2, which equals 2. Thus, the overall cost is $10 + 5 + 4 + 2 = 21$.

2.43 Shortest Path in an Acyclic Graph (SPA)

The problem of finding a shortest path from a single-source s to a single-target t in a weighted, directed graph $G = (V, E)$ with no cycles is considered here. This problem is discussed in detail in Chap. 1. Without loss of generality, assume that $V = \{0, \ldots, n-1\}$, $s = 0$, and $t = n - 1$. The edges are given in the form of an adjacency matrix $C = (c_{i,j})$.

By modeling a state x as the current node, a decision d is the next-state, and the cost (reward) of the decision is $c_{x,d}$ the DP functional equation can be expressed as

$$f(x) = \begin{cases} \min_{d \in V}\{f(d) + c_{x,d}\} & \text{if } x < n-1 \\ 0 & \text{if } x = n-1. \end{cases} \tag{2.44}$$

In this formulation $f(x)$ is the length of the shortest path from a node $x \in V$ to the target node $n - 1$. The goal becomes to compute $f(0)$.

Consider the following problem instance of finding the shortest path from node 0 to node 3 in the graph of figure 2.7 defined by the weighted adjacency matrix

$$C = \begin{pmatrix} \infty & 3 & 5 & \infty \\ \infty & \infty & 1 & 8 \\ \infty & \infty & \infty & 5 \\ \infty & \infty & \infty & \infty \end{pmatrix}$$

The shortest path length is computed as $f(0) = 9$ and the shortest path is $(0, 1, 2, 3)$. Note that a large class of optimization problems can be solved by formulating them in terms of finding shortest paths.

2.44 Shortest Path in an Cyclic Graph (SPC)

In this variant of the shortest path problem there may be cycles in the graph. Also, it is permissible to have edges with negative weights. However, in order

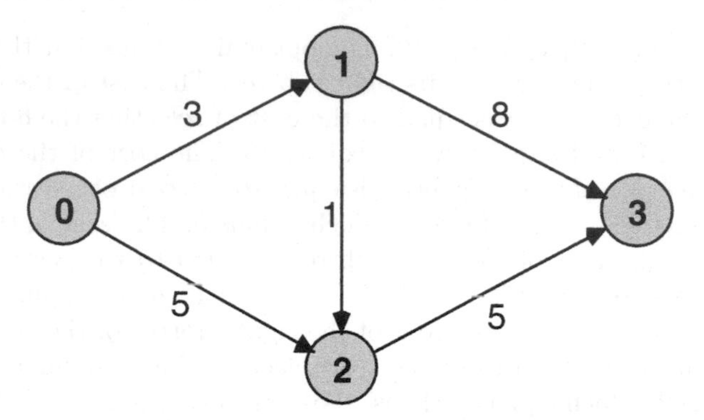

Fig. 2.7. Shortest Path Problem in an Acyclic Graph

for the problem to remain well defined there must not be a negative-weight cycle reachable from the source s. Methods for solving such problems are discussed in Chap. 1.

Without loss of generality, assume that $V = \{0, \ldots, n-1\}$, $s = 0$, and $t = n - 1$. The edges are given in the form of an adjacency matrix $C = (c_{i,j})$. We will give two different DP models along with their DP functional equation that solve this problem. Note that the DPFE (2.44) that we used in the acyclic case in Sect. 2.43 no longer works if there are cycles in the graph.

The first model utilizes the approach taken for TSP (section 2.47) and codes a state as a pair (x, S) where x is the current vertex and S is the set of vertices already visited.

$$f(x,S) = \begin{cases} \min_{d \notin S}\{f(d, S \cup \{d\}) + c_{x,d}\} & \text{if } x < n-1 \\ 0 & \text{if } x = n-1. \end{cases} \tag{2.45}$$

The length of the shortest path is computed as $f(0, \{0\})$.

The second model utilizes the "relaxation approach" discussed in Chap. 1, and codes a state as a pair (x, i) where x is the current vertex and i is an integer upper bound for the number of edges on the path to the target node.

$$f(x,i) = \begin{cases} \min_{d}\{f(d, i-1) + c_{x,d}\} & \text{if } x < n-1 \text{ and } i > 0 \\ \infty & \text{if } x < n-1 \text{ and } i = 0 \\ 0 & \text{if } x = n-1. \end{cases} \tag{2.46}$$

Since during the course of the computation $f(x, i)$ gives the length of the shortest path from vertex x to the target node $n - 1$ having at most i edges, the length of the shortest path from the source to the target node is computed as $f(0, n-1)$. Note that the shortest path consists of at most $n - 1$ edges, because n or more path segments would imply a circuit in the path and since this circuit has by assumption nonnegative length, the path could be improved by cutting out the circuit (or at least simplified in case the circuit length is 0).

Consider the following problem instance [35, p.335] of finding the shortest path from node 0 to node 3 in the graph of Fig. 2.8 defined by the weighted adjacency matrix

$$\begin{pmatrix} \infty & 3 & 5 & \infty \\ \infty & \infty & 1 & 8 \\ \infty & 2 & \infty & 5 \\ \infty & \infty & \infty & \infty \end{pmatrix}$$

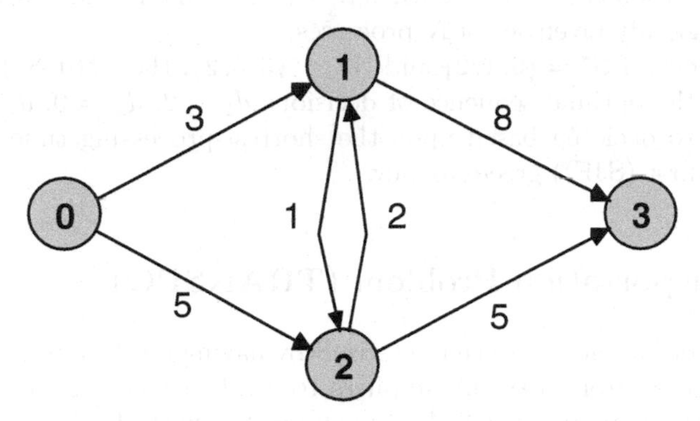

Fig. 2.8. Shortest Path Problem in a Cyclic Graph

In the first model, the shortest path length is computed as $f(0, \{0\}) = 9$, in the second model it is computed as $f(0, 3) = 9$. The shortest path is $(0, 1, 2, 3)$.

2.45 Process Scheduling Problem (SPT)

Suppose we have a set of processes whose weights are their processing or execution times, 3, 5, and 2, respectively. If the processes are executed in that given order (on a single processor), they would complete execution at times 3, 8, and 10, respectively, for a total of 21. If the processes are executed in shortest-processing-time (SPT) order, 2,3,5, they would complete execution at times 2,5,10, for a total of 17. The optimal process scheduling problem is that of determining the optimal sequence in which processes should be executed on a single processor so as to minimize the total (or average) of the completion times for a given set of processes. This problem is equivalent to the linear search problem (LINSRC) and related permutation problems.

One way to solve this problem is to use a staged formulation, or what we called virtual-stages in Chap. 1. Observe that the cost of the next decision is the sum of the costs of the earlier decisions plus the individual cost for this next decision. Thus, we define the state as (k, S) where k is the stage number and S is the set of processes that have not yet been scheduled. A decision d is a member of S. The cost of choosing d in state (k, S) is the sum of the execution times $\sum_{i \notin S} w_i$ of processes already executed (i.e., for processes not in S) plus the execution time w_d of process d. The sum $\sum_{i \notin S} w_i$ can also be incorporated into the definition of state by introducing it as a virtual stage parameter k. Then

$$f(k, S) = \min_{d \in S}\{(k + w_d) + f(k + w_d, S - \{d\})\}. \tag{2.47}$$

The base-condition is $f(k, S) = 0$ when $S = \emptyset$. The goal is find $f(0, S^*)$, where S^* is the originally given set of N processes.

For instance, if $S^* = \{0, 1, 2\}$ and $W^* = \{3, 5, 2\}$, then $f(0, S^*) = 2 + 5 + 10 = 17$ for the optimal sequence of decisions $d_1 = 2$, $d_2 = 0$, $d_3 = 1$. This corresponds to deciding based upon the shortest-processing time (SPT) (or shortest-job-first (SJF)) greedy policy.

2.46 Transportation Problem (TRANSPO)

Transportation problems are characterized by having a set of objects that are to be distributed from a set of suppliers to a set of receivers in an optimal fashion, where there are specified costs associated with destination q receiving objects from source p. In a graph model, suppliers and receivers would be represented by nodes, and a flow by a branch from p to q, with a label that specifies the cost $c(p, q, i)$ of transporting i objects from p to q. A transportation problem is equivalent to finding a minimum cost flow in this graph, but can be solved by formulating it as a production problem.

For example, the transportation problem of [8, problem 19.22] can be formulated as a production problem where the production costs C (per unit) are, for three stages

$$C = \begin{pmatrix} 0.0 & 35.0 & 74.0 & 113.0 & \infty & \infty & \infty \\ 0.0 & 43.0 & 86.0 & 133.0 & 180.0 & \infty & \infty \\ 0.0 & 40.0 & 80.0 & 120.0 & 165.0 & 210.0 & \infty \end{pmatrix}$$

the inventory cost I is 3 per unit, the demand D is 2 in all stages, and the production capacity (i.e. the maximum number that can be produced in a stage) is 6. The DPFE is of the same form as (2.36) and reads

$$f(k, s) = \min_{x}\{C(k, x) + I(k, s) + f(k + 1, s + x - D(k))\}, \tag{2.48}$$

where k is the stage number, s is the current inventory, and the decision x is the amount to produce.

The goal is to find $f(0, 0)$ with base condition $f(k, s) = 0$ when $k = n$. For the given example, for initial stage $k = 0$ and initial inventory $s = 0$,

$f(0,0) = \min_x\{C(0,x) + I(0,0) + f(1, x - D(0))\}$. $C(0,x)$ is given by the zero-th row of the cost matrix, $I(0,0) = 0$, and $D(0) = 2$; the decision x is constrained by the capacity, $x \leq 6$, but must also be great enough to meet the initial demand (minus the initial inventory), $x \geq 2$. Continuing these calculations, we have $f(0,0) = 239$, and $x_1 = 3$, $x_2 = 1$, and $x_3 = 2$.

2.47 Traveling Salesman Problem (TSP)

Given a complete weighted directed graph $G = (V, E)$ with distance matrix $C = (c_{i,j})$ the optimization version of the traveling saleman problem asks to find a minimal Hamiltonian cycle (visiting each of the $n = |V|$ vertices exactly once). Without loss of generality assume $V = \{0, \ldots, n-1\}$. The DP functional equation can be expressed as

$$f(v, S) = \begin{cases} \min\limits_{d \notin S}\{f(d, S \cup \{d\}) + c_{v,d}\} & \text{if } |S| < n \\ c_{v,s} & \text{if } |S| = n \end{cases} \tag{2.49}$$

where the length of the minimal cycle is computed as $f(s, \{s\})$ where $s \in V$. (The choice of the starting vertex s is irrelevant, since we are looking for a cycle, so arbitrarily pick $s = 0$.) In the above DP functional equation (2.49), a state is a pair (v, S) where v can be interpreted as the current vertex and S as the set of vertices already visited.

In contrast, the following equivalent DP functional equation (2.50) maintains S as the set of vertices not yet visited.

$$f(v, S) = \begin{cases} \min\limits_{d \in S}\{f(d, S - \{d\}) + c_{v,d}\} & \text{if } |S| > 1 \\ c_{v,s} & \text{if } S = \emptyset \end{cases} \tag{2.50}$$

The length of the minimal cycle is computed as $f(s, V - \{s\})$ where $s \in V$.

For instance, let

$$C = \begin{pmatrix} 0 & 1 & 8 & 9 & 60 \\ 2 & 0 & 12 & 3 & 50 \\ 7 & 11 & 0 & 6 & 14 \\ 10 & 4 & 5 & 0 & 15 \\ 61 & 51 & 13 & 16 & 0 \end{pmatrix}$$

This instance is shown in Fig. 2.9.

We apply the DP functional equation (2.49) to compute $f(0, \{0\}) = 39$ or we apply the DP functional equation (2.50) to compute $f(0, \{1, 2, 3, 4\}) = 39$. By keeping track of which arguments minimize the min-expressions in each case, we find that the minimal cycle is $(0, 1, 3, 4, 2)$.

We remark in conclusion that TSP is one of many "intractable" problems that cannot be solved efficiently (see Sect. 1.1.13) using any known algorithm, using DP or otherwise, at least for large n. On the other hand, for small n,

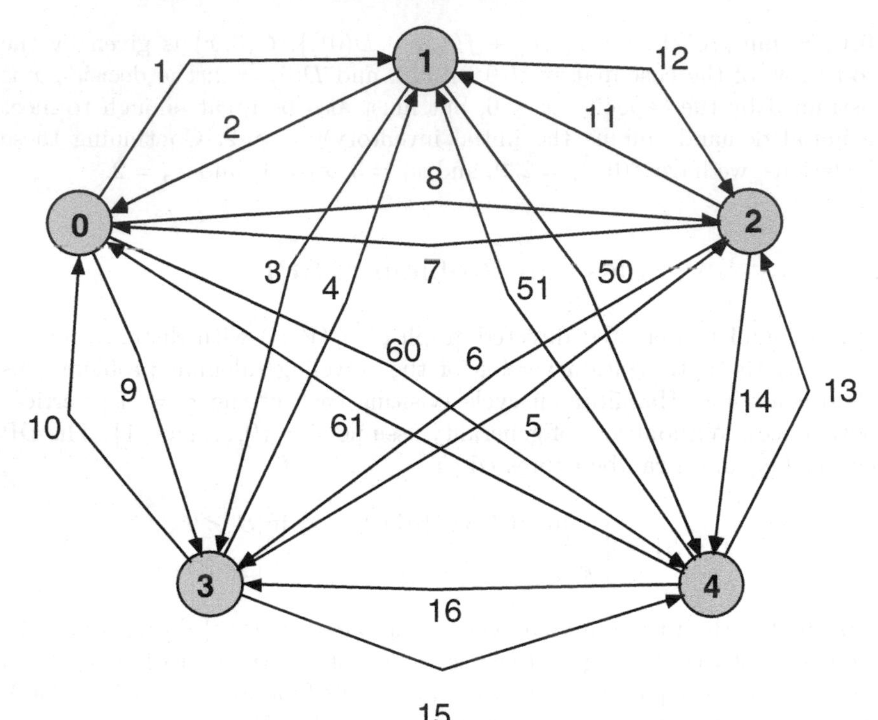

Fig. 2.9. Instance of a TSP Problem

TSP can be solved in practice using the above DP formulation and using our software tool. (In this book, we have emphasized generality, not necessarily practicality.)

Part II

Modeling of DP Problems

3

The DP Specification Language gDPS

This chapter gives an overview of the "general DP specification" (gDPS) language as the result of our design efforts for the DP2PN2Solver software [44, 45]. Section 3.1 gives a brief introduction to gDPS. In Sect. 3.2 we describe the design principles of gDPS. Section 3.3 describes the syntactical details of gDPS. The grammar of gDPS is given in Sect. 3.4.

3.1 Introduction to gDPS

The gDPS source language combines features of common computer programming languages, such as Java, C++, and Visual Basic, and of mathematical programming languages, such as AMPL, GAMS, and LINGO. We note that the latter are verbal (text-based) languages, which are commonly considered more "user-friendly" than symbolic languages such as APL [56]. Such text-based languages have been more readily accepted in the mathematical programming community, and it is our model for gDPS as well.

We view gDPS as a *specification* rather than *procedural* language, although it has optional procedural elements that allow users the flexibility to code their own "subroutines" to handle unusual cases. For these procedural elements, we adopt Java syntax for specificity, since some choice had to be made in order to implement an actual software tool. We chose Java in large measure because of its universal availability, at no cost in all major platforms, and also because of its suitability for a web-based package (see `http://java.sun.com`). Brief introductions to Java can be found in numerous books, such as [5].

A gDPS source consists of several structured sections, which reflect the standard parts of every DP problem. To increase writability, additional information such as helper functions that ease the DP model formulation can be included into a gDPS source. Of particular importance are the following sections.

- The optional `GENERAL_VARIABLES` section allows the definition of variables (e.g. of type `int`, `double`, or `String`) in a Java syntax.

A. Lew and H. Mauch: *The DP Specification Language gDPS*, Studies in Computational Intelligence (SCI) **38**, 103–123 (2007)
www.springerlink.com

- The optional `SET_VARIABLES` section allows the definition of variables of type `Set` (who can take on a set of integers as value) in a convenient enumerative fashion the way it is written in mathematics. Ellipsis is supported with a subrange syntax.
- The optional `GENERAL_FUNCTIONS` section allows the definition of arbitrary functions in a Java style syntax. For set types the use of type `NodeSet` (which is subclassed from `java.util.TreeSet`) is recommended in order to be most compatible with sets specified elsewhere in the gDPS source.
- The mandatory `STATE_TYPE` section. A state in a DP problem can be viewed as an ordered tuple of variables. The variables can be of heterogeneous types. The "stage" of a DP problem would, as an integer coordinate of its state, be one part of the ordered tuple.
- The mandatory `DECISION_VARIABLE` and `DECISION_SPACE` sections declare the type of a decision and the set of decisions from which to choose (the space of alternatives), which is given as a function of the state.
- The mandatory DP functional equation (DPFE). The recursive equation is described in the `DPFE` section and its base cases must either be expressed in conditional fashion in the `DPFE_BASE_CONDITIONS` or in a enumerative way in the `DPFE_BASE` section.
- The computational goal of the DP instance is given in the mandatory `GOAL` section.
- The reward function (cost function) is given as a function of the state and the decision in the mandatory `REWARD_FUNCTION` section.
- The transformation functions (one or more) are given as functions of the state and the decision in the mandatory `TRANSFORMATION_FUNCTION` section.
- Probabilistic DP problems (see Sect. 2.35 and Sect. 2.19 for examples) require a probability weight to be multiplied to every recursive functional in the DPFE. Discounted DP problems (see Sect. 2.9 for an example) require a discount factor to be multiplied to every recursive functional in the DPFE. Since these weights (or discount factors) can be arbitrary functions of the state and the decision taken, the optional `TRANSITION_WEIGHTS` section provides the means to specify these weights that are used during the transition from one state to the next. If the weights are simply constants, this is expressed by defining constant functions independent of the state and the decision taken. Transition weights can also be used to model a 0-1-binary choice (see the `intvl3` formulation of the INTVL problem (Sect. 2.16)).

Set theoretic features in the gDPS language simplify the specification and improve its readability. A variable of the type `Set` can be declared in the `SET_VARIABLES` section. Such variables and set literals (described using complete enumeration, or a subrange syntax) can be operands of set theoretic operations like `SETUNION`, `SETINTERSECTION` or `SETMINUS` and it is possible to compare sets for equality using the keyword `SETEQUALS`.

Since the `GENERAL_VARIABLES` section allows the definition of arbitrary variables and the `GENERAL_FUNCTIONS` section allows the definition of arbitrary functions in a Java style syntax, the gDPS language is powerful and flexible. Despite this power and flexibility, the other sections require a very structured format, which makes a gDPS very readable and produces a specification that resembles the mathematical formulation closely. The hybrid approach of flexible Java style elements and strictly structured sections makes it easy to learn the gDPS language.

A gDPS source file is a plain text file ending in the suffix `.dp`. It can be created and modified with a simple text editor. Throughout a gDPS source it is legal to document it using Java style block comments (between `/*` and `*/`) and line comments (starting with `//`).

3.2 Design Principles of gDPS

DP problems can take on various forms and shapes. This is why it is hard to specify a strict and fixed specification format as can be done for LP. But there are common themes across all DP problems and those are captured in the gDPS language.

A special purpose language should allow both a convenient description of a DP problem instance and efficient parsing of the specification. The gDPS language has been designed as a general source language that can describe a variety of DP problems. It offers the flexibility needed for the various types of DP problems that arise in reality. All sample problems from Chapter 2 have been successfully modelled with gDPS and solved by DP2PN2Solver. Sources (and snippets of sources) in gDPS for some of these sample problems are presented in this chapter with the purpose of illustrating special features of gDPS. For a complete listing of all gDPS sources, the reader is referred to Chapter 4.

Input data that is specific to a DP instance is hardcoded into the gDPS source in order to have a single gDPS source file that contains the complete specification of the DP instance. If it seems more desirable to keep separate files for the instance specific data and the problem specific data, this would require a conceptually straightforward modification (which we leave as a future extension).

In the worst-case it is necessary that the decision variable loops over all values of the current decision set [48, p.70]. Thus in the `DECISION_SPACE` section the current decision set (which is dependent on the current state) needs to be defined as a set-valued function.

Reward and transformation functions are expressed either in closed-form expressions or as tabular data [48, p.48,67]. The same applies to transition weight functions. When a closed-form expression exists, it can usually be expressed directly within the `REWARD_FUNCTION`, the `TRANSFORMATION_FUNCTION`, or the `TRANSITION_WEIGHTS` section. If these functions are given as tabular data (as is frequently the case in real-world problems) then the table can be stored as an array in the GENERAL_VARIABLES section and table entries can be referred to as array entries (using the usual Java style syntax) from the REWARD_FUNCTION, TRANSFORMATION_FUNCTION and TRANSITION_WEIGHTS sections. If necessary, helper functions for table access can be defined in the GENERAL_FUNCTIONS section using Java syntax.

3.3 Detailed Description of the gDPS Sections

This section focuses on the details of how to generate a specification file in the "general DP specification" (gDPS) language. Such a file constitutes one possible input format for the DP2PN2Solver software. Section 3 already gave an overview of the gDPS language. Here we examine each language construct in more detail. A gDPS source file always begins with key word BEGIN and ends with the keyword END and its body consists of several structured sections, each of which we will consider in more detail now.

At any point in the body of the gDPS file, one may have block comments that start with /* and end with */ and one-line comments that start with // and extend to the end of the line. Comments and white-space characters such as blanks, tabs and newlines (CR and LF) do not have any functionality associated with them, except that they act as token separators for the parser of the DP2PN module.

3.3.1 Name Section

The mandatory Name section begins with the keyword NAME followed by the name you would like to assign to your particular model, followed by a semicolon. This is a logical name, which does not have to match the file name of your gDPS source; however it seems good practice to match them. For example, you might have a gDPS file named tsp2005.dp and your name section might read

```
NAME TravelingSalesperson;
```

Since the logical model name is used for naming intermediate files and output files, it is prudent to use only names that do not cause trouble for your operating system's file name conventions (i.e. avoid special symbols and blanks).

3.3.2 General Variables Section

The optional General Variables section begins with the keyword GENERAL_VARIABLES_BEGIN and ends with the keyword GENERAL_VARIABLES_END. It allows the definition of variables and constants in a C++/Java style syntax.

A legal variable name starts with a character from the alphabet $\{a, \ldots, z, A, \ldots, Z\}$ and may be followed by any number of alphanumerical characters from the alphabet $\{a, \ldots, z, A, \ldots, Z, 0, 1, \ldots, 9\}$. Variable names are case sensitive.

As a general rule, if a variable definition is legal in Java, then it is also legal in gDPS. In order to emphasize the global scope of the variables defined in this section, we adopted the Java convention to declare variables as `static`, and in order to emphasize that the variables defined in this section are to be accessed exclusively within a gDPS file, we adopted the convention to declare variables as `private`. Constants carry the qualifier `final`.

Integer Type

In order to define and initialize a variable of integral type, the `int` keyword is used, followed by a legal variable name, the equation (or equal-sign) character, an integer literal and a semicolon. For example,

```
private static int n=5;
```

defines an integer variable `n` which is initialized to the value 5.

Constants of Java classes can be used as well, as can be seen in the example

```
private static final int infty=Integer.MAX_VALUE;
```

where we define an infinity constant `infty` to be the maximally representable integer as provided in `java.lang.Integer.MAX_VALUE`.

Floating Point Type

In order to define and initialize a variable of floating point type, the `double` keyword is used, followed by a legal variable name, the equal-sign character, a floating point literal and a semicolon. For example,

```
private static double x3=3.14;
```

defines a floating point variable `x3` which is initialized to the value 3.14.

String Type

In order to define and initialize a variable of String type, the `String` keyword is used, followed by a legal variable name, the equal-sign character, a string literal (in quotes) and a semicolon. For example,

```
private static String s="Hello";
```

defines a string variable `s` which is initialized to “Hello”.

Array Types

Array variables can be defined and initialized by placing a pair of brackets for each dimension after the type declaration, and by using (nested) lists, enclosed in braces, for initialization purposes. For example,

```
private static double[][] myMatrix=
  {
   { -2.5, 13  , -0.2 },
   { 3  ,  4.0,  1.234},
  };
```

defines and initializes a 2-dimensional array of doubles, representing a 2×3 matrix. We can reference the value 1.234 via `myMatrix[1][2]` in the usual way. As is typical for Java arrays, the index count or subscript in each dimension starts with 0, not 1.

The length of an array dimension can be determined using the `length` construct. Thus, the number of rows and columns of `myMatrix` can be determined as follows:

```
private static int m=myMatrix.length; //m=2
private static int n=myMatrix[0].length; //n=3
```

3.3.3 Set Variables Section

The optional Set Variables section begins with the keyword `SET_VARIABLES_BEGIN` and ends with the keyword `SET_VARIABLES_END`. It provides a compact and flexible way to define and initialize set variables and constants.

A legal set variable name follows the same rules as a general variable name, it cannot start with a numerical chracter. Set definitions do not take `private` and `static` modifiers. Those are added internally by the compiler, which generates `NodeSet` objects for each variable or constant defined as `Set` in this section. The `NodeSet` type is a subclass of `java.util.TreeSet`; we discuss this further in the next section.

A variable or constant defined in this section starts with the keyword `Set`, an equals character, and then a variety of notations are available to specify the set.

1. To resemble the common mathematical notation, all elements may be enumerated within braces. Duplicate elements are ignored. The order in which elements are enumerated is irrelevant.
2. Ellipsis is supported with the following subrange syntax: an opening brace, followed by the smallest element to be in the set, followed by two dots, followed by the largest element to be in the set, followed by a closing brace.
3. The empty set $\emptyset$ is defined as `{}`.

4. If $s1$ and $s2$ are correctly defined sets, then the set union $s1 \cup s2$ is expressed using the keyword SETUNION instead of the mathematical symbol $\cup$.
5. If $s1$ and $s2$ are correctly defined sets, then the set intersection $s1 \cap s2$ is expressed using the keyword SETINTERSECTION instead of the mathematical symbol $\cap$.
6. If $s1$ and $s2$ are correctly defined sets, then the set difference $s1 - s2$ is expressed using the keyword SETMINUS instead of the mathematical symbol $-$.
7. The binary set operators mentioned above can be nested, the default hierarchy (set difference has highest priority, then set intersection, then set union) can be overridden by using parenthesization.

All sets are assumed to contain a finite number of integer typed elements. Each set variable definition is terminated with a semicolon. For example,

```
Set enumeratedSet={5,2,2,3,7,2};
```

defines and initializes the set $\{2, 3, 5, 7\}$.

In another example,

```
Set subrangeSet={1,..,n};
```

defines and initializes a set that contains all positive integers up to (and including) n, where n must be defined in the General Variables section.

3.3.4 General Functions Section

The optional General Functions section begins with the keyword GENERAL_FUNCTIONS_BEGIN and ends with the keyword GENERAL_FUNCTIONS_END. It provides a flexible way to define functions in Java syntax.

As a general rule, if a function (method) definition is legal in Java, then it is also legal in gDPS. In order to emphasize the global, non-object oriented character of the functions defined in this section, we adopted the Java convention to declare functions as "static", and in order to emphasize that the functions defined in this section are to be accessed exclusively within a gDPS file, we adopted the convention to declare functions as "private".

For set types the use of class type NodeSet (which is subclassed from the standard Java class java.util.TreeSet and named so for legacy reason) is recommended in order to be most compatible with sets specified in the set variables section. Since java.util.TreeSet implements the standard Java interface java.util.SortedSet all sets are actually understood as sorted sets, which is sometimes advantageous. For instance, in the gDPS source of the BST problem (Sect. 2.6) we can conveniently split a set using the headSet() method. Additional useful methods such as tailSet(), subSet(), etc., are documented in the Java 2 SDK 1.4.2 API specification (available at http://java.sun.com).

For example, assume we are given the adjacency matrix distance[][] of a directed graph in the general variables section, where ∞-entries represent edges that are not present in the graph. Given a node node we can compute the set of adjacent nodes using the function

```
private static NodeSet possibleNextNodes(int node) {
  NodeSet result = new NodeSet();
  for (int i=0; i<distance[node].length; i++) {
    if (distance[node][i]!=infty) {
      result.add(new Integer(i));
    }
  }
  return result;
}
```

where infty is a constant declared in the general variables section. This function uses a conventional Java for-loop and an if-statement to detect whether an index should be included in the resulting set result which is eventually returned. The NodeSet object result inherits the methods from java.util.TreeSet, so we use the add() method to add integers to the resulting set result. Since in Java sets are containers of objects, we need to wrap the primitive int type into an Integer object.

3.3.5 State Type Section

The mandatory State Type section begins with the keyword STATE_TYPE, followed by a colon, the parameter list, and ends with a semicolon. For our definition of DP, a state is considered to be an ordered, fixed-length tuple of parameters. Each parameter has a type and a name. Permissible parameter types are int and Set. The parameter list is enclosed in a pair of parentheses and parameters are separated by commas; this resembles the Java style parameter list for functions. For example,

```
STATE_TYPE: (int stage, Set s1, int x, Set s2);
```

defines a DP problem where a state is an ordered quadruple (stage, s_1, x, s_2) consisting of an integer, a set, an integer, and another set, respectively.

3.3.6 Decision Variable Section

The mandatory Decision Variable section begins with the keyword DECISION_VARIABLE, followed by a colon, the decision variable type, the decision variable name, and ends with a semicolon. All decisions are considered to be choices from a finite set, so at this point we can assume without loss of generality that the decision variable type be int. The decision variable name follows the usual requirement for Java identifiers (cannot start with a digit). For example,

```
DECISION_VARIABLE: int d;
```

declares the decision variable for this DP problem to be d.

3.3.7 Decision Space Section

The mandatory Decision Space section begins with the keyword DECISION_SPACE, followed by a colon, the decision set function name and parameter list, the equal-sign symbol, the decision set (which can expressed in various ways), and ends with a semicolon. Since the current decision set under consideration depends on the current state, the parameter list is populated with precisely those variables that make up a state (see state type section). The decision set is the space of alternatives, one of which gets to be chosen to be in the optimal policy; it has to be of type Set and can be expressed in one of the following ways.

1. Explicitly build a set using complete enumeration or subrange syntax, possibly using variables from the parameter list. For example, if a state is a pair of integers, then

   ```
   DECISION_SPACE: decisionSet(i,j)={i,..,j - 1};
   ```

   characterizes the decision set as the set containing all integers between the first index i and the second index j, including i but excluding j.
2. Delegate the task of building the decision set, or parts of it, to a function specified in the general functions section, which provides a set-typed return value. For example, if a state consists of a single integer representing the current node, then

   ```
   DECISION_SPACE: possibleAlternatives(currentNode)
                   =possibleNextNodes(currentNode);
   ```

   delegates the task of computing the decision set possibleAlternatives to the function possibleNextNodes() defined in the General Functions section.
3. If S_1 and S_2 are correctly defined sets, it is possible to form the union $S_1 \cup S_2$, the intersection $S_1 \cap S_2$ and the set difference $S_1 - S_2$ using the keywords SETUNION, SETINTERSECTION and SETMINUS, respectively, as was the case in the Set Variables section.

3.3.8 Goal Section

The mandatory Goal section describes the computational goal of the DP instance at hand. It begins with the keyword GOAL, followed by a colon, the identifier of the DPFE functional, and then, within parentheses, a goal state is declared by providing a suitable value for each parameter of the state. This can be done by either providing an explicit numerical value, or by providing a symbolic variable or constant defined in the general variables section. The section ends with a semicolon. For example, if a state is a triple consisting of two integers, followed by a set, then

```
GOAL: f(7,n,goalSet);
```

assumes that n is predefined integer variable or constant and that `goalSet` is a predefined set variable or constant. The goal of our computation has been reached once we have evaluated $f(7, n, \text{goalSet})$.

3.3.9 DPFE Base Section

The mandatory DPFE Base section is used to define base cases or conditions to terminate the recursive evaluation of a DPFE. There are two alternative ways of specifying the base of a DPFE. Exactly one of these two possibilities must be present in a gDPS file.

The first possibility is to express the base cases in a conditional fashion. The section begins with the keyword `DPFE_BASE_CONDITIONS`, followed by a colon, followed by one or more base condition statements. Each base condition statement starts with the DPFE functional and its parameter list (see DPFE section), followed by an equal-sign character and an arithmetic expression that evaluates to a floating point number (see the reward function section for details about arithmetic expressions). This is followed by the keyword `WHEN`, a conditional expression in parentheses and a semicolon, which designates the end of a base condition statement. The conditional expression may consist of numerical conditions involving Java style operators to compare numbers (or numerical variables or constants), such as `<`, `<=`, `>`, `>=`, `!=`, `==`. Sets (or set variables) can be compared for equality using the keyword `SETEQUALS`. Atomic conditions can be combined using the logical boolean operators `&&` ("and"), `||` ("or"), `!` ("not"); conditions can be nested, if desired, and the default hierarchy for evaluating the logical operations can be influenced by using parentheses. Every conditional expression evaluates to either "true" or "false". For example, if a state is a pair of integers, then

```
DPFE_BASE_CONDITIONS:
  f(i,j)=0.0 WHEN (i==j);
  f(i,j)=1.0 WHEN (i>j) && (i<9) && (j>5);
```

designates states such as $(1,1)$, $(2,2)$, etc. as base states which get assigned the value 0.0 (by the first base condition statement). The second base condition statement designates the states $(8,6)$, $(8,7)$, and $(7,6)$ as base states, which get assigned the value 1.0. If a state is a set, then

```
DPFE_BASE_CONDITIONS:
  f(currentSet)=4.0 WHEN (currentSet SETEQUALS {3,4});
  f(currentSet)=8.0 WHEN (currentSet SETEQUALS {5,..,8});
```

designates the state $(\{3,4\})$ as a base state, which gets assigned the value 4.0 and designates the state $(\{5,6,7,8\})$ as a base state, which gets assigned the value 8.0.

The second possibility is to express the base cases in an enumerative way. The section begins with the keyword `DPFE_BASE`, followed by a colon, followed by one or more DPFE base statements. A DPFE base statement can be an assignment statement, or a block of possibly nested for-loops that contain assignment statements within its body. Each assignment statement starts with the DPFE functional and an argument list, enclosed in parentheses, that provides suitable values for each of the components that make up a state. This is followed by an equal-sign symbol, an expression that evaluates to a floating point number, and a semicolon, which designates the end of an assignment statement. For example, assuming the set variable `setOfAll` defined in the set variables section is valued $\{0, 1, 2, 3\}$ and the array `distance` is defined and initialized in the general variables section with `distance[1][0]` equaling 7.5, then

```
DPFE_BASE:
  f(0,setOfAll)=0.0 ;
  f(1,setOfAll)=distance[1][0];
```

designates the state $(0, \{0, 1, 2, 3\})$ to be a base state, which gets assigned the value 0.0 (by the first assignment statement). The second assignment statement designates the state $(1, \{0, 1, 2, 3\})$ as a base state, which gets assigned the value 7.5.

The for-loop notation allows a convenient shortcut when consecutive integers are involved in assignment statements:

```
DPFE_BASE:
  FOR(i=2;i<=10;i++) {
    f(emptySet,i)=0.0;
  }
```

In a DP model, one has to be careful to make sure that every DPFE base case that can occur during the computation is actually covered in this section. No damage is done, if a base case is covered more than once, provided that the same value is assigned.

3.3.10 DPFE Section

The mandatory DPFE section describes the recursive equation that is at the center of a DP model. It begins with the keyword `DPFE`, followed by a colon, the DP functional name and parameter list which is populated with precisely those variables that make up a state (see state type section). Then there is an equal-sign symbol, followed by either the `MAX_` or `MIN_` operator indicating the direction of optimization, followed by an opening brace, the decision variable as defined in the decision variable section, the keyword `IN`, the decision set identifier as defined in the decision space section, followed by a closing brace. Then, within the next pair of braces, will be the DP functional, possibly more than once, performing the recursive call(s) after applying a transformation function to it, and also the call to the reward function. All these functionals must be connected by either the addition or multiplication operator. The parameter lists of the transformation functions and of the reward function is populated with precisely those variables that make up a state (see state type section) and in addition with the decision variable, since the transformation function value and the reward function value are in general dependent on both the state and the decision. The DPFE section ends with a semicolon. Optionally, each recursively invoked functional may be multiplied with a weight, which must itself be expressed as a function in the transition weight section. The multiplication symbol for these weights is "." (dot), to distinguish it from the "*" (star) symbol used in other contexts.

As a first example, if a state consists of a pair (i, j) of integers, then

```
DPFE: f(i,j)
      =MIN_{k IN decisionSet}
                              { f(t1(i,j,k))
                               +f(t2(i,j,k))
                               +r(i,j,k)
                              };
```

specifies a DPFE whose functional is named f. There are two recursive calls via the two transformation functions t_1 and t_2. The reward function is named r, and all functionals are connected by "+" resulting in an additive DPFE.

In the following example

```
DPFE: fun(a,b)
      =MAX_{m IN myAlternatives}
                              { fun(t(a,b,m))
                               *cost(a,b,m)
                              };
```

there is only a single recursive call to `fun` via the transformation function t. It is connected to the reward functional `cost` by the multiplication operator "*" resulting in a multiplicative DPFE.

Assuming there are transition weights p_1 and p_2 specified in the transition weight section, the following example

```
DPFE: f(i,j)
      =MAX_{d IN decisionSet}
                              { p1.f(t1(i,j,d))
                               +p2.f(t2(i,j,d))
                               +r(i,j,d)
                              };
```

shows an additive DPFE with two recursive calls via the transformation functions t_1 and t_2 the result of each of which is weighted with the floating point values p_1 and p_2 respectively. For details about the calculation of the weights, please refer to the Transition Weight section.

3.3.11 Cost/Reward Function Section

The mandatory Reward Function section defines the cost or profit function referenced in the DPFE. If a DP model does not need a reward function, it can be defined to be a constant function, equaling the identity element of addition (i.e. 0) or multiplication (i.e. 1), depending on whether we have an additive or multiplicative DPFE. The section begins with the keyword REWARD_FUNCTION, followed by a colon, the identifier of the reward function as used in the DPFE section and the parameter list which contains the identifiers of the state components, and the decision variable. Then there is an equal-sign symbol followed by an arithmetic expression that evaluates to a floating point number. This expression may involve the parameters. A semicolon denotes the end of the Reward Function section. For example, in

```
REWARD_FUNCTION: rew(i,j,k)
                 =myArr[i]*myArr[j]*myArr[k];
```

the reward function is named rew and the expression it evaluates to is a product of three array variables from the array myArr.

In addition to the usual arithmetic operators such as addition (+), subtraction (-), multiplication (*), integer or real-vauled division (/), and integer remainder (%) there is the possibility to delegate more complicated arithmetic to a helper function defined in the general functions section. For example, in

```
REWARD_FUNCTION: r(stage,remainingMoney,m)
                 =reliabilityOfStage(stage,m);
```

the calculation of the reward is performed by a helper function named reliabilityOfStage, which happens to require only two of the three parameters as arguments (i.e. the reward is independent of the parameter remainingMoney). The helper function is assumed to return a floating-point value of type double.

3.3.12 Transformation Function Section

The mandatory Transformation Function section defines the one or more next-state transformation (or transition) functions referenced in the DPFE. It begins with the keyword TRANSFORMATION_FUNCTION, followed by a colon, and a semicolon separated list of function definitions.

Since a transformation function computes the next-state when provided with the current state and the decision taken, a suitable definition starts with the function identifier as used in the DPFE section and the parameter list, which contains the identifiers of the state components and the decision variable. Then there is an equal-sign symbol followed by a parenthesized list of expressions, each of which evaluates to an integer or to a set, depending on the definition of what constitutes a state. Any expression may involve the parameters. A semicolon denotes the end of a transformation function definition. In the example

```
TRANSFORMATION_FUNCTION: t1(i,j,k)
                          =(i,k);
                         t2(i,j,k)
                          =(k+1,j);
```

there are two transformation functions t_1 and t_2, each of which maps a state, represented by a pair of integers (i, j) and a decision k to a new state. The new state is (i, k) in case of t_1 and it is $(k+1, j)$ in case of t_2. The arithmetic operators `+,-,*,/,%` can be used for state components that are integers.

For state components that are sets it is possible to use the set operator keywords `SETUNION`, `SETINTERSECTION` and `SETMINUS`. For example, if the state is a pair of an integer x and a set `nSet`, and y is the decision variable, then

```
TRANSFORMATION_FUNCTION: t(x,nSet,y)
                          =(y, nSet SETUNION {y});
```

defines the transformation function t in the following way: the new state's first component is the integer y and the second component is the set that results from adding the element y to the set `nSet`. More complex transformation functions may require use of helper functions. For example, exponentiation, maximization, minimization, etc. have to be delegated to a helper function in the General Functions section. There, methods of `java.lang.Math` such as `abs()`, `ceil()`, `floor()`, `exp()`, `max()`, `min()`, `pow()`, etc., can be used.

3.3.13 Transition Weight Section

The optional Transition Weight section describes the transition weights that may precede the recursive function calls in the DPFE. It is assumed that the number of transition weights and the order in which the transition weights are defined in this section matches the number and order of the corresponding transformation functions, according to the DPFE.

The section begins with the keyword `TRANSITION_WEIGHTS`, followed by a colon, and a semicolon separated list of real-valued transition weight function definitions.

A transition weight function definition begins with the identifier of the transition weight as used in the DPFE section, followed by the parameter list which contains the identifiers of the state components, and the decision variable. Then there is an equal-sign symbol followed by an arithmetic expression that evaluates to a floating point number (see the reward function section for details about legal arithmetic expressions). A semicolon denotes the end of the transition weight function definition. For example,

```
TRANSITION_WEIGHTS: p1(n,sn,xn)=1.0-winProbability;
                    p2(n,sn,xn)=winProbability;
```

defines the two transition weight functions p_1 and p_2 which in this case are both constant functions independent of the state and the decision.

In the example

```
TRANSITION_WEIGHTS: p(x,y,d)=myProbFunction(x,d);
```

the computation of the transition weight p is delegated to the helper function `myProbFunction` which returns a floating-point value of type `double` and which is defined in the general functions section. The function `myProbFunction` happens to require only two of the three parameters as arguments (i.e. the transition weight is independent of the state component y).

3.4 BNF Grammar of the gDPS language

The Backus-Naur form (BNF) for the gDPS language is given in this section. The extended form is used, where "*" denotes "zero or more occurence", "+" denotes "one or more occurence", and "?" denotes "zero or one occurence (option)." Terminal symbols are delimited by an opening angled bracket ("<") and a closing angled bracket (">").

```
dpSpecification ::= <BEGIN> sectionList <END> <EOF>
sectionList ::= nameSection ( generalVariablesSection )?
   ( setVariablesSection )? ( generalFunctionsSection )?
   stateTypeSection decisionVariableSection
   decisionSpaceSection goalSection
   ( dbfeBaseConditionsSection | dpfeBaseSection ) dpfeSection
   rewardFunctionSection transformationFunctionSection
   ( transitionWeightSection )?
nameSection ::= <NAME> <IDENTIFIER> <SEMICOLON>
generalVariablesSection ::= <GENERAL_VARIABLES>
setVariablesSection ::= <SET_VARIABLES_BEGIN>
   ( setVariableAssignment )* <SET_VARIABLES_END>
setVariableAssignment ::= <SET> <IDENTIFIER> <EQUALS>
   setUnionExpression <SEMICOLON>
setUnionExpression ::= setIntersectionExpression
   ( <SETUNION> setIntersectionExpression )*
setIntersectionExpression ::= setDifferenceExpression
   ( <SETINTERSECTION> setDifferenceExpression )*
setDifferenceExpression ::= setPrimaryExpression
   ( <SETMINUS> setPrimaryExpression )*
setPrimaryExpression ::= setGlobalFunctional
 | setArrayVariable
 | explicitSet
 | <LPAREN> setUnionExpression <RPAREN>
setGlobalFunctional ::= <IDENTIFIER> <LPAREN>
   setGlobalFunctionalArgumentList <RPAREN>
setGlobalFunctionalArgumentList ::= ( globalFunctionalArgument
```

```
   ( <COMMA> globalFunctionalArgument )* )?
setArrayVariable ::= <IDENTIFIER>
   ( <LBRACKET> additiveExpression <RBRACKET> )*
explicitSet ::= explicitSetInDoubleDotNotation
 | explicitSetEnumeration
explicitSetInDoubleDotNotation ::= <LBRACE> additiveExpression
   <COMMA> <DOUBLEDOT> <COMMA> additiveExpression <RBRACE>
explicitSetEnumeration ::= <LBRACE> ( additiveExpression
   ( <COMMA> additiveExpression )* )? <RBRACE>
additiveExpression ::= multiplicativeExpression
   ( ( <PLUS> | <MINUS> ) multiplicativeExpression )*
multiplicativeExpression ::= primaryExpression
   ( ( <MULT> | <DIV> | <MOD> ) primaryExpression )*
primaryExpression ::= globalFunctional
 | arrayVariable
 | <INTEGER_LITERAL>
 | <LPAREN> additiveExpression <RPAREN>
globalFunctional ::= <IDENTIFIER> <LPAREN>
   globalFunctionalArgumentList <RPAREN>
globalFunctionalArgumentList ::= ( globalFunctionalArgument
   ( <COMMA> globalFunctionalArgument )* )?
globalFunctionalArgument ::= arrayVariable
arrayVariable ::= <IDENTIFIER> ( <LBRACKET> additiveExpression
   <RBRACKET> )*
generalFunctionsSection ::= <GENERAL_FUNCTIONS>
stateTypeSection ::= <STATE_TYPE> <COLON>
   stateTypeParameterList <SEMICOLON>
stateTypeParameterList ::= <LPAREN> stateTypeFormalParameter
   ( <COMMA> stateTypeFormalParameter )* <RPAREN>
stateTypeFormalParameter ::= stateTypeType
   stateTypeVariableDeclaratorId
stateTypeType ::= <INT> | <SET>
stateTypeVariableDeclaratorId ::= <IDENTIFIER>
decisionVariableSection ::= <DECISION_VARIABLE> <COLON>
   decisionVariableType <IDENTIFIER> <SEMICOLON>
decisionVariableType ::= <INT> | <STRING>
decisionSpaceSection ::= <DECISION_SPACE> <COLON> <IDENTIFIER>
   <LPAREN> argumentList <RPAREN> <EQUALS>
   decisionSetUnionExpression <SEMICOLON>
argumentList ::= <IDENTIFIER> ( <COMMA> <IDENTIFIER> )*
decisionSetUnionExpression ::= decisionSetIntersectionExpression
   ( <SETUNION> decisionSetIntersectionExpression )*
decisionSetIntersectionExpression ::=
   decisionSetDifferenceExpression ( <SETINTERSECTION>
   decisionSetDifferenceExpression )*
```

```
decisionSetDifferenceExpression ::= decisionSetPrimaryExpression
   ( <SETMINUS> decisionSetPrimaryExpression )*
decisionSetPrimaryExpression ::= decisionSetGlobalFunctional
 | decisionSetArrayVariable
 | decisionSetExplicit
 | <LPAREN> decisionSetUnionExpression <RPAREN>
decisionSetGlobalFunctional ::= <IDENTIFIER> <LPAREN>
   decisionSetGlobalFunctionalArgumentList <RPAREN>
decisionSetGlobalFunctionalArgumentList ::=
   ( decisionSetGlobalFunctionalArgument ( <COMMA>
   decisionSetGlobalFunctionalArgument )* )?
decisionSetGlobalFunctionalArgument ::= <IDENTIFIER>
decisionSetArrayVariable ::= <IDENTIFIER> ( <LBRACKET>
   decisionAdditiveExpression <RBRACKET> )*
decisionSetExplicit ::= decisionSetExplicitInDoubleDotNotation
 | decisionSetExplicitEnumeration
decisionSetExplicitInDoubleDotNotation ::= <LBRACE>
   decisionAdditiveExpression <COMMA> <DOUBLEDOT> <COMMA>
   decisionAdditiveExpression <RBRACE>
decisionSetExplicitEnumeration ::= <LBRACE>
   decisionAdditiveExpression ( <COMMA>
   decisionAdditiveExpression )* <RBRACE>
decisionAdditiveExpression ::= decisionMultiplicativeExpression
   ( ( <PLUS> | <MINUS> ) decisionMultiplicativeExpression )*
decisionMultiplicativeExpression ::= decisionPrimaryExpression
   ( ( <MULT> | <DIV> | <MOD> ) decisionPrimaryExpression )*
decisionPrimaryExpression ::= decisionArrayVariable
 | <INTEGER_LITERAL>
 | <LPAREN> decisionAdditiveExpression <RPAREN>
decisionArrayVariable ::= <IDENTIFIER> ( <LBRACKET>
   decisionAdditiveExpression <RBRACKET> )*
goalSection ::= <GOAL> <COLON> <IDENTIFIER> <LPAREN>
   primaryExpression ( <COMMA> primaryExpression )* <RPAREN>
   <SEMICOLON>
singleState ::= <LPAREN> numberOrConstant ( <COMMA>
   numberOrConstant )* <RPAREN>
numberOrConstant ::= <INTEGER_LITERAL> | <IDENTIFIER>
dbfeBaseConditionsSection ::= <DPFE_BASE_CONDITIONS> <COLON>
   ( dpfeBaseConditionStatement )+
dpfeBaseConditionStatement ::= <IDENTIFIER> <LPAREN>
   argumentList <RPAREN> <EQUALS>
   rewardFunctionAdditiveExpression <WHEN> <LPAREN>
   conditionalOrCExpression <RPAREN> <SEMICOLON>
conditionalOrCExpression ::= conditionalAndCExpression
   ( <COND_OR> conditionalAndCExpression )*
```

```
conditionalAndCExpression ::= equalityCExpression
   ( <COND_AND> equalityCExpression )*
equalityCExpression ::= setEqualityCExpression
 | numericalEqualityCExpression
setEqualityCExpression ::=
   transformationFunctionSetUnionExpression <SETEQUALS>
   transformationFunctionSetUnionExpression
numericalEqualityCExpression ::= relationalCExpression
   ( ( <EQ> | <NE> ) relationalCExpression )*
relationalCExpression ::= negatingCExpression
   ( ( <LT> | <GT> | <LTE> | <GTE> ) negatingCExpression )*
negatingCExpression ::= ( <EXCLAMATION_MARK> )?
   primaryCExpression
primaryCExpression ::= transformationFunctionAdditiveExpression
 | <LPAREN> conditionalOrCExpression <RPAREN>
dpfeBaseSection ::= <DPFE_BASE> <COLON> ( dpfeBaseStatement )+
dpfeBaseStatement ::= dpfeBaseBlock
 | dpfeBaseForStatement
 | dpfeBaseAtomicStatement
dpfeBaseBlock ::= <LBRACE> ( dpfeBaseStatement )* <RBRACE>
dpfeBaseForStatement ::= <FOR> <LPAREN> <IDENTIFIER> <EQUALS>
   numberOrConstant <SEMICOLON> <IDENTIFIER>
   ( <LT> | <LTE> | <GT> | <GTE> ) numberOrConstant <SEMICOLON>
   <IDENTIFIER> ( <INC> | <DEC> ) <RPAREN> dpfeBaseStatement
dpfeBaseAtomicStatement ::= <IDENTIFIER> singleState <EQUALS>
   rewardFunctionAdditiveExpression <SEMICOLON>
dpfeSection ::= <DPFE> <COLON> <IDENTIFIER> <LPAREN>
   argumentList <RPAREN> <EQUALS> dpfeMinOrMax dpfeDecisionLoop
   dpfeExpression <SEMICOLON>
dpfeMinOrMax ::= <MIN_> | <MAX_>
dpfeDecisionLoop ::= <LBRACE> <IDENTIFIER> <IN> <IDENTIFIER>
   <RBRACE>
dpfeExpression ::= <LBRACE> ( dpfeAdditiveExpression |
   dpfeMultiplicativeExpression ) <RBRACE>
dpfeAdditiveExpression ::= dpfeFunctional
   ( <PLUS> dpfeFunctional )+
dpfeMultiplicativeExpression ::= dpfeFunctional
   ( <MULT> dpfeFunctional )+
dpfeFunctional ::= dpfeFunctionalAtom
 | dpfeDoublyNestedFunctional
dpfeFunctionalAtom ::= <IDENTIFIER> <LPAREN> argumentList
   <RPAREN>
dpfeDoublyNestedFunctional ::= ( <IDENTIFIER> <DOT> )?
   <IDENTIFIER> <LPAREN> <IDENTIFIER> <LPAREN> argumentList
   <RPAREN> <RPAREN>
```

```
rewardFunctionSection ::= <REWARD_FUNCTION> <COLON>
   <IDENTIFIER> <LPAREN> argumentList <RPAREN> <EQUALS>
   rewardFunctionBody <SEMICOLON>
rewardFunctionBody ::= rewardFunctionAdditiveExpression
rewardFunctionAdditiveExpression ::=
   rewardFunctionMultiplicativeExpression ( ( <PLUS> | <MINUS> )
   rewardFunctionMultiplicativeExpression )*
rewardFunctionMultiplicativeExpression ::=
   rewardFunctionPrimaryExpression ( ( <MULT> | <DIV> | <MOD> )
   rewardFunctionPrimaryExpression )*
rewardFunctionPrimaryExpression ::=
   rewardFunctionGlobalFunctional
  | rewardFunctionArrayVariable
  | rewardFunctionLiteral
  | <LPAREN> rewardFunctionAdditiveExpression <RPAREN>
rewardFunctionGlobalFunctional ::= <IDENTIFIER> <LPAREN>
   rewardFunctionGlobalFunctionalArgumentList <RPAREN>
rewardFunctionGlobalFunctionalArgumentList ::=
   ( rewardFunctionGlobalFunctionalArgument ( <COMMA>
   rewardFunctionGlobalFunctionalArgument )* )?
rewardFunctionGlobalFunctionalArgument ::= <IDENTIFIER>
rewardFunctionArrayVariable ::= <IDENTIFIER> ( <LBRACKET>
   rewardFunctionAdditiveExpression <RBRACKET> )*
rewardFunctionLiteral ::= <INTEGER_LITERAL>
  | <FLOATING_POINT_LITERAL>
transformationFunctionSection ::= <TRANSFORMATION_FUNCTION>
   <COLON> ( <IDENTIFIER> <LPAREN> argumentList <RPAREN>
   <EQUALS> transformationFunctionNewState <SEMICOLON> )+
transformationFunctionNewState ::= <LPAREN>
   transformationFunctionNewStateCoordinateList <RPAREN>
transformationFunctionNewStateCoordinateList ::=
   transformationFunctionNewStateCoordinate ( <COMMA>
   transformationFunctionNewStateCoordinate )*
transformationFunctionNewStateCoordinate ::=
   transformationFunctionAdditiveExpression
  | transformationFunctionSetUnionExpression
transformationFunctionAdditiveExpression ::=
   transformationFunctionMultiplicativeExpression
   ( ( <PLUS> | <MINUS> )
   transformationFunctionMultiplicativeExpression )*
transformationFunctionMultiplicativeExpression ::=
   transformationFunctionPrimaryExpression
   ( ( <MULT> | <DIV> | <MOD> )
   transformationFunctionPrimaryExpression )*
transformationFunctionPrimaryExpression ::=
```

```
   transformationFunctionGlobalFunctional
 | transformationFunctionArrayVariable
 | <INTEGER_LITERAL>
 | <LPAREN> transformationFunctionAdditiveExpression <RPAREN>
transformationFunctionGlobalFunctional ::= <IDENTIFIER> <LPAREN>
   transformationFunctionGlobalFunctionalArgumentList <RPAREN>
transformationFunctionGlobalFunctionalArgumentList ::=
   ( transformationFunctionGlobalFunctionalArgument ( <COMMA>
   transformationFunctionGlobalFunctionalArgument )* )?
transformationFunctionGlobalFunctionalArgument ::= <IDENTIFIER>
transformationFunctionArrayVariable ::= <IDENTIFIER>
   ( <LBRACKET> transformationFunctionAdditiveExpression
   <RBRACKET> )*
transformationFunctionSetUnionExpression ::=
   transformationFunctionSetIntersectionExpression ( <SETUNION>
   transformationFunctionSetIntersectionExpression )*
transformationFunctionSetIntersectionExpression ::=
   transformationFunctionSetDifferenceExpression
   ( <SETINTERSECTION>
   transformationFunctionSetDifferenceExpression )*
transformationFunctionSetDifferenceExpression ::=
   transformationFunctionSetPrimaryExpression ( <SETMINUS>
   transformationFunctionSetPrimaryExpression )*
transformationFunctionSetPrimaryExpression ::=
   transformationFunctionSetGlobalFunctional
 | transformationFunctionSetArrayVariable
 | transformationFunctionExplicitSet
 | <LPAREN> transformationFunctionSetUnionExpression <RPAREN>
transformationFunctionSetGlobalFunctional ::= <IDENTIFIER>
   <LPAREN>
   transformationFunctionSetGlobalFunctionalArgumentList
   <RPAREN>
transformationFunctionSetGlobalFunctionalArgumentList ::=
   ( transformationFunctionSetGlobalFunctionalArgument ( <COMMA>
   transformationFunctionSetGlobalFunctionalArgument )* )?
transformationFunctionSetGlobalFunctionalArgument ::=
   <IDENTIFIER>
transformationFunctionSetArrayVariable ::= <IDENTIFIER>
   ( <LBRACKET> transformationFunctionAdditiveExpression
   <RBRACKET> )*
transformationFunctionExplicitSet ::=
   transformationFunctionExplicitSetInDoubleDotNotation
 | transformationFunctionExplicitSetEnumeration
transformationFunctionExplicitSetInDoubleDotNotation ::=
   <LBRACE> transformationFunctionAdditiveExpression <COMMA>
```

```
   <DOUBLEDOT> <COMMA> transformationFunctionAdditiveExpression
   <RBRACE>
transformationFunctionExplicitSetEnumeration ::= <LBRACE>
   ( transformationFunctionAdditiveExpression ( <COMMA>
   transformationFunctionAdditiveExpression )* )? <RBRACE>
transitionWeightSection ::= <TRANSITION_WEIGHTS> <COLON>
   ( <IDENTIFIER> <LPAREN> argumentList <RPAREN> <EQUALS>
   rewardFunctionAdditiveExpression <SEMICOLON> )+
```

4

DP Problem Specifications in gDPS

In this chapter, gDPS source files for all of the DP examples from Chap. 2 are listed. We elaborate on some of the files to illustrate some of the finer points on how to code with gDPS.

4.1 gDPS source for ALLOT

In the ALLOTt problem allotment decisions and their costs are defined in separate tables.

```
BEGIN
  NAME ALLOTt; //Optimal Allotment Problem (multiplicative)
               // cost defined by tables
               // REF: Hillier,pp.549-552

  GENERAL_VARIABLES_BEGIN
    private static final int num=3; //no.of items
    private static final int lim=2; //capacity
    private static final double infty=Double.MAX_VALUE;
    private static int[][] allotment =
      { {0,1,2},
        {0,1,2},
        {0,1,2}
      };
    private static double [][] probFail =
      { {.40,.20,.15},
        {.60,.40,.20},
        {.80,.50,.30}
      };
  GENERAL_VARIABLES_END
```

A. Lew and H. Mauch: *DP Problem Specifications in gDPS*, Studies in Computational Intelligence (SCI) **38**, 125–203 (2007)
www.springerlink.com

```
  STATE_TYPE: (int stage, int total);

  DECISION_VARIABLE: int d;

  DECISION_SPACE: decisionSet(stage,total)={0,..,lim};

  GOAL: f(0,0);

  DPFE_BASE_CONDITIONS:
    f(stage,total)=infty WHEN ((stage==num)&&(total>lim));
    f(stage,total)=1.00 WHEN ((stage==num)&&(total<=lim));

  DPFE:
    f(stage,total)=MIN_{d IN decisionSet}
        {cost(stage,total,d) * f(t(stage,total,d))};

  REWARD_FUNCTION:
    cost(stage,total,d)=probFail[stage][d];

  TRANSFORMATION_FUNCTION:
    t(stage,total,d)=(stage+1,total+allotment[stage][d]);

END
```

In the ALLOTf problem the costs are defined nontabularly, i.e., by general functions.

```
BEGIN
  NAME ALLOTf; // Optimal Allotment Problem (additive)
               // cost defined by function
               // REF: Winston04,pp.975-977

  GENERAL_VARIABLES_BEGIN
    private static int N = 3; // no. of users
    private static int MAX = 6; // max amt of resourse
    private static final double negInfty=Double.MIN_VALUE;
  GENERAL_VARIABLES_END

  GENERAL_FUNCTIONS_BEGIN
    private static double cost(int k, int d) {
       double result=negInfty;
       if (d==0) {
         result=0.0;
       }
       else {
         switch(k) {
```

```
          case 1: {
            result=7.0*d+2.0;
            break;
          }
          case 2: {
            result=3.0*d+7.0;
            break;
          }
          case 3: {
            result=4.0*d+5.0;
            break;
          }
          default: {
            result=negInfty;
            break;
          }
        }
      }
      return result;
    }
  GENERAL_FUNCTIONS_END

  STATE_TYPE: (int k, int m); //k=stage; m=remaining

  DECISION_VARIABLE: int d;   //amt allotted to user k

  DECISION_SPACE: decisionSet(k,m)={0,..,m};

  GOAL: f(1,MAX);

  DPFE_BASE_CONDITIONS:
      f(k,m)=0.0 WHEN ((k>N)&&(m>=0));

  DPFE: f(k,m)=MAX_{d IN decisionSet}
          { r(k,m,d) + f(t(k,m,d)) };

  REWARD_FUNCTION: r(k,m,d)=cost(k,d);

  TRANSFORMATION_FUNCTION: t(k,m,d)=(k+1,m-d);

END
```

In the ALLOTm problem costs are multiplicative rather than additive.

```
BEGIN
  NAME ALLOTm; // Optimal Allotment Problem--multiplicative
```

```
            // REF: Winston04,pp.998-999

  GENERAL_VARIABLES_BEGIN
    private static double[][] cost={
                     {0.6, 0.8, 0.85},
                     {0.5, 0.7, 0.85},
                     {0.3, 0.55, 0.7}
                   };
    private static int N = 3; // no. of users
    private static int MAX = 2; // max amt of resourse
  GENERAL_VARIABLES_END

  STATE_TYPE: (int k, int m); //k=stage; m=remaining

  DECISION_VARIABLE: int d;    //amt allotted to user k

  DECISION_SPACE: decisionSet(k,m)={0,..,m};

  GOAL: f(1,MAX);

  DPFE_BASE_CONDITIONS:
      f(k,m)=1.0 WHEN ((k>N)&&(m>=0));

  DPFE: f(k,m)=MAX_{d IN decisionSet}
          { r(k,m,d) * f(t(k,m,d)) };

  REWARD_FUNCTION: r(k,m,d)=cost[k - 1][d];

  TRANSFORMATION_FUNCTION: t(k,m,d)=(k+1,m-d);

END
```

4.2 gDPS source for APSP

The first DP model uses the relaxation approach from (2.2).

```
BEGIN
  NAME APSP;

  GENERAL_VARIABLES_BEGIN
    private static final int infty=Integer.MAX_VALUE;
    private static int[][] distance =
      {
         {infty,     3,     5, infty},
```

```
        {infty, infty,     1,     8},
        {infty,     2, infty,     5},
        {infty, infty, infty, infty}
      };
    private static int N=distance.length; // number of nodes
    private static int N1=N-1;
    private static int s=0;                 // source
    private static int t=3;                 // target
  GENERAL_VARIABLES_END

  GENERAL_FUNCTIONS_BEGIN
    private static int cost(int p, int q) {
      if (p==q) {
        return 0;
      }
      else {
        return distance[p][q];
      }
    }
  GENERAL_FUNCTIONS_END

  STATE_TYPE: (int k, int p, int q);

  DECISION_VARIABLE: int d;
  DECISION_SPACE: possibleSuccessors(k,p,q)={0,..,N1};

  GOAL:  f(N1,s,t); //at most N1=N-1 edges in solution path

  DPFE_BASE_CONDITIONS:
    f(k,p,q)=0.0   WHEN ((k==0)&&(p==q));
    f(k,p,q)=infty WHEN ((k==0)&&(p!=q));

  DPFE: f(k,p,q)
        =MIN_{d IN possibleSuccessors}
          { r(k,p,q,d) + f(t(k,p,q,d))};

  REWARD_FUNCTION: r(k,p,q,d)=cost(p,d);

  TRANSFORMATION_FUNCTION: t(k,p,q,d)=(k - 1, d, q);

END
```

The second DP model uses the Floyd-Warshall DPFE (2.4).

```
BEGIN
  NAME APSPFW;  //Floyd-Warshall
```

```
GENERAL_VARIABLES_BEGIN
  private static final int infty=Integer.MAX_VALUE;
  private static int[][] distance =
    {
     {infty,     3,     5, infty},
     {infty, infty,     1,     8},
     {infty,     2, infty,     5},
     {infty, infty, infty, infty}
    };
  private static int maxNodeIndex=distance.length-1;
GENERAL_VARIABLES_END

STATE_TYPE: (int k, int p, int q);

DECISION_VARIABLE: int d;
DECISION_SPACE: possibleSuccessors(k,p,q)={0,1};

GOAL:  f(maxNodeIndex,0,3); //maxNodeIndex=3,
                            //source=0, target=3

DPFE_BASE_CONDITIONS:
  f(k,p,q)=0.0            WHEN ((k==0)&&(p==q));
  f(k,p,q)=distance[p][q] WHEN ((k==0)&&(p!=q));

DPFE: f(k,p,q)
      =MIN_{d IN possibleSuccessors}
      { r(k,p,q,d) + p1.f(t1(k,p,q,d))
                   + p2.f(t2(k,p,q,d)) + p3.f(t3(k,p,q,d)) };

REWARD_FUNCTION: r(k,p,q,d)=0;

TRANSFORMATION_FUNCTION:
    t1(k,p,q,d)=(k - 1, p, q);
    t2(k,p,q,d)=(k - 1, p, k);
    t3(k,p,q,d)=(k - 1, k, q);

TRANSITION_WEIGHTS:
    p1(k,p,q,d)=1-d;
    p2(k,p,q,d)=d;
    p3(k,p,q,d)=d;

END
```

4.3 gDPS source for ARC

```
BEGIN
  NAME ARC; //alphabetic radix code;
            //uses top-down formulation as for MCM

  GENERAL_VARIABLES_BEGIN
    //one instance:
    //private static double[] weight = {1, 2, 3, 4};
    //another instance:
    private static double[] weight = {2, 3, 3, 4};
    //N is the no. of partition points:
    private static int N = weight.length-1;
  GENERAL_VARIABLES_END

  GENERAL_FUNCTIONS_BEGIN
    private static double sum(int p, int q) {
      double result=0.0;
      for (int k=p; k<=q; k++) {
        result+=weight[k];
      }
      return result;
    }
  GENERAL_FUNCTIONS_END

  STATE_TYPE: (int firstIndex, int secondIndex);

  DECISION_VARIABLE: int d;
  DECISION_SPACE: decisionSet(firstIndex,secondIndex)
                 ={firstIndex,..,secondIndex - 1};

  GOAL: f(0,N);

  DPFE_BASE_CONDITIONS:
    f(firstIndex,secondIndex)=0.0
        WHEN (firstIndex==secondIndex);

  DPFE: f(firstIndex,secondIndex)
          =MIN_{d IN decisionSet}
            { f(t1(firstIndex,secondIndex,d))
             +f(t2(firstIndex,secondIndex,d))
             +r(firstIndex,secondIndex,d)
            };

  REWARD_FUNCTION: r(firstIndex,secondIndex,d)
```

```
                    =sum(firstIndex,secondIndex);

  TRANSFORMATION_FUNCTION: t1(firstIndex,secondIndex,d)
                            =(firstIndex,d);
                           t2(firstIndex,secondIndex,d)
                            =(d+1,secondIndex);


END
```

4.4 gDPS source for ASMBAL

We present two solutions, corresponding to the two different DPFEs given in Sect. 2.4. The first formulation uses the staged DPFE (2.6). The `arccost` function is used to set the cost to zero for staying in the same line, as opposed to adding such zero costs to the `cost` array.

```
BEGIN
  NAME asmbals; //assembly line balancing problem
                //staged version

  GENERAL_VARIABLES_BEGIN
    private static int[][] cost =           // (stage,line)
      {
       {2,4},                               // to next line
       {2,2},{1,3},{2,1},{2,3},{1,4},       // to next line
       {3,2}                                // from line
      };
    private static int[][] vertexcost =     // (stage,line)
      {
       {0,0}, {7,8},{9,5},{3,6},{4,4},{8,5},{4,7}
      };
     private static int N = vertexcost.length-1;
  GENERAL_VARIABLES_END

  GENERAL_FUNCTIONS_BEGIN
    private static int arccost(int g, int x, int d) {
      if (g==0) return cost[g][d];          // to next line d
      else if (g==N) return cost[g][x];     // from line x
      else if (x==d) return 0;              // stay same line
      else return cost[g][d];               // to next line d
    }
  GENERAL_FUNCTIONS_END
```

```
  STATE_TYPE: (int g, int x);  // g is stage, x is line

  DECISION_VARIABLE: int d;    // d is next line

  DECISION_SPACE: possibleSuccessors(g,x)={0,1}; // line numbers

  GOAL: f(0,0);

  DPFE_BASE_CONDITIONS: f(g,x)=0.0 WHEN (g>N);

  DPFE: f(g,x) = MIN_{d IN possibleSuccessors}
                   { r(g,x,d) + f(t(g,x,d)) };

  REWARD_FUNCTION:
      r(g,x,d) = vertexcost[g][x] + arccost(g,x,d);

  TRANSFORMATION_FUNCTION: t(g,x,d) = (g+1,d);

END
```

The second formulation uses the DPFE (2.7). To represent infinity entries in a matrix, we may use

```
    private static final int infty=Integer.MAX_VALUE;
```

where `Integer.MAX_VALUE` is the largest representable integer.

```
BEGIN
  NAME asmbala; //assembly line balancing problem
                //(based upon SPA)

  GENERAL_VARIABLES_BEGIN
    private static final int infty=Integer.MAX_VALUE;
    private static int[][] arccost =
      {
       {infty,     2,     4, infty, infty, infty, infty,
        infty, infty, infty, infty, infty, infty, infty},
       {infty, infty, infty,     0,     2, infty, infty,
        infty, infty, infty, infty, infty, infty, infty},
       {infty, infty, infty,     2,     0, infty, infty,
        infty, infty, infty, infty, infty, infty, infty},
       {infty, infty, infty, infty, infty,     0,     3,
        infty, infty, infty, infty, infty, infty, infty},
       {infty, infty, infty, infty, infty,     1,     0,
        infty, infty, infty, infty, infty, infty, infty},
       {infty, infty, infty, infty, infty, infty, infty,
            0,     1, infty, infty, infty, infty, infty},
```

```
      {infty, infty, infty, infty, infty, infty, infty,
           2,     0, infty, infty, infty, infty, infty},
      {infty, infty, infty, infty, infty, infty, infty,
       infty, infty,     0,     3, infty, infty, infty},
      {infty, infty, infty, infty, infty, infty, infty,
       infty, infty,     2,     0, infty, infty, infty},
      {infty, infty, infty, infty, infty, infty, infty,
       infty, infty, infty, infty,     0,     4, infty},
      {infty, infty, infty, infty, infty, infty, infty,
       infty, infty, infty, infty,     1,     0, infty},
      {infty, infty, infty, infty, infty, infty, infty,
       infty, infty, infty, infty, infty, infty,     3},
      {infty, infty, infty, infty, infty, infty, infty,
       infty, infty, infty, infty, infty, infty,     2},
      {infty, infty, infty, infty, infty, infty, infty,
       infty, infty, infty, infty, infty, infty, infty}
    };
   private static int[] vertexcost =
      {    0,     7,     8,     9,     5,     3,     6,
           4,     4,     8,     5,     4,     7,     0};
   //max value of subscript for 0-indexing:
   private static int N = vertexcost.length-1;
 GENERAL_VARIABLES_END

 STATE_TYPE: (int x);

 DECISION_VARIABLE: int d;

 DECISION_SPACE: possibleSuccessors(x)={x+1,..,N};

 GOAL: f(0);

 DPFE_BASE: f(N)=0.0;

 DPFE: f(x) = MIN_{d IN possibleSuccessors}
                 { cost(x,d) + f(t(x,d)) };

 REWARD_FUNCTION: cost(x,d) = vertexcost[x] + arccost[x][d];

 TRANSFORMATION_FUNCTION: t(x,d) = (d);

END
```

4.5 gDPS source for ASSIGN

This program can be modified to solve the linear search (LINSRC) problem and other permutation problems (PERM, SPT) by simply changing the `cost` function. It is easy to loop over all elements in a set, as is done in the `sumofwgts` functions using an `Iterator` object.

```
BEGIN
  NAME assign; // Optimal Assignment Problem

  GENERAL_VARIABLES_BEGIN
    private static int[] weight = {3,5,2};
    private static int ttlwgt = 10;
    private static int N = weight.length;
  GENERAL_VARIABLES_END

  SET_VARIABLES_BEGIN
    Set setOfAllItems={0,..,N - 1};
    Set emptySet={};
  SET_VARIABLES_END

  GENERAL_FUNCTIONS_BEGIN
    private static int sumofwgts(SortedSet items) {
      int result=0;
      Iterator i=items.iterator();
      while(i.hasNext()) {
        int j=((Integer) i.next()).intValue();
        result+=weight[j];
      }
      return result;
    }
    private static int cost(SortedSet items, int k, int d) {
      //return k*weight[d]; //LINSRCS: min at end;
                            //k=N+1-size => 1..N
      //return sumofwgts(items); //LINSRCW: max at front; sum(S)
      //return items.size()*weight[d]; //PERM: min at front;
                                       //size => N..1

      return (ttlwgt-sumofwgts(items)+weight[d]);
        // SPT: min at front; sum(~S)+nextitem
        // ttlwgt-sumofwgts(itemsToBeChosen)
        //     =sumofwgts(itemsChosenSoFar)=sum(~S)
    }
  GENERAL_FUNCTIONS_END
```

```
  STATE_TYPE: (Set items, int k);

  DECISION_VARIABLE: int d;

  DECISION_SPACE: decisionSet(items, k)=items;

  GOAL: f(setOfAllItems, 1);

  DPFE_BASE_CONDITIONS: f(items,k)=0.0 WHEN (k>N);

  DPFE: f(items,k)=MIN_{d IN decisionSet}
             { r(items,k,d) + f(t(items,k,d)) };

  REWARD_FUNCTION: // cost of ASSIGNing item d at stage k
    r(items,k,d) = cost(items,k,d);

  TRANSFORMATION_FUNCTION: // omit item in next stage
    t(items,k,d) = (items SETMINUS {d},k+1);

END
```

4.6 gDPS source for BST

```
BEGIN
  NAME BST; //OptimalBinarySearchTree;

  GENERAL_VARIABLES_BEGIN
    //assume the n data items are the ints {0,1,..,n-1}
    //this instance: n=5, i.e. {  0 ,   1 ,  2 ,  3 ,  4 }
    //corresponding to strings { "A",  "B", "C", "D", "E"}
    //now specify the probabilities of access:
    private static double[] probability
                                = {0.25, 0.05, 0.2, 0.4, 0.1};
    private static int n = probability.length; //n=5 data items
  GENERAL_VARIABLES_END

  SET_VARIABLES_BEGIN
    Set setOfAllItems={0,..,n - 1};
    Set emptySet={};
  SET_VARIABLES_END

  GENERAL_FUNCTIONS_BEGIN
    private static double sumOfProbabilitiesOfItems
        (SortedSet items) {
```

```
      double result=0.0;
      for (int i=((Integer) items.first()).intValue();
           i<=((Integer) items.last()).intValue();
           i++) {
        result+=probability[i];
      }
      return result;
    }
    private static NodeSet setOfItemsLessThanPivot
        (NodeSet items, int alpha) {
      //conveniently use method headSet() from SortedSet
      //headSet() DOES NOT include alpha
      return new NodeSet(items.headSet(new Integer(alpha)));
    }
  GENERAL_FUNCTIONS_END

  STATE_TYPE: (Set items);

  DECISION_VARIABLE: int splitAtAlpha;
  DECISION_SPACE: decisionSet(items)
                  =items;

  GOAL: f(setOfAllItems);
  DPFE_BASE:
    f(emptySet)=0.0 ;

  DPFE: f(items)
        =MIN_{splitAtAlpha IN decisionSet}
                                   { f(tLeft(items,splitAtAlpha))
                                    +f(tRight(items,splitAtAlpha))
                                    +cost(items,splitAtAlpha)
                                   };

  REWARD_FUNCTION: cost(items,splitAtAlpha)
                   =sumOfProbabilitiesOfItems(items);

  TRANSFORMATION_FUNCTION:
    tLeft(items,splitAtAlpha)
         =( setOfItemsLessThanPivot(items,splitAtAlpha) );
    tRight(items,splitAtAlpha)
         =( (items SETMINUS {splitAtAlpha}) SETMINUS
               setOfItemsLessThanPivot(items,splitAtAlpha) );

END
```

4.7 gDPS source for COV

```
BEGIN
  NAME COV; //optimal COVering problem

  GENERAL_VARIABLES_BEGIN
    //cost of cover sizes 0,1,..,9 in dollar
    private static int[] cost
        = {1, 4, 5, 7, 8, 12, 13, 18, 19, 21};
  GENERAL_VARIABLES_END

  STATE_TYPE: (int numberOfCoverSizes, int largestSize);
   //numberOfCoverSizes denotes the stage number in this problem

  DECISION_VARIABLE: int nextCoverSize;
  DECISION_SPACE: decisionSet(numberOfCoverSizes,largestSize)
                 ={numberOfCoverSizes - 2,..,largestSize - 1};

  GOAL: f(3,9); //choose 3 cover sizes to cover
                //shrubs sized 0 through 9

  DPFE_BASE_CONDITIONS:
    f(numberOfCoverSizes,largestSize)
        =(largestSize+1)*cost[largestSize]
        WHEN (numberOfCoverSizes==1);

  DPFE:
    f(numberOfCoverSizes,largestSize)
      = MIN_{nextCoverSize IN decisionSet}
            { f(t(numberOfCoverSizes,largestSize,nextCoverSize))
             +r(numberOfCoverSizes,largestSize,nextCoverSize)
            };

  REWARD_FUNCTION:
    r(numberOfCoverSizes,largestSize,nextCoverSize)
      =(largestSize - nextCoverSize)*cost[largestSize];

  TRANSFORMATION_FUNCTION:
    t(numberOfCoverSizes,largestSize,nextCoverSize)
      =(numberOfCoverSizes - 1, nextCoverSize);

END
```

4.8 gDPS source for DEADLINE

```
BEGIN
  NAME deadline;  //deadline scheduling of unit-time jobs

  GENERAL_VARIABLES_BEGIN
    private static int[] profit = {10,15,20,1,5};
    private static int[] deadline = {1,2,2,3,3};  //sorted
    private static int N = profit.length; //no.of jobs
  GENERAL_VARIABLES_END

  SET_VARIABLES_BEGIN
    Set setOfAllJobs={0,..,N - 1};
    Set emptySet={};
  SET_VARIABLES_END

  GENERAL_FUNCTIONS_BEGIN
    private static boolean feasible(Set jobs, int k, int d) {
      int j=0;
      for (int i=0; i<N; i++) {
        if (!(jobs.contains(new Integer(i)))||i==d) {
        //if i already chosen or next (and is j-th),
        //does it meet its deadline?
          if (deadline[i]<++j) {
            return false;
          }
        }
      }
      return true;
    }
    private static int cost(SortedSet jobs, int k, int d) {
      if (feasible(jobs,k,d)) {
        return profit[d];
      }
      else {
        return 0;
      }
    }
  GENERAL_FUNCTIONS_END

  STATE_TYPE: (Set jobs, int k);
      //jobs not yet chosen at stage k

  DECISION_VARIABLE: int d; //next job
```

```
  DECISION_SPACE: decisionSet(jobs,k)=jobs;

  GOAL: f(setOfAllJobs, 1);

  DPFE_BASE_CONDITIONS: f(jobs,k)=0.0 WHEN (k>N);

  DPFE: f(jobs,k)=MAX_{d IN decisionSet}
            { r(jobs,k,d) + f(t(jobs,k,d)) };

  REWARD_FUNCTION: r(jobs,k,d) = cost(jobs,k,d);

  TRANSFORMATION_FUNCTION: t(jobs,k,d)=(jobs SETMINUS {d},k+1);

END
```

4.9 gDPS source for DPP

```
BEGIN
  NAME DPP; //Discounted Profits Problem
  //A discounted DP problem from Winston/Venkataramanan
  //pp.779--780
  //Used a reproductionFactor of 2 instead of 1.2 so number of
  //fish takes on integral values, without the need to use a
  //round or a floor function.

  GENERAL_VARIABLES_BEGIN
    private static int T=2;  //planning horizon T=2 years
    private static double interestRate=0.05; //assume 5%
                                             //interest rate
    private static int initialFishAmount=10; //initially 10,000
                                             //fish in lake
    private static int reproductionFactor=2; //in 1 year 100%
                                             //more fish
  GENERAL_VARIABLES_END

  GENERAL_FUNCTIONS_BEGIN
    private static double revenue(int xt) {
      //simply assume that the sale price
      //is $3 per fish, no matter what
      return 3.0*xt;
    }
    private static double cost(int xt, int b) {
      //simply assume that it costs $2 to catch (and process)
      //a fish, no matter what
```

```
    return 2.0*xt;
  }
GENERAL_FUNCTIONS_END

STATE_TYPE: (int t, int b);
  //t is stage number, each stage represents one year
  //b is current number of fish in lake (scaled to thousands)

DECISION_VARIABLE: int xt;
  //xt is the number of fish to catch and sell
  //during year t

DECISION_SPACE: decisionSet(t,b)={0,..,b};
  //can't sell more than what is in lake

GOAL: f(1,initialFishAmount);

DPFE_BASE_CONDITIONS:
  f(t,b)=0.0 WHEN (t==T+1); //T+1 is outside planning horizon

DPFE: f(t,b)
      =MAX_{xt IN decisionSet}
                            { yearlyNetProfit(t,b,xt)
                             +beta.f(trans(t,b,xt))
                            };

REWARD_FUNCTION: yearlyNetProfit(t,b,xt)
                   = revenue(xt) - cost(xt,b);

TRANSFORMATION_FUNCTION: trans(t,b,xt)
                          =(t+1,reproductionFactor*(b-xt));

TRANSITION_WEIGHTS:
  beta(t,b,xt)=1/(1+interestRate); //discount factor

END
```

4.10 gDPS source for EDP

```
BEGIN
  NAME EDPalt; //EditDistanceProblem;

  GENERAL_VARIABLES_BEGIN
    //example from Gusfield p.223
```

```
  private static String s1 = "CAN";
  private static String s2 = "ANN";

  private static final int insertionCost=1;
  private static final int deletionCost=1;
  private static final int replacementCost=1;

  //it is useful to have the string lengths as
  //symbolic constants
  private static int s1Length = s1.length();
  private static int s2Length = s2.length();
GENERAL_VARIABLES_END

GENERAL_FUNCTIONS_BEGIN
  //costOfOperation()
  //returns 0 if the specified characters in the 2 strings
  //            match and the decision is to perform
  //            "replacement" operation for matching chars
  //            (whose cost is usually defined as 0)
  //returns 1 otherwise (if delete, insert, or a real
                replacement operation happens)
  private static int costOfOperation(int i, int j, int dec) {
    if(dec==12) { //dec==12 means decision is to replace
      if (  s1.charAt(i-1) //note: subtract 1 because array
                           //      starts at index 0
          ==s2.charAt(j-1)) { //matching chars, cost is 0
        return 0;
      }
      else { //real replacement
        return replacementCost; //cost of real replacement
      }
    }
    if(dec==1) { //dec==1 means decision is to delete
      return deletionCost;
    }
    //otherwise it must be that dec==2, decision to insert
    return insertionCost;
  }
  private static int s1Adjustment(int dec) {
    if(dec==2) { //insert
      return 0;
    }
    return 1;
  }
  private static int s2Adjustment(int dec) {
```

```
      if(dec==1) { //delete
        return 0;
      }
      return 1;
    }
  GENERAL_FUNCTIONS_END

  STATE_TYPE: (int i, int j);

  DECISION_VARIABLE: int dec;

  DECISION_SPACE: decisionSet(i,j)
                  ={1,2,12};
   //The decision space is constant in this example.
   //It does not depend on the current state.
   //We chose to code the 3 allowed operations
   //as follows:
   //1 stands for delete
   //2 stands for insert
   //12 stands for replace

  GOAL: d(s1Length,s2Length);
    //the lengths of s1 and s2, respectively

  DPFE_BASE_CONDITIONS:
    d(i,j)=j WHEN (i==0);
    d(i,j)=i WHEN (j==0);

  DPFE: d(i,j)
        =MIN_{dec IN decisionSet}
                              { d(t(i,j,dec))
                               +r(i,j,dec)
                              };

  REWARD_FUNCTION: r(i,j,dec)
                   =costOfOperation(i,j,dec);

  TRANSFORMATION_FUNCTION: t(i,j,dec)
                           =(i-s1Adjustment(dec),
                             j-s2Adjustment(dec));

END
```

4.11 gDPS source for FIB

```
BEGIN
  NAME FIB; //Fibonacci numbers

  //Interesting feature: The size of the Bellman Net for this
  //example is linear, but the size of the solution tree output
  //for this example (something we are not interested anyway)
  //is exponential.
  //A PN simulator or a spreadsheet solver is more suitable
  //than the Java Solver!

  STATE_TYPE: (int m);

  DECISION_VARIABLE: int dummy;
  DECISION_SPACE: dummyDecisionSet(m)
                 ={777}; //any singleton works

  GOAL: f(7); //calculate 7th Fibonacci number

  DPFE_BASE_CONDITIONS:
    f(m)=1.0 WHEN ((m==1)||(m==2));

  DPFE: f(m)
        =MIN_{dummy IN dummyDecisionSet}
                                  { f(t1(m,dummy))
                                   +f(t2(m,dummy))
                                   +r(m,dummy)
                                  };

  REWARD_FUNCTION: r(m,dummy)=0.0;

  TRANSFORMATION_FUNCTION: t1(m,dummy)
                            =(m - 1);
                           t2(m,dummy)
                            =(m - 2);

END
```

4.12 gDPS source for FLOWSHOP

```
BEGIN
  NAME FLOWSHOP; //Flowshop Problem
```

```
  GENERAL_VARIABLES_BEGIN
    private static int[] first  = {3,4,8,10}; //sum=25
    private static int[] second = {6,2,9,15};  //sum=32
    //upper bound on final completion time is 25+32:
    private static int sum=57;
    private static int m=first.length;
  GENERAL_VARIABLES_END

  SET_VARIABLES_BEGIN
    Set setOfAllItems={0,..,m - 1};
    Set emptySet={};
  SET_VARIABLES_END

  GENERAL_FUNCTIONS_BEGIN
    private static int fct(int t, int d) {
      return Math.max(t - first[d],0) + second[d] ;
    }
  GENERAL_FUNCTIONS_END

  STATE_TYPE: (Set S, int t);

  DECISION_VARIABLE: int d;
  DECISION_SPACE: decset(S,t) = S;

  GOAL: f(setOfAllItems,0);

  DPFE_BASE_CONDITIONS:
    f(S,t)=t WHEN (S SETEQUALS emptySet);

  DPFE:  f(S,t) = MIN_{d IN decset} {cost(S,t,d) + f(g(S,t,d))};

  REWARD_FUNCTION:  cost(S,t,d) = first[d];

  TRANSFORMATION_FUNCTION: g(S,t,d)
      = ( S SETMINUS {d}, fct(t,d) );

END
```

4.13 gDPS source for HANOI

```
BEGIN
  NAME hanoi; //tower of hanoi problem
    //Compute the number of moves when the recursive
    //strategy is used. Shows that DP2PN2Solver can be used to
```

```
    //solve basic recurrences that do not require optimization.
    //Model does not return the sequence of moves.

  STATE_TYPE: (int m, int i, int j, int k);
    //m: number of disks to move
    //i: index of source tower
    //j: index of destination tower
    //k: index of temporary tower

  DECISION_VARIABLE: int dummy;
  DECISION_SPACE: dummyDecisionSet(m)
                  ={-1}; //any singleton works

  GOAL: f(3,1,2,3); //move 3 disks from tower 1 to tower 2
                    //using tower 3 as temporary storage

  DPFE_BASE_CONDITIONS:
        f(m,i,j,k)=1.0 WHEN (m==1); // counts actual moves

  DPFE: f(m) = MIN_{dummy IN dummyDecisionSet}
        {  f(t1(m,i,j,k,dummy))
         + f(t2(m,i,j,k,dummy))
         + f(t3(m,i,j,k,dummy))
         + r(m,i,j,k,dummy)
        };

  REWARD_FUNCTION:
        r(m,i,j,k,dummy) = 0.0; // cost deferred to base case

  TRANSFORMATION_FUNCTION:
        t1(m,i,j,k,dummy) = (m - 1, i, k, j);
        t2(m,i,j,k,dummy) = (  1   , i, j, k);
        t3(m,i,j,k,dummy) = (m - 1, k, j, i);

  END
```

4.14 gDPS source for ILP

```
BEGIN
  NAME ILP; //IntegerLinearProgramming;

  GENERAL_VARIABLES_BEGIN
    // max c'x=c1*x1+..+cn*xn
    // s.t.  Ax<=b
```

```
  //        xi>=0 and integer
  // For this problem formulation,
  // must assume A,b,c have all nonnegative integer entries.
  // E.g. if a constraint like 3x1-2x2<=18 were allowed,
  // we could NOT limit the decision set for variable x1
  // to be {0,..,6}.

  //objective function coefficients
  private static int[] c = {3, 5};
  //right hand side of constraints vector
  private static int[] b = {4, 12, 18};
  //constraint matrix
  private static int[][] a=
    {
     { 1, 0},
     { 0, 2},
     { 3, 2}
    };
  private static int n = c.length;
  private static int m = b.length;
  private static final int infty=Integer.MAX_VALUE;
GENERAL_VARIABLES_END

GENERAL_FUNCTIONS_BEGIN
  private static NodeSet calculateDecisionSet(int stage,
      int y1, int y2, int y3) {
    NodeSet result = new NodeSet();
    //maxPossibleChoiceBecauseOfResourceiRestriction, i=1,2,3
    int mpc1=infty;
    int mpc2=infty;
    int mpc3=infty;
    if(a[0][stage]!=0){
      mpc1=y1/a[0][stage];
    }
    if(a[1][stage]!=0){
      mpc2=y2/a[1][stage];
    }
    if(a[2][stage]!=0){
      mpc3=y3/a[2][stage];
    }
    for (int i=0; i<=Math.min(mpc1,Math.min(mpc2,mpc3)); i++){
      result.add(new Integer(i));
    }
    return result;
  }
```

```
  GENERAL_FUNCTIONS_END

  //here: yi denotes how much of resource i is still available,
  //(in other words how much slack is still available)
  STATE_TYPE: (int stage, int y1, int y2, int y3);

  DECISION_VARIABLE: int d;
  DECISION_SPACE: decisionSet(stage,y1,y2,y3)
                 =calculateDecisionSet(stage,y1,y2,y3);

  GOAL: f(0,b[0],b[1],b[2]);
        //convenient notation for f(0,4,12,18)


  DPFE_BASE_CONDITIONS:
    f(stage,y1,y2,y3)=0.0 WHEN (stage == n);

  DPFE: f(stage,y1,y2,y3)
        =MAX_{d IN decisionSet}
                               { f(t1(stage,y1,y2,y3,d))
                                +r(stage,y1,y2,y3,d)
                               };

  REWARD_FUNCTION: r(stage,y1,y2,y3,d)
                   =c[stage]*d;

  TRANSFORMATION_FUNCTION: t1(stage,y1,y2,y3,d)
                           =(stage+1,
                             y1-a[0][stage]*d,
                             y2-a[1][stage]*d,
                             y3-a[2][stage]*d);

END
```

4.15 gDPS source for ILPKNAP

```
BEGIN
  NAME ILPKNAP; //IntegerLinearProgramming formulation
                //of 0-1 Knapsack problem;
    // max c'x=c1*x1+..+cn*xn
    // s.t.  Ax<=b
    //       xi>=0 and integer

  GENERAL_VARIABLES_BEGIN
```

```
  //objective function coefficients:
  private static int[] c = {15, 25, 24};
  //right hand side constraint vector:
  private static int[] b = {22, 1, 1, 1};
  //constraint matrix:
  private static int[][] a =
    {
     { 10, 18, 15},
     { 1, 0, 0},
     { 0, 1, 0},
     { 0, 0, 1},
    };
  private static int n = c.length;
  private static int m = b.length;
  private static final int infty=Integer.MAX_VALUE;
GENERAL_VARIABLES_END

GENERAL_FUNCTIONS_BEGIN
  private static NodeSet calculateDecisionSet(int stage,
      int y1, int y2, int y3, int y4) {
    NodeSet result = new NodeSet();
    //maxPossibleChoiceBecauseOfResourceiRestriction,i=1,2,3,4
    int mpc1=infty;
    int mpc2=infty;
    int mpc3=infty;
    int mpc4=infty;
    if(a[0][stage]!=0){ mpc1=y1/a[0][stage]; }
    if(a[1][stage]!=0){ mpc2=y2/a[1][stage]; }
    if(a[2][stage]!=0){ mpc3=y3/a[2][stage]; }
    if(a[3][stage]!=0){ mpc4=y4/a[3][stage]; }
    for (int i=0;
         i<=Math.min(mpc1,Math.min(mpc2,Math.min(mpc3,mpc4)));
         i++) {
      result.add(new Integer(i));
    }
    return result;
  }
GENERAL_FUNCTIONS_END

// the "slack" yi denotes how much of resource i is still
// available
STATE_TYPE: (int stage, int y1, int y2, int y3, int y4);

DECISION_VARIABLE: int d;
DECISION_SPACE: decisionSet(stage,y1,y2,y3,y4)
```

```
                =calculateDecisionSet(stage,y1,y2,y3,y4);

  GOAL: f(0,b[0],b[1],b[2],b[3]);

  DPFE_BASE_CONDITIONS:
    f(stage,y1,y2,y3,y4)=0.0 WHEN (stage == n);

  DPFE: f(stage,y1,y2,y3,y4) = MAX_{d IN decisionSet}
        { f(t1(stage,y1,y2,y3,y4,d)) + r(stage,y1,y2,y3,y4,d) };

  REWARD_FUNCTION: r(stage,y1,y2,y3,y4,d)=c[stage]*d;

  TRANSFORMATION_FUNCTION: t1(stage,y1,y2,y3,y4,d)
                            =(stage+1,
                              y1-a[0][stage]*d,
                              y2-a[1][stage]*d,
                              y3-a[2][stage]*d,
                              y4-a[3][stage]*d);

END
```

4.16 gDPS source for INTVL

The interval scheduling problem can be solved by the following program, which uses DPFE (2.21).

```
BEGIN
  NAME INTVL1; //interval scheduling or
               //activity selection problem;

  GENERAL_VARIABLES_BEGIN
    private static int[] begintime = {1, 2, 5, 3, 8, 9};
    private static int[] endtime   = {4, 6, 7,10,11,12};
        //endtime sorted!!
    private static int[] weight    = {2, 4, 4, 7, 2, 1};
    private static int[] pred      = {0, 0, 1, 0, 3, 3};
    private static int N = weight.length;
    private static int L = 20;  //>= max endtime
  GENERAL_VARIABLES_END

  GENERAL_FUNCTIONS_BEGIN
    private static int cost(int k, int d) {
      if (d==1) {
        return weight[k-1];
```

```
      }
      else {
        return 0;
      }
    }
    private static int next(int k, int d) {
       if (d==1) {
         return pred[k-1];
       }
       else {
         return k-1;
       }
    }
  GENERAL_FUNCTIONS_END

  STATE_TYPE: (int k); //stage number

  DECISION_VARIABLE: int d; //decision is whether to
                            //include k-th or not

  DECISION_SPACE: decisionSet(k)={1,0}; //boolean values
                                        //(1=include k-th)

  GOAL: f(N);

  DPFE_BASE: f(0)=0.0;

  DPFE: f(k) = MAX_{d IN decisionSet} { r(k,d) + f(t(k,d)) };

  REWARD_FUNCTION:            r(k,d) =  cost(k,d);

  TRANSFORMATION_FUNCTION:    t(k,d) = (next(k,d));

END
```

An alternate version, which illustrates use of nonprobabilistic transition weights, which uses DPFE (2.22), is as follows.

```
BEGIN
  NAME INTVL3; //interval scheduling or
               //activity selection problem;
               //reformulation of INTVL1

  GENERAL_VARIABLES_BEGIN
    private static int[] begintime = {1, 2, 5, 3, 8, 9};
    private static int[] endtime   = {4, 6, 7,10,11,12};
```

```
        //endtime sorted!!
    private static int[] weight    = {2, 4, 4, 7, 2, 1};
    private static int[] pred      = {0, 0, 1, 0, 3, 3};
    private static int N = weight.length;
    private static int L = 20;  //>= max endtime
  GENERAL_VARIABLES_END

  STATE_TYPE: (int k); //stage number

  DECISION_VARIABLE: int d; //decision is whether to
                            //include k-th or not

  DECISION_SPACE: decisionSet(k)={0,1};
  //boolean values: (1=include k-th), (0=don't include)

  GOAL: f(N);

  DPFE_BASE: f(0)=0.0;

  DPFE: f(k) = MAX_{d IN decisionSet}
          { r(k,d) + p1.f(takeItTrans(k,d))
                   + p2.f(leaveItTrans(k,d))};

  REWARD_FUNCTION:  r(k,d) = d*weight[k - 1];

  TRANSFORMATION_FUNCTION:
    takeItTrans(k,d)  = (pred[k - 1]);
    leaveItTrans(k,d) = (k - 1);

  TRANSITION_WEIGHTS:
    p1(k,d)=d;
    p2(k,d)=1-d;

END
```

The problem can also be solved by the following nonserial version, which uses DPFE (2.20).

```
BEGIN
  NAME INTVL2; //weighted interval scheduling
               //or activity selection problem;

  GENERAL_VARIABLES_BEGIN
    private static int[] begintime = { 9, 8, 3,5,2,1};
    private static int[] endtime   = {12,11,10,7,6,4};
        //unordered!
```

```
  private static int[] weight    = { 1, 2, 7,4,4,2};
  private static int L = 20;             // L >= max endtime
  private static int N = weight.length; //N=no. of activities
GENERAL_VARIABLES_END

SET_VARIABLES_BEGIN
  Set setOfAllJobs={0,..,N - 1};
  Set emptySet={};
SET_VARIABLES_END

GENERAL_FUNCTIONS_BEGIN
  private static boolean feasible(int i, int j, int d) {
     // activity d in interval (i,j) ?
     if (i<=begintime[d] && endtime[d]<=j)
       return true;
     else
       return false;
  }
  private static int cost(int i, int j, int d) {
     // return weight if activity d in interval (i,j)
     if (feasible(i,j,d)) return weight[d];
     else return 0;
  }
  private static NodeSet leftFeasibles(NodeSet S, int i,
      int j, int d) {
    NodeSet result = new NodeSet();
    int j2=begintime[d];
    // only add those activities of S that fit into
    // interval [i,j2]
    Iterator iter=S.iterator();
    while(iter.hasNext()) {
      Integer element=(Integer) iter.next();
      if ((feasible(i,j2,element.intValue()))) {
        result.add(element);
      }
    }
    return result;
  }
  private static NodeSet rightFeasibles(NodeSet S, int i,
      int j, int d) {
    NodeSet result = new NodeSet();
    int i2=endtime[d];
    // only add those activities of S that fit into
    // interval [i2,j]
    Iterator iter=S.iterator();
```

```
      while(iter.hasNext()) {
        Integer element=(Integer) iter.next();
        if ((feasible(i2,j,element.intValue()))) {
          result.add(element);
        }
      }
      return result;
    }
  GENERAL_FUNCTIONS_END

  STATE_TYPE: (Set S, int i, int j);
    // S=decision set; (i,j)=time interval

  DECISION_VARIABLE: int d;

  DECISION_SPACE: decisionSet(S,i,j)=S;

  GOAL: f(setOfAllJobs,0,L);

  DPFE_BASE_CONDITIONS:
    f(S,i,j)=0.0 WHEN (i>=j);
    f(S,i,j)=0.0 WHEN (S SETEQUALS emptySet);

  DPFE: f(S,i,j) = MAX_{d IN decisionSet}
                   { r(S,i,j,d) + f(tLeft(S,i,j,d))
                                + f(tRight(S,i,j,d)) };

  REWARD_FUNCTION:   r(S,i,j,d) = cost(i,j,d);

  TRANSFORMATION_FUNCTION:
    tLeft(S,i,j,d) = (leftFeasibles(S,i,j,d), i, begintime[d]);
    tRight(S,i,j,d) = (rightFeasibles(S,i,j,d), endtime[d], j);

END
```

4.17 gDPS source for INVENT

```
BEGIN
  NAME INVENT; // Inventory Production Problem;
               // Winston02, p.758--763

  GENERAL_VARIABLES_BEGIN
    private static int[] demand={1,3,2,4};
    private static int N=demand.length; //no. of stages
```

```
    private static final int CAP= 5; // production capacity
    private static final int LIM= 4; // inventory limit
  GENERAL_VARIABLES_END

  GENERAL_FUNCTIONS_BEGIN
    private static double holdcost(int k, int s, int x) {
      return 0.5*(s+x-demand[k]); // holding cost
    }
    private static double prodcost(int x) {
      if (x==0) {
        return 0.0; //no production cost for 0 production
      }
      else {
        return 3.0+x; //production cost
      }
    }
    private static NodeSet possibleDecisions(int k, int s) {
      NodeSet result = new NodeSet();
      int p=Math.max(0, demand[k] - s);
      int q=Math.min(LIM + demand[k] - s, CAP);
      for (int i=p; i<=q; i++) {
        result.add(new Integer(i));
      }
      return result;
    }
  GENERAL_FUNCTIONS_END

  STATE_TYPE: (int k, int s); //stage k, inventory s

  DECISION_VARIABLE: int x; //amount to produce

  DECISION_SPACE: decisionSet(k,s) = possibleDecisions(k,s);

  GOAL:  f(0,0);

  DPFE_BASE_CONDITIONS: f(k,s) = 0. WHEN (k == N);

  DPFE: f(k,s) = MIN_{x IN decisionSet}
                   { r(k,s,x) + f(t(k,s,x)) };

  REWARD_FUNCTION: r(k,s,x) = prodcost(x) + holdcost(k,s,x);

  TRANSFORMATION_FUNCTION: t(k,s,x) = (k+1,s+x-demand[k]);

END
```

4.18 gDPS source for INVEST

```
BEGIN
  NAME INVEST; // Bronson97 Investment Problem 19.29;

  GENERAL_VARIABLES_BEGIN
    private static final int infty=Integer.MAX_VALUE;
    private static double[] prob
      = {1.0, 0.2, 0.4};  // gain prob. as fct of decision
    private static int n=4;         //time limit
    private static int bankroll=2; //initial amount
  GENERAL_VARIABLES_END

  STATE_TYPE: (int g, int s);

  DECISION_VARIABLE: int d;

  DECISION_SPACE: decisionSet(g,s)={0,..,2};

  GOAL: f(1,bankroll);

  DPFE_BASE_CONDITIONS:
    f(g,s)=s WHEN (g==n);

  DPFE: f(g,s)
        =MAX_{d IN decisionSet}
                                  { p1.f(t1(g,s,d))   //gain
                                   +p2.f(t2(g,s,d))   //loss
                                   +cost(g,s,d)
                                  };

  REWARD_FUNCTION: cost(g,s,d)=0.0;

  TRANSFORMATION_FUNCTION: t1(g,s,d)
                            =(g+1,s+d);
                           t2(g,s,d)
                            =(g+1,s-d+1);

  TRANSITION_WEIGHTS: p1(g,s,d)=prob[d];
                      p2(g,s,d)=1.0-prob[d];

END
```

4.19 gDPS source for INVESTWLV

The INVESTWLV problem instance from Section 2.19 illustrates the use of probability weights in the DPFE section and shows how these weights are defined as functions in the TRANSITION_WEIGHTS section. For INVESTWLV there are two probability weights p1 and p2 within the DPFE section. They are both defined in the TRANSITION_WEIGHTS section, in this case simply as constants. The following gDPS source code shows these gDPS features.

```
BEGIN
  NAME INVESTWLV; //INVESTment : Winning in LV;
  //probabilistic problem from Hillier/Lieberman p.423

  GENERAL_VARIABLES_BEGIN
    private static int startAmount=3; //start with 3 chips
    private static int targetAmount=5; //need 5 chips
    private static int maxNumberOfPlays=3;
    private static double winProbability=2.0/3.0;
  GENERAL_VARIABLES_END

  STATE_TYPE: (int n, int sn);
    //n is stage number, sn is number of chips

  DECISION_VARIABLE: int xn; //xn is number of chips
                             //to bet at stage n
  DECISION_SPACE: decisionSet(n,sn)={0,..,sn};

  GOAL: f(1,startAmount);

  DPFE_BASE_CONDITIONS:
    f(n,sn)=0.0 WHEN ((n>maxNumberOfPlays)&&(sn<targetAmount));
    f(n,sn)=1.0 WHEN ((n>maxNumberOfPlays)&&(sn>=targetAmount));

  DPFE: f(n,sn)
        =MAX_{xn IN decisionSet}
                                 { p1.f(t1(n,sn,xn))
                                  +p2.f(t2(n,sn,xn))
                                  +r(n,sn,xn)
                                 };

  REWARD_FUNCTION: r(n,sn,xn)=0;
```

```
  TRANSFORMATION_FUNCTION: t1(n,sn,xn)
                             =(n+1,sn-xn);
                           t2(n,sn,xn)
                             =(n+1,sn+xn);

  TRANSITION_WEIGHTS: p1(n,sn,xn)=1.0-winProbability;
                      p2(n,sn,xn)=winProbability;


END
```

4.20 gDPS source for KS01

```
BEGIN
  NAME KS01; //Knapsack01Problem;

  GENERAL_VARIABLES_BEGIN
    private static int knapsackCapacity = 22;
    private static int[] value = {25, 24, 15};
    private static int[] weight = {18, 15, 10};
    private static int n = value.length; //number of objects n=3
    private static int highestIndex=n-1; //items are indexed
                                         //from 0 through n-1
  GENERAL_VARIABLES_END

  SET_VARIABLES_BEGIN
    Set setOfAllNodes={0,..,highestIndex};
    Set emptySet={};
    Set goalSet={0};
  SET_VARIABLES_END

  GENERAL_FUNCTIONS_BEGIN
    private static NodeSet calculateDecisionSet(int objInd,
        int w) {
      NodeSet decSet = new NodeSet();
      decSet.add(new Integer(0)); //decision to not take object
                                  //is always feasible
      if(w>=weight[objInd]) { //check if there is enough space
                              //to take object
        decSet.add(new Integer(1));
      }
      return decSet;
    }
  GENERAL_FUNCTIONS_END
```

```
  STATE_TYPE: (int currentObjectIndex, int weightToGive);

  DECISION_VARIABLE: int d;
  DECISION_SPACE: decisionSet(currentObjectIndex,weightToGive)
      =calculateDecisionSet(currentObjectIndex, weightToGive);

  GOAL:
    f(highestIndex, knapsackCapacity); //f(2,22)

  DPFE_BASE_CONDITIONS:
    f(currentObjectIndex,weightToGive)=0.0
        WHEN (currentObjectIndex == -1);

  DPFE: f(currentObjectIndex,weightToGive)
        =MAX_{d IN decisionSet}
          { profit(currentObjectIndex,weightToGive,d)
            +f(t(currentObjectIndex,weightToGive,d))};

  REWARD_FUNCTION: profit(currentObjectIndex,weightToGive,d)
                   =d*value[currentObjectIndex];

  TRANSFORMATION_FUNCTION: t(currentObjectIndex,weightToGive,d)
      =(currentObjectIndex - 1,
        weightToGive - d*weight[currentObjectIndex]);

END
```

4.21 gDPS source for KSCOV

```
BEGIN
  NAME KSCOV; // Integer Knapsack model for COV problem;

  GENERAL_VARIABLES_BEGIN
    private static int[] value = {1,4,5,7,8,12,13,18,19,21};
    private static int n = value.length; //number of objects
    private static int M = 3;            //number of stages
    private static final double infty=Double.MAX_VALUE;
  GENERAL_VARIABLES_END

  //state=(stage,remaining largest object)
  STATE_TYPE: (int k, int s);

  DECISION_VARIABLE: int d; //d=number of objects covered
```

```
  DECISION_SPACE: decisionSet(k,s) = {1,..,s}; //cover 1 to all

  GOAL:  f(1,n);

  DPFE_BASE_CONDITIONS:
    f(k,s)=0.0 WHEN (s==0);
    f(k,s)=infty WHEN ((k>M)&&(s>0));

  DPFE:  f(k,s) = MIN_{d IN decisionSet}
                        { r(k,s,d) + f(t(k,s,d)) };

  REWARD_FUNCTION:          r(k,s,d) = d*value[s - 1];

  TRANSFORMATION_FUNCTION: t(k,s,d) = (k + 1, s - d);

END
```

4.22 gDPS source for KSINT

```
BEGIN
  NAME KSINT; // Integer Knapsack Problem;

  GENERAL_VARIABLES_BEGIN
    private static int cap = 22;
    private static int[] value = {24, 15, 25};
    private static int[] weight = {15, 10, 18};
    private static int n = value.length; //number of objects
    private static int highestIndex=n-1; //items indexed
                                          //from 0 to n-1
  GENERAL_VARIABLES_END

  SET_VARIABLES_BEGIN
    Set setOfAllNodes={0,..,highestIndex};
    Set emptySet={};
    Set goalSet={0};
  SET_VARIABLES_END

  GENERAL_FUNCTIONS_BEGIN
    private static NodeSet calculateDecisionSet(int objInd,
        int w) {
      NodeSet decSet = new NodeSet();
      int limit=w/weight[objInd];
      for (int i=0; i<=limit; i++)
        decSet.add(new Integer(i));
```

```
      return decSet;
    }
  GENERAL_FUNCTIONS_END

  STATE_TYPE: (int currentObjectIndex, int weightToGive);

  DECISION_VARIABLE: int d;
  DECISION_SPACE: decisionSet(currentObjectIndex,weightToGive)
      =calculateDecisionSet(currentObjectIndex, weightToGive);

  GOAL:
    f(highestIndex, cap);  //f(n-1,cap)

  DPFE_BASE_CONDITIONS:
    f(currentObjectIndex,weightToGive)=0.0
        WHEN (currentObjectIndex == -1);

  DPFE: f(currentObjectIndex,weightToGive)
        =MAX_{d IN decisionSet}
          { profit(currentObjectIndex,weightToGive,d)
            +f(t(currentObjectIndex,weightToGive,d))};

  REWARD_FUNCTION: profit(currentObjectIndex,weightToGive,d)
                   =d*value[currentObjectIndex];

  TRANSFORMATION_FUNCTION: t(currentObjectIndex,weightToGive,d)
       =(currentObjectIndex - 1,
         weightToGive - d*weight[currentObjectIndex]);

END
```

4.23 gDPS source for LCS

Two different gDPS sources are given. The first one is for the DP functional equation (2.30).

```
BEGIN
  NAME LCSalt;
  //alternative formulation using indices
  //into global strings
  //DPFE based on Gusfield, p.227,228
  GENERAL_VARIABLES_BEGIN
    //the 2 globally defined input strings
    private static String x = "abcbdab";
```

```
    private static String y = "bdcaba";
    //it is useful to have the string lengths
    //as symbolic constants
    private static int xLength = x.length(); //here = 7
    private static int yLength = y.length(); //here = 6
  GENERAL_VARIABLES_END

  GENERAL_FUNCTIONS_BEGIN
    //returns 1 if the specified characters in the 2 strings
    //          match and the decision is to prune both strings
    //returns 0 otherwise
    private static int matchingCharactersAndPruneBoth
        (int xIndex, int yIndex, int d) {
      if((d==12) &&  //d=12 means decision is to prune both
         (x.charAt(xIndex-1)==y.charAt(yIndex-1))) {
         //above, subtract 1 because array starts at index 0
        return 1;
      }
      return 0;
    }
    private static int xAdjustment(int d) {
      if(d==2) {
        return 0;
      }
      return 1;
    }
    private static int yAdjustment(int d) {
      if(d==1) {
        return 0;
      }
      return 1;
    }
  GENERAL_FUNCTIONS_END

  STATE_TYPE: (int xIndex, int yIndex);

  DECISION_VARIABLE: int pruneD;

  DECISION_SPACE: decisionSet(xIndex,yIndex)
                    ={1,2,12};
   //The decision space is constant in this example
   //It does not depend on the current state.
   //1 stands for x is pruned    and y is unchanged
   //2 stands for x is unchanged and y is pruned
   //12 stands for prune both x and y
```

```
  GOAL: f(xLength,yLength);
      //the lengths of x and y, respectively


  DPFE_BASE_CONDITIONS:
    f(xIndex,yIndex)=0.0 WHEN ((xIndex==0)||(yIndex==0));

  DPFE: f(xIndex,yIndex)
        =MAX_{pruneD IN decisionSet}
                                { f(t(xIndex,yIndex,pruneD))
                                 +r(xIndex,yIndex,pruneD)
                                };

  REWARD_FUNCTION: r(xIndex,yIndex,pruneD)
      =matchingCharactersAndPruneBoth(xIndex,yIndex,pruneD);

  TRANSFORMATION_FUNCTION: t(xIndex,yIndex,pruneD)
                              =(xIndex-xAdjustment(pruneD),
                                yIndex-yAdjustment(pruneD));


END
```

The second one is the improved model based on the DP functional equation (2.29) that produces fewer states is given in the following.

```
BEGIN
  NAME LCSaltShort;
  //alternative formulation using indices
  //into global strings
  //try to capture the shorter version
  //from Cormen et al., p.354
  GENERAL_VARIABLES_BEGIN
    //the 2 globally defined input strings
    private static String x = "abcbdab";
    private static String y = "bdcaba";
    //the string lengths as symbolic constants
    private static int xLength = x.length(); //here = 7
    private static int yLength = y.length(); //here = 6
  GENERAL_VARIABLES_END

  GENERAL_FUNCTIONS_BEGIN
    private static NodeSet calculateDecisionSet(int xIndex,
        int yIndex) {
      NodeSet result = new NodeSet();
```

```
    //subtract 1 because array starts at index 0
    if(x.charAt(xIndex-1)==y.charAt(yIndex-1)) {
      //characters match, so the only decision possible is to
      //prune both
      result.add(new Integer(12));
    }
    else {
      //characters do not match, so either prune x or prune y
      result.add(new Integer(1));
      result.add(new Integer(2));
    }
    return result;
  }

  private static int calculateReward(int d) {
    if(d==12) { //decision==prune both
      return 1;
    }
    //other decisions have a reward of 0
    return 0;
  }

  //shorter alternatives:
  private static int xAdjustment(int d) {
    if(d==2) {
      return 0;
    }
    return 1;
  }
  private static int yAdjustment(int d) {
    if(d==1) {
      return 0;
    }
    return 1;
  }

GENERAL_FUNCTIONS_END

//a state is a vector/list of primitive or Set types
STATE_TYPE: (int xIndex, int yIndex);

DECISION_VARIABLE: int d;

//the set of alternatives is a function of the state
DECISION_SPACE: decisionSet(xIndex,yIndex)
```

```
                =calculateDecisionSet(xIndex,yIndex);
  //The decision space is calculated by
  //a global function here.
  //1 stands for x is pruned    and y is unchanged
  //2 stands for x is unchanged and y is pruned
  //12 stands for prune both x and y

 GOAL: f(xLength,yLength); //the lengths of x, y, respectively

 DPFE_BASE_CONDITIONS:
   f(xIndex,yIndex)=0.0 WHEN ((xIndex==0)||(yIndex==0));

 DPFE: f(xIndex,yIndex)
       =MAX_{d IN decisionSet}
                               { f(t(xIndex,yIndex,d))
                                +r(xIndex,yIndex,d)
                               };

 REWARD_FUNCTION: r(xIndex,yIndex,d)
                  =calculateReward(d);

 TRANSFORMATION_FUNCTION: t(xIndex,yIndex,d)
                          =(xIndex-xAdjustment(d),
                            yIndex-yAdjustment(d));

END
```

4.24 gDPS source for LINSRC

Two different gDPS sources, both using DPFE (1.23), are given. The difference is in the formulation of the cost function. The first model uses method W (see Sects. 1.1.4 and 1.1.5).

```
BEGIN
  NAME LINSRCW; //OptimalLinearSearch-W;

  GENERAL_VARIABLES_BEGIN
    private static double[] prob= {.2,.5,.3};
    private static int n = prob.length;
  GENERAL_VARIABLES_END

  SET_VARIABLES_BEGIN
    Set setOfAllItems={0,..,n - 1};
    Set emptySet={};
```

```
  SET_VARIABLES_END

  GENERAL_FUNCTIONS_BEGIN
    private static double sumOfProbabilities(SortedSet items) {
      SortedSet iter=items;
      double result=0.0; int j;
          int n=items.size();
      for (int i=0; i<=n-1; i++) {
        j=((Integer) iter.first()).intValue();
        result+=prob[j];
        iter.remove(iter.first());
      }
      return result;
    }

  GENERAL_FUNCTIONS_END

  STATE_TYPE: (Set items);

  DECISION_VARIABLE: int d;  // item to place next AT FRONT

  DECISION_SPACE: decisionSet(items)=items;

  GOAL: f(setOfAllItems);

  DPFE_BASE: f(emptySet)=0.0 ;

  DPFE: f(items)=MIN_{d IN decisionSet}
            { cost(items,d)+f(t(items,d))};

  REWARD_FUNCTION: cost(items,d) = sumOfProbabilities(items);

  TRANSFORMATION_FUNCTION: t(items,d)=(items SETMINUS {d});

END
```

The second model uses method S (see Sects. 1.1.4 and 1.1.5).

```
BEGIN
  NAME LINSRCS; //OptimalLinearSearch-S;

  GENERAL_VARIABLES_BEGIN
    private static double[] prob= {.2,.5,.3};
    private static int N = prob.length;
  GENERAL_VARIABLES_END
```

```
  SET_VARIABLES_BEGIN
    Set setOfAllItems={0,..,N - 1};
    Set emptySet={};
  SET_VARIABLES_END

  GENERAL_FUNCTIONS_BEGIN
     private static int size(Set items) {
      return items.size();
    }
  GENERAL_FUNCTIONS_END

  STATE_TYPE: (Set items);

  DECISION_VARIABLE: int d;  // item to place next at front

  DECISION_SPACE: decisionSet(items)=items;

  GOAL: f(setOfAllItems);

  DPFE_BASE: f(emptySet)=0.0 ;

  DPFE: f(items)=MIN_{d IN decisionSet}
             { cost(items,d)+f(t(items,d))};

  REWARD_FUNCTION: cost(items,d) = (N+1-size(items)) * prob[d];
        // N+1-size = no. compares to find item (=1 if at front)

    // Note: cost=size*prob if d is item to place next AT END!

  TRANSFORMATION_FUNCTION: t(items,d)=(items SETMINUS {d});

END
```

4.25 gDPS source for LOT

```
BEGIN
  NAME LOT; // Lot Size Problem (Wagner-Whitin);
            // Winston04, example 15

  GENERAL_VARIABLES_BEGIN
    private static double[] demand={220,280,360,140,270};
    private static int N=demand.length;  //N=5
  GENERAL_VARIABLES_END
```

```
  GENERAL_FUNCTIONS_BEGIN
    private static double prodcost(int k, int x) {
      double kFix = 250.0;
      double c = 2.0;
      double ttl=0.0;
      for (int i=k; i<=k+x; i++) {
        ttl=ttl+demand[i - 1];
      }
      return kFix+c*ttl;
    }
    private static double holdcost(int g, int x) {
      double h = 1.0;
      double ttl=0.0;
      for (int i=x; i>=1; i--) {
        ttl=ttl+i*demand[g + i - 1];
      }
      return h*ttl;
    }
  GENERAL_FUNCTIONS_END

  STATE_TYPE: (int k);  //stage

  DECISION_VARIABLE: int x;

  DECISION_SPACE: decisionSet(k) = {0,..,N - k};

  GOAL:  f(1);

  DPFE_BASE_CONDITIONS: f(k)=0.0 WHEN (k>N);

  DPFE:  f(k) = MIN_{x IN decisionSet}
                       { f(t(k,x)) + r(k,x) };

  REWARD_FUNCTION: r(k,x) = prodcost(k,x) + holdcost(k,x);

  TRANSFORMATION_FUNCTION: t(k,x) = (k+x+1);

END
```

4.26 gDPS source for LSP

```
BEGIN
  NAME LSP;
```

```
GENERAL_VARIABLES_BEGIN
  private static final int negInfty=Integer.MIN_VALUE;
  //adjacency matrix of graph from CLRS01, p.343
  private static int[][] distance =
    {
     {negInfty,        1, negInfty,        1},
     {       1, negInfty,        1, negInfty},
     {negInfty,        1, negInfty,        1},
     {       1, negInfty,        1, negInfty}
    };
  //note: negInfty entry represents the fact that there is no
  //      edge between nodes
  //      (negative infinity, since we are maximizing...)

  private static int n = distance.length;//number of nodes n=4
GENERAL_VARIABLES_END

SET_VARIABLES_BEGIN
  Set goalSet={0};
SET_VARIABLES_END

GENERAL_FUNCTIONS_BEGIN
  private static NodeSet possibleNextNodes(int node,
      NodeSet dummy) {
    NodeSet result = new NodeSet();
    for (int i=0; i<distance[node].length; i++) {
      if (distance[node][i]!=negInfty) {
        result.add(new Integer(i));
      }
    }
    return result;
  }
GENERAL_FUNCTIONS_END

STATE_TYPE: (Set nodesVisited, int currentNode);

DECISION_VARIABLE: int alpha;
DECISION_SPACE: decisionSet(currentNode,nodesVisited)
                = possibleNextNodes(currentNode,nodesVisited)
                  SETMINUS nodesVisited;

GOAL:
  f(goalSet,0); //that is f({0},0);
```

```
  DPFE_BASE_CONDITIONS:
    f(nodesVisited,currentNode)=0.0 WHEN (currentNode==3);

  DPFE: f(nodesVisited,currentNode)
        =MAX_{alpha IN decisionSet}
          { cost(nodesVisited,currentNode,alpha)
            +f(t(nodesVisited,currentNode,alpha))};

  REWARD_FUNCTION: cost(nodesVisited,currentNode,alpha)
                   =distance[currentNode][alpha];

  TRANSFORMATION_FUNCTION:
    t(nodesVisited,currentNode,alpha)
      =(nodesVisited SETUNION {alpha}, alpha);

END
```

4.27 gDPS source for MCM

The MCM problem instance from section 2.27 can be coded in gDPS as follows.

```
BEGIN
  NAME MCM; //MatrixChainMultiplication;

  GENERAL_VARIABLES_BEGIN
    //dimensions in MatMult problem
    private static int[] dimension = {3, 4, 5, 2, 2};
    private static int n = dimension.length-1; //n=4 matrices
  GENERAL_VARIABLES_END

  STATE_TYPE: (int firstIndex, int secondIndex);

  DECISION_VARIABLE: int k;
  DECISION_SPACE: decisionSet(firstIndex,secondIndex)
                 ={firstIndex,..,secondIndex - 1};

  GOAL: f(1,n);

  DPFE_BASE_CONDITIONS:
    f(firstIndex,secondIndex)=0.0
      WHEN (firstIndex==secondIndex);

  DPFE: f(firstIndex,secondIndex)
```

```
        =MIN_{k IN decisionSet}
                                  { f(t1(firstIndex,secondIndex,k))
                                   +f(t2(firstIndex,secondIndex,k))
                                   +r(firstIndex,secondIndex,k)
                                  };

  REWARD_FUNCTION: r(firstIndex,secondIndex,k)
                    =dimension[firstIndex - 1]
                     *dimension[k]*dimension[secondIndex];

  TRANSFORMATION_FUNCTION: t1(firstIndex,secondIndex,k)
                            =(firstIndex,k);
                           t2(firstIndex,secondIndex,k)
                            =(k+1,secondIndex);

END
```

The MCM problem exemplifies that the TRANSFORMATION_FUNCTION section allows the definition of an arbitrary number of successor states. Here, MCM has two successor states, computed by t1 and t2.

4.28 gDPS source for MINMAX

```
BEGIN
  NAME minmax;

  GENERAL_VARIABLES_BEGIN
    private static final int infty=Integer.MIN_VALUE;
    private static int[][] distance =
      { // 0     1     2     3     4     5
        // 6     7     8     9
       {infty,   10,    7,    6,infty,infty,
        infty,infty,infty,infty},               //0
       {infty,infty,infty,infty,    9,infty,
        infty,infty,infty,infty},               //1
       {infty,infty,infty,infty,    7,infty,
        infty,infty,infty,infty},               //2
       {infty,infty,infty,infty,   11,    7,
        infty,infty,infty,infty},               //3
       {infty,infty,infty,infty,infty,infty,
            8,    7,   10,infty},               //4
       {infty,infty,infty,infty,infty,infty,
            8,    6,    7,infty},               //5
       {infty,infty,infty,infty,infty,infty,
```

```
       infty,infty,infty,   13},              //6
      {infty,infty,infty,infty,infty,infty,
       infty,infty,infty,  8},                //7
      {infty,infty,infty,infty,infty,infty,
       infty,infty,infty,    9},              //8
      {infty,infty,infty,infty,infty,infty,
       infty,infty,infty,infty}               //9
    };
  private static int numStages=5;  // number of stages
GENERAL_VARIABLES_END

SET_VARIABLES_BEGIN
  Set goalSet={0};          //0 is start node
SET_VARIABLES_END

GENERAL_FUNCTIONS_BEGIN
  private static NodeSet possibleNextNodes(int node) {
    NodeSet result = new NodeSet();
    for (int i=0; i<distance[node].length; i++) {
      if (distance[node][i]!=infty) {
        result.add(new Integer(i));
      }
    }
    return result;
  }

 // maxlink finds maximum branch label in path connecting
 // given set of nodes.
 // Assumes nodes in path are topologically ordered.
 private static int maxlink(SortedSet nodes) {
    int result=Integer.MIN_VALUE;
    int p=0;
    Iterator i=nodes.iterator();
    while (i.hasNext()) {
      int q=((Integer) i.next()).intValue();
      result=Math.max(result,distance[p][q]);
      p=q;
    }
    return result;
  }
GENERAL_FUNCTIONS_END

STATE_TYPE: (int stage, Set nodesVisited, int currentNode);

DECISION_VARIABLE: int alpha;
```

```
  DECISION_SPACE:
    possibleSuccessors(stage, nodesVisited, currentNode)
      = possibleNextNodes(currentNode) SETMINUS nodesVisited;

  GOAL:  f(1,goalSet,0);

  DPFE_BASE_CONDITIONS:
    f(stage,nodesVisited,currentNode) = maxlink(nodesVisited)
       WHEN (stage>=numStages); //base-condition cost
                                //is attached to final decision

  DPFE: f(stage,nodesVisited,currentNode)
       =MIN_{alpha IN possibleSuccessors}
         { r(stage,nodesVisited,currentNode,alpha)
          +f(t(stage,nodesVisited,currentNode,alpha)) };

  REWARD_FUNCTION:
      r(stage,nodesVisited,currentNode,alpha)=0.0;
      //decision-cost=0

  TRANSFORMATION_FUNCTION:
      t(stage,nodesVisited,currentNode,alpha)
        =(stage + 1,nodesVisited SETUNION {alpha},alpha);

END
```

4.29 gDPS source for MWST

```
BEGIN
  NAME mwst;  //minimum weight spanning tree

  GENERAL_VARIABLES_BEGIN
    private static final int infty=Integer.MAX_VALUE;
    private static int[][] w =  // weighted adjacency matrix of
                                // graph (not used)
      {
       {infty,      1,      3,  infty},
       {    1,  infty,      2,      4},
       {    3,      2,  infty,      5},
       {infty,      4,      5,  infty}
      };
    private static int N = 4; // no. of nodes in graph
    private static int B = 5; // no. of branches in graph
```

```
  private static int[] wgt = {5,4,3,2,1};
    // branch weights, unsorted
  private static int N1 = N-1;
    // no. of branches in spanning tree

  private static int[][] cycle = // set of basic cycles
                                 // (len<=N-1)
    { {0,0,1,1,1},{1,1,0,1,0} }; // 321, 542
  private static int NC = 2;     // no. of basic cycles
GENERAL_VARIABLES_END

SET_VARIABLES_BEGIN
  Set setOfAllbranches={0,..,B - 1};
  Set emptySet={};
SET_VARIABLES_END

GENERAL_FUNCTIONS_BEGIN
  private static boolean match(int[] a, int[] b, int n) {
   for (int i=0; i<n; i++){
     if (a[i]!=b[i]) {
       return false;
     }
   }
   return true;
  }
  private static int[] invertvec(int[] a, int n) {
    int[] vec = new int[n];
    for (int i=0; i<n; i++) {
      vec[i]=1-a[i];
    }
    return vec;
  }
  private static int[] characteristicVector(Set s, int n) {
    int [] vec = new int[n];
    for (int i=0; i<n; i++) {
      if (s.contains(new Integer(i))) {
        vec[i]=1;
      }
      else {
        vec[i]=0;
      }
    }
    return vec;
  }
  private static int cost(Set branches, int k, int d) {
```

```
    //char.vector, dim=B
    int [] vec = characteristicVector(branches, B);
    vec=invertvec(vec,B);  //TreeSoFar=branches not in state
    vec[d]=1;              //vec=TSF+d
    //use given cycles, vs. computing cycles from d
    for (int j=0; j<=NC-1; j++) {
      if (match(vec,cycle[j],B)) { //TreeSoFar is cyclic!
        return(infty);
      }
    }
    //cost=wgt[d] of adding branch d to tree at stage k
    return wgt[d];
  }
GENERAL_FUNCTIONS_END

STATE_TYPE: (Set branches, int k); //branches to choose from
                                   //(=~TSF) at stage k

DECISION_VARIABLE: int d; //choice of k-th branch
                          //to add to TreeSoFar

DECISION_SPACE: decisionSet(branches, k)=branches;

GOAL: f(setOfAllbranches, 0); //TreeSoFar (TSF) initially
                              //empty, at stage 0

DPFE_BASE_CONDITIONS: f(branches,k)=0.0 WHEN (k==N1);
               //since no. of branches in sp.tree=N-1

DPFE: f(branches,k)=MIN_{d IN decisionSet}
         { r(branches,k,d) + f(t(branches,k,d)) };

REWARD_FUNCTION:
  r(branches,k,d) = cost(branches,k,d);
  //cost of adding branch d to TSF at stage k

TRANSFORMATION_FUNCTION:
  t(branches,k,d) = (branches SETMINUS {d},k+1);
  //omit branch d in next stage

END
```

4.30 gDPS source for NIM

```
BEGIN
  NAME NIM;

  GENERAL_VARIABLES_BEGIN
    private static int m = 10; //a winning state (small example)
    //private static int m = 9; //a losing state (small example)
    //private static int m = 30; //winning state (large example)
    //private static int m = 29; //losing state (large example)
  GENERAL_VARIABLES_END

  STATE_TYPE: (int s);

  DECISION_VARIABLE: int d;
  DECISION_SPACE: decisionSet(d)={1,..,3};

  GOAL: f(m);

  DPFE_BASE_CONDITIONS:
    f(s) = 1.0 WHEN (s<0);  //win!
                            //(adversary removed last sticks)
    f(s) = 0.0 WHEN (s==1); //loss!
                            //(my turn with one remaining stick)
    //Note:  much time can be saved by adding f(2)=f(3)=f(4)=1.0

  DPFE: f(s)
        =MAX_{d IN decisionSet}
             { r(s,d) * f(t1(s,d)) * f(t2(s,d)) * f(t3(s,d)) };

  REWARD_FUNCTION: r(s,d)=1.0;

  TRANSFORMATION_FUNCTION:
      t1(s,d) = (s - d - 1);
      t2(s,d) = (s - d - 2);
      t3(s,d) = (s - d - 3);

END
```

4.31 gDPS source for ODP

```
BEGIN
  NAME ODP; //Optimal Distribution Problem;
```

```
GENERAL_VARIABLES_BEGIN
  private static final int infty=Integer.MAX_VALUE;
  private static int[][] cashFlow =
    { {0,1,2,3},        //creditor 0
      {0,1,2,3,4},      //creditor 1
      {0,3,4}           //creditor 2
    };
  private static int[][] netPresentValue =
    { {0,4,12,21},        //creditor 0
      {0,6,11,16,20},     //creditor 1
      {0,16,22}           //creditor 2
    };
GENERAL_VARIABLES_END

GENERAL_FUNCTIONS_BEGIN
  //create the index set of possible alternatives
  //which may vary from stage to stage
  private static NodeSet possibleAlternatives(int stage) {
    NodeSet result = new NodeSet();
    for (int i=0; i<cashFlow[stage].length; i++) {
      result.add(new Integer(i));
    }
    return result;
  }
GENERAL_FUNCTIONS_END

STATE_TYPE: (int stage, int moneySecured);

DECISION_VARIABLE: int d;
DECISION_SPACE: decisionSet(stage,moneySecured)
                =possibleAlternatives(stage);

GOAL:
  f(0,0); //stage 0, no money secured

DPFE_BASE_CONDITIONS:
  f(stage,moneySecured)=0.0
    WHEN ((stage==3)&&(moneySecured>=6));
  f(stage,moneySecured)=infty
    WHEN ((stage==3)&&(moneySecured<6));

DPFE: f(stage,moneySecured)
      =MIN_{d IN decisionSet}
        { futureNPV(stage,moneySecured,d)
          +f(t(stage,moneySecured,d))};
```

```
  REWARD_FUNCTION: futureNPV(stage,moneySecured,d)
                   =netPresentValue[stage][d];

  TRANSFORMATION_FUNCTION:
    t(stage,moneySecured,d)
      =(stage+1,moneySecured+cashFlow[stage][d]);

END
```

4.32 gDPS source for PERM

```
BEGIN
  NAME PERM; //optimal PERMutation problem;

  GENERAL_VARIABLES_BEGIN
    private static int[] programLength = {5,3,2};
    private static int n = programLength.length;
      //n=number of programs
  GENERAL_VARIABLES_END

  SET_VARIABLES_BEGIN
    Set setOfAllItems={0,..,n - 1}; //the n programs are indexed
                                    //from 0 to n-1
    Set emptySet={};
  SET_VARIABLES_END

  GENERAL_FUNCTIONS_BEGIN
    //need a function that returns the cardinality of a set
    private static int cardinality(Set items) {
      return items.size();
        //size() method from java.util.Set interface
    }
  GENERAL_FUNCTIONS_END

  STATE_TYPE: (Set items);

  DECISION_VARIABLE: int d;
  DECISION_SPACE: decisionSet(items)
                 =items;

  GOAL:  f(setOfAllItems);

  DPFE_BASE:  f(emptySet)=0.0;
```

```
  DPFE: f(items) = MIN_{d IN decisionSet}
        { cost(items,d)
          +f(t(items,d))};

  REWARD_FUNCTION:
      cost(items,d) = cardinality(items)*programLength[d];

  TRANSFORMATION_FUNCTION:
      t(items,d) = (items SETMINUS {d});

END
```

4.33 gDPS source for POUR

```
BEGIN
  NAME pour;  //"wine pouring" example
              //Winston02 pp.750--751, example 2

  GENERAL_VARIABLES_BEGIN
    private static int P = 9;                //capacity of I
    private static int Q = 4;                //capacity of J
    private static int TGT = 6;              //target amount
    private static int R = P+Q;
    //private static int LIM = (P+1)*(Q+1); //bound on no.stages
    private static int LIM = R; //LIM is used to prevent
                                //out-of-memory error
  GENERAL_VARIABLES_END

  GENERAL_FUNCTIONS_BEGIN
    private static int nexti(int i, int j, int k, int d) {
      if (d==1)      return P;                         //fill i
      else if (d==2) return i;                         //fill j
      else if (d==3) return 0;                         //empty i
      else if (d==4) return i;                         //empty j
      else if (d==5) return Math.max(0,i - (Q - j));   //i2j
      else           return Math.min(P,i+j);           //j2i
    }
    private static int nextj(int i, int j, int k, int d) {
      if (d==1)      return j;                         //fill i
      else if (d==2) return Q;                         //fill j
      else if (d==3) return j;                         //empty i
      else if (d==4) return 0;                         //empty j
      else if (d==5) return Math.min(i+j,Q);           //i2j
      else           return Math.max(0,j - (P - i));   //j2i
```

```
    }
    private static int nextk(int i, int j, int k, int d) {
      if (d==1)       return k - (P - i);                  //fill i
      else if (d==2) return k - (Q - j);                   //fill j
      else if (d==3) return k+i;                           //empty i
      else if (d==4) return k+j;                           //empty j
      else if (d==5) return k;                             //i2j
      else            return k;                            //j2i
    }
    private static NodeSet possibleDec(int i, int j, int k) {
      NodeSet result = new NodeSet();
      for (int d=1; d<=6; d++) {
        result.add(new Integer(d));
      }
////    NOTE: The following constraints can be added to reduce
////      the size of the decision set, but are unnecessary.
      if (i==0) result.remove(new Integer(3));             //empty i
      if (i==0) result.remove(new Integer(5));             //i2j
      if (j==0) result.remove(new Integer(4));             //empty j
      if (j==0) result.remove(new Integer(6));             //j2i
      if (i==P) result.remove(new Integer(1));             //fill i
      if (i==P) result.remove(new Integer(6));             //j2i
      if (j==Q) result.remove(new Integer(2));             //fill j
      if (j==Q) result.remove(new Integer(5));             //i2j
      if (k==0) result.remove(new Integer(1));             //fill i
      if (k==0) result.remove(new Integer(2));             //fill j
      return result;
    }
GENERAL_FUNCTIONS_END

  STATE_TYPE: (int s, int i, int j, int k);
    //s: stage number
    //i: amount in glass 1
    //j: amount in glass 2
    //k: amount in carafe

  DECISION_VARIABLE: int d;
      //d=1: fill i (from k)
      //d=2: fill j (from k)
      //d=3: empty i (into k)
      //d=4: empty j (into k)
      //d=5: i2j (pour contents from i to j)
      //d=6: j2i (pour contents from j to i)

  DECISION_SPACE: decSet(s,i,j,k) //={1,..,6};
```

```
      =possibleDec(i,j,k); //use possibleDec if you want to
                            //prune the decision set

  GOAL: f(1,0,0,R);                     //initial state

  DPFE_BASE_CONDITIONS:
        f(s,i,j,k)=0.0 WHEN (i==TGT);   //target state
        f(s,i,j,k)=0.0 WHEN (j==TGT);   //target state
        f(s,i,j,k)=999.0 WHEN (s>LIM);  //cycle!

  DPFE: f(s,i,j,k) = MIN_{d IN decSet}
                     { r(s,i,j,k,d) + f(t(s,i,j,k,d)) };

  REWARD_FUNCTION: r(s,i,j,k,d) = 1.0;

  TRANSFORMATION_FUNCTION: t(s,i,j,k,d)
      = (s+1, nexti(i,j,k,d), nextj(i,j,k,d), nextk(i,j,k,d));

END
```

4.34 gDPS source for PROD

```
BEGIN
  NAME PROD; // Production Problem Bronson97 19.14;

  GENERAL_VARIABLES_BEGIN
    private static final int infty=Integer.MAX_VALUE;
    private static double prob = .6;
    private static int n=5;
  GENERAL_VARIABLES_END

  GENERAL_FUNCTIONS_BEGIN
    private static double p(int amount) {
      double[] productioncost = {0.0,10.0,19.0};
      if (amount<0) return infty;
      else return productioncost[amount];
    }
    private static double q(int inventory) {
      if (inventory<0) return -1.5 * inventory;
      else return 1.1 * inventory;
    }
  GENERAL_FUNCTIONS_END

  //g: stage number; s: size of inventory
```

```
  STATE_TYPE: (int g, int s);

  DECISION_VARIABLE: int d;
  DECISION_SPACE: decisionSet(g,s)={0 - s,..,2};
    //s=-1 or s=-2 denotes a shortfall in inventory
    //which must be made up!
  GOAL: f(1,0);

  DPFE_BASE_CONDITIONS:
    f(g,s)=0.0 WHEN (g==n);

  DPFE: f(g,s)
        =MIN_{d IN decisionSet}
                                { p1.f(t1(g,s,d))
                                 +p2.f(t2(g,s,d))
                                 +cost(g,s,d)
                                };

  REWARD_FUNCTION: cost(g,s,d) = p(d)+q(s);

  TRANSFORMATION_FUNCTION: t1(g,s,d)
                            =(g+1,s+d - 1);
                           t2(g,s,d)
                            =(g+1,s+d - 2);

  TRANSITION_WEIGHTS: p1(g,s,d)=prob;
                      p2(g,s,d)=1.0-prob;

END
```

4.35 gDPS source for PRODRAP

The PRODRAP problem instance from Section 2.35 illustrates the use of probability weights in the `DPFE` section and shows how these weights are defined as functions in the `TRANSITION_WEIGHTS` section.

For PRODRAP there appears only a single probability weight `p1` within the `DPFE` section. Note that in order to assign a weight to a functional the dot notation "." is used as the multiplication symbol, so it can easily be distinguished by the parser and the modeler from the star notation "*" which is used for multiplicative DP problems like the RDP problem from section 2.36. The probability weight `p1` is then defined in the `TRANSITION_WEIGHTS` section, where the helper function `probabilityThatAllDefect` is used, since only the basic arithmetic operators (but not exponentiation) can be used in the TRANSITION_WEIGHTS section.

```
BEGIN
  NAME PRODRAP; //PRODuction --- Reject Allowance Problem
  //probabilistic problem from Hillier/Lieberman p.421

  GENERAL_VARIABLES_BEGIN
    private static double defectProbability=0.5;
    private static int productionRuns=3;
    private static double marginalProductionCost=1.0;
    private static double setupCost=3.0;
    private static double penaltyCosts=16.0;
    private static int maxLotSize=5;
  GENERAL_VARIABLES_END

  GENERAL_FUNCTIONS_BEGIN
    //calculate probability that all xn items
    //produced are defect
    private static double probabilityThatAllDefect(int xn) {
      return Math.pow(defectProbability,xn);
    }
    //function K calculates the setup cost
    private static double K(int xn) {
      if (xn==0) {  //if nothing is produced
        return 0;   //there is no setup cost
      }
      //otherwise we encounter a fix setup cost of $300
      return setupCost;
    }
  GENERAL_FUNCTIONS_END

  STATE_TYPE: (int n); //n is stage number
                       //(production run number)

  DECISION_VARIABLE: int xn; //xn is lot size for stage n
  //Hillier/Lieberman do not explicitly set an upper bound for
  //the lot size in the general problem formulation. However in
  //the example instance they use an upper bound of 5, making
  //the decision set {0,..,5}
  DECISION_SPACE: decisionSet(n)={0,..,maxLotSize};

  GOAL: f(1);

  DPFE_BASE_CONDITIONS:
```

```
    f(n)=penaltyCosts WHEN (n==productionRuns+1);
    //penalty of $1600 if no acceptable item after
    //3 production runs

  DPFE: f(n) = MIN_{xn IN decisionSet}
                 {    r(n,xn)
                  +p1.f(t1(n,xn))
                 };

  REWARD_FUNCTION: r(n,xn)=xn*marginalProductionCost+K(xn);

  TRANSFORMATION_FUNCTION: t1(n,xn)
                             =(n+1);

  TRANSITION_WEIGHTS: p1(n,xn)=probabilityThatAllDefect(xn);

END
```

4.36 gDPS source for RDP

```
BEGIN
  NAME RDP; //ReliabilityDesignProblem
  GENERAL_VARIABLES_BEGIN
    //cost for each device type
    //(we can use ints instead of doubles)
    private static int[] cost = {30, 15, 20};
    //reliability of each device type
    private static double[] reliability = {0.9, 0.8, 0.5};
    //total budget
    private static int budget = 105;
    //This instance has 3 stages. Number them 0, 1 and 2.
    private static int numberOfStages=reliability.length;
  GENERAL_VARIABLES_END

  GENERAL_FUNCTIONS_BEGIN
    //calculate reliability of stage i given the number of
    //components m_i
    private static double reliabilityOfStage(int stage,
        int noOfComponents) {
      return 1-Math.pow(1-reliability[stage],noOfComponents);
    }

    //calculate upper bound for number of components
    //(Horowitz p.297 introduces a different attempt for this)
```

```
  //the following improved version disallows that we run out
  //of funds too early
  private static int upperBoundForNumberOfComponents
      (int stage,int remainingMoney) {
    int totalCostForOneDeviceAtEachLowerStage=0;
    for(int i=0; i<stage; i++) {
      totalCostForOneDeviceAtEachLowerStage+=cost[i];
    }
    //note: integer division makes the floor superfluous
    //in the following equation
    int result =
        (remainingMoney-totalCostForOneDeviceAtEachLowerStage)
        /cost[stage];
    return result;
  }

  //produce the set {1,..,upperBoundForNumberOfComponents}
  private static NodeSet calculateDecisionSet(int stage,
      int remainingMoney) {
    NodeSet result = new NodeSet();
    int upperBound
      = upperBoundForNumberOfComponents(stage,remainingMoney);
    for(int i=1; i<=upperBound; i++) {
      result.add(new Integer(i));
    }
    return result;
  }
GENERAL_FUNCTIONS_END

STATE_TYPE: (int stage, int remainingMoney);

DECISION_VARIABLE: int m;
DECISION_SPACE: decisionSet(stage,remainingMoney)
              =calculateDecisionSet(stage,remainingMoney);

GOAL: f(2,budget);

DPFE_BASE_CONDITIONS:
  f(stage,remainingMoney)=1.0 WHEN (stage== -1);

DPFE: f(stage,remainingMoney)
     =MAX_{m IN decisionSet}
                              { f(t(stage,remainingMoney,m))
                               *r(stage,remainingMoney,m)
                              };
```

```
  REWARD_FUNCTION: r(stage,remainingMoney,m)
                          =reliabilityOfStage(stage,m);

  TRANSFORMATION_FUNCTION: t(stage,remainingMoney,m)
      =(stage - 1,remainingMoney - m*cost[stage]);

END
```

4.37 gDPS source for REPLACE

```
BEGIN
  NAME replace; // Replacement Problem;

  GENERAL_VARIABLES_BEGIN
    private static double priceOfNewMachine=1000.0;
    private static int[] tm={60,140,260}; //total maintenance
                                          //cost for d stages
    private static int[] v={800,600,500}; //salvage value
                                          //after d stages
    private static int L=tm.length;       //max. lifetime
    private static int N=5;               //no. of stages
    private static double infty=Double.MAX_VALUE;
  GENERAL_VARIABLES_END

  GENERAL_FUNCTIONS_BEGIN
    private static double cost(int d) {
      if (d>L) {
        return infty;
      }
      else {
        return priceOfNewMachine + tm[d - 1] - v[d - 1];
      }
    }
  GENERAL_FUNCTIONS_END

  STATE_TYPE: (int k); // stage number

  DECISION_VARIABLE: int d;
      //d = no. of stages before replacement = usage time

  DECISION_SPACE: decisionSet(k) = {1,..,L};

  GOAL:  f(0);
```

```
  DPFE_BASE_CONDITIONS:
        f(k) = 0.0 WHEN (k>=N);

  DPFE: f(k) = MIN_{d IN decisionSet}
                    { r(k,d) + f(trans(k,d)) };

  REWARD_FUNCTION: r(k,d) = cost(d);

  TRANSFORMATION_FUNCTION:
      trans(k,d) = (k+d);  //stage of next replacement

END
```

4.38 gDPS source for SCP

```
BEGIN
  NAME SCPwS; //StagecoachProblem with explicit stages;
  //uses explicit stages numbered 0 through 4 in DPFE

  GENERAL_VARIABLES_BEGIN
    private static final int infty=Integer.MAX_VALUE;
    //adjacency matrix for stagecoach problem (Winston03, p.753)
    //The 10 nodes are numbered from 0 through 9 here,
    //not 1 to 10.
    private static int[][] distance =
      {
       {infty,  550,  900,  770,infty,
        infty,infty,infty,infty,infty},
       {infty,infty,infty,infty,  680,
          790, 1050,infty,infty,infty},
       {infty,infty,infty,infty,  580,
          760,  660,infty,infty,infty},
       {infty,infty,infty,infty,  510,
          700,  830,infty,infty,infty},
       {infty,infty,infty,infty,infty,
        infty,infty,  610,  790,infty},
       {infty,infty,infty,infty,infty,
        infty,infty,  540,  940,infty},
       {infty,infty,infty,infty,infty,
        infty,infty,  790,  270,infty},
       {infty,infty,infty,infty,infty,
        infty,infty,infty,infty, 1030},
```

```
      {infty,infty,infty,infty,infty,
       infty,infty,infty,infty, 1390},
      {infty,infty,infty,infty,infty,
       infty,infty,infty,infty,infty}
    };
  GENERAL_VARIABLES_END

  GENERAL_FUNCTIONS_BEGIN
    private static NodeSet possibleNextNodes(int node) {
      NodeSet result = new NodeSet();
      for (int i=0; i<distance[node].length; i++) {
        if (distance[node][i]!=infty) {
          result.add(new Integer(i));
        }
      }
      return result;
    }
  GENERAL_FUNCTIONS_END

  STATE_TYPE: (int g, int x);
    //g is stage number, x is current node

  DECISION_VARIABLE: int d;
  //the set of alternatives is a function of the state
  DECISION_SPACE: possibleSuccessors(g,x)
                  =possibleNextNodes(x);

  GOAL:
    f(0,0); //We start from node number zero in stage 0, so
            //the goal is to compute f(0).

  DPFE_BASE_CONDITIONS:  f(g,x)=0.0 WHEN (x==9);
  DPFE: f(g,x)=MIN_{d IN possibleSuccessors}
          { cost(g,x,d)
            +f(t(g,x,d))};

  REWARD_FUNCTION: cost(g,x,d)=distance[x][d];

  TRANSFORMATION_FUNCTION: t(g,x,d)=(g+1,d);

END
```

4.39 gDPS source for SEEK

```
BEGIN
  NAME seek; // optimal total file seek time

  GENERAL_VARIABLES_BEGIN
    private static int[] track= {100,50,190};
    private static int start = 140;
    private static int N = track.length;
  GENERAL_VARIABLES_END

  SET_VARIABLES_BEGIN
    Set setOfAllItems={0,..,N - 1};
    Set emptySet={};
  SET_VARIABLES_END

  GENERAL_FUNCTIONS_BEGIN
    private static int cost(int x, int d) {
       return Math.abs(x-track[d]);
     }
  GENERAL_FUNCTIONS_END

  STATE_TYPE: (Set items, int x);

  DECISION_VARIABLE: int d;

  DECISION_SPACE: decisionSet(items, x)=items;

  GOAL: f(setOfAllItems,start);

  DPFE_BASE_CONDITIONS:
    f(items, x)=0.0 WHEN (items SETEQUALS emptySet);

  DPFE: f(items,x)=MIN_{d IN decisionSet}
            { r(items,x,d)+f(t(items,x,d)) };

  REWARD_FUNCTION: r(items,x,d) = cost(x,d);

  TRANSFORMATION_FUNCTION: t(items,x,d)
      =(items SETMINUS {d}, track[d]);

END
```

4.40 gDPS source for SEGLINE

Two different gDPS sources are given. The first model uses the DP functional equation (2.40).

```
BEGIN

  NAME SEGline;    // segmented curve fitting

  GENERAL_VARIABLES_BEGIN
    private static double[][] e = // least-squares error
      {
        { 9999.,    0.0, 0.0556,   1.45},
        { 9999.,  9999.,    0.0, 1.3889},
        { 9999.,  9999.,  9999.,    0.0},
        { 9999.,  9999.,  9999.,  9999.}
      };
    private static int N1 = e.length-1;
      // no. of partitions (segments)
    //private static double K = 10.0;
      // segmentation cost => 1 part
    private static double K = 1.0;
      // segmentation cost => 2 parts
    //private static double K = 0.01;
      // segmentation cost => 3 parts
  GENERAL_VARIABLES_END

  STATE_TYPE: (int s); // s=state=current break location

  DECISION_VARIABLE: int d; // d=decision=next break location
  DECISION_SPACE: decisionSet(s)={s+1,..,N1};
      // d = current+1,..,end

  GOAL: f(0); // initial break at location 0

  DPFE_BASE: f(N1)=0.0; // final break at location N-1
  DPFE: f(s) = MIN_{d IN decisionSet} { r(s,d) + f(t(s,d)) };

  REWARD_FUNCTION:
      r(s,d) = e[s][d] + K; // cost of line + cost of segment

  TRANSFORMATION_FUNCTION:
      t(s,d) = (d); // set current state to d

END
```

The second model is an alternative formulation using the DP functional equation (2.41).

```
BEGIN

  NAME SEGlineAlt; // segmented curve fitting
                   // with LIMit on no. of segments

  GENERAL_VARIABLES_BEGIN
    private static double[][] e = // least-squares error
      {
        { 9999.,    0.0, 0.0556,   1.45},
        { 9999., 9999.,     0.0, 1.3889},
        { 9999., 9999.,  9999.,     0.0},
        { 9999., 9999.,  9999.,  9999.}
      };
    private static int N1 = e.length-1;
        // max. no. of partitions (segments)
    private static double K =  10.0;
        // segmentbreak cost (0.01,..,10.0)
    private static int LIM = 2;
        // constraint on no. of segments (1..N1)
  GENERAL_VARIABLES_END

  STATE_TYPE: (int i, int s);
      // i=no. segmentbreaks allowed, s=current break location

  DECISION_VARIABLE: int d; // d=decision=next break location
  DECISION_SPACE: decisionSet(i,s)={s+1,..,N1};
      // d = current+1,..,end

  GOAL: f(LIM,0); // initial break at location 0
                  // up to LIM segmentbreaks

  DPFE_BASE_CONDITIONS:
    f(i,s)=0.0 WHEN ((i==0)&&(s==N1));
        // final break at location N-1
    f(i,s)=9999.9 WHEN ((i==0)&&(s<N1));//no more breaks allowed
    //f(i,s)=0.0 WHEN ((i>0)&&(s==N1));// no. breaks .le. LIM
    f(i,s)=9999.9 WHEN ((i>0)&&(s==N1));// no. breaks .eq. LIM

  DPFE: f(i,s) = MIN_{d IN decisionSet}
                    { r(i,s,d) + f(t(i,s,d)) };

  REWARD_FUNCTION:
```

```
      r(i,s,d) = e[s][d] + K; // cost of line + cost of break

  TRANSFORMATION_FUNCTION:
      t(i,s,d) = (i - 1, d); // one fewer break for next segment

END
```

4.41 gDPS source for SEGPAGE

```
BEGIN
  NAME SEGPAGE;

  GENERAL_VARIABLES_BEGIN
    private static final int inf=Integer.MAX_VALUE; //infinity
    private static int[][] distance =
      {
       {inf,   0,   2,  82,  82,  82,  82,   2, 202,
        202,   2,  42,  42,   2,   2,   0,   0},
       {inf, inf,   2,  82,  82,  82,  82,   2, 202,
        202,   2,  42,  42,   2,   2,   0,   0},
       {inf, inf, inf,  82,  82,  82,  82,   2, 202,
        202,   2,  42,  42,   2,   2,   0,   0},
       {inf, inf, inf, inf,  42,  82,  82,   2, 202,
        202,   2,  42,  42,   2,   2,   0,   0},
       {inf, inf, inf, inf, inf,  42,  42,   2, 202,
        202,   2,  42,  42,   2,   2,   0,   0},
       {inf, inf, inf, inf, inf, inf,  41,   1, 201,
        201,   2,  42,  42,   2,   2,   0,   0},
       {inf, inf, inf, inf, inf, inf, inf,   1, 201,
        201,   2,  42,  42,   2,   2,   0,   0},
       {inf, inf, inf, inf, inf, inf, inf, inf, 201,
        201,   2,  42,  42,   2,   2,   0,   0},
       {inf, inf, inf, inf, inf, inf, inf, inf, inf,
        101,   2,  42,  42,   2,   2,   0,   0},
       {inf, inf, inf, inf, inf, inf, inf, inf, inf,
        inf,   2,  42,  42,   2,   2,   0,   0},
       {inf, inf, inf, inf, inf, inf, inf, inf, inf,
        inf, inf,  42,  42,   2,   2,   0,   0},
       {inf, inf, inf, inf, inf, inf, inf, inf, inf,
        inf, inf, inf,  22,   2,   2,   0,   0},
       {inf, inf, inf, inf, inf, inf, inf, inf, inf,
        inf, inf, inf, inf,   2,   2,   0,   0},
       {inf, inf, inf, inf, inf, inf, inf, inf, inf,
        inf, inf, inf, inf, inf,   2,   0,   0},
```

```
      {inf, inf, inf, inf, inf, inf, inf, inf, inf,
       inf, inf, inf, inf, inf, inf,   0,   0},
      {inf, inf, inf, inf, inf, inf, inf, inf, inf,
       inf, inf, inf, inf, inf, inf, inf,   0},
      {inf, inf, inf, inf, inf, inf, inf, inf, inf,
       inf, inf, inf, inf, inf, inf, inf, inf}
    };
  private static int N1=distance.length-1;
  private static final int m=4; // page size
GENERAL_VARIABLES_END

GENERAL_FUNCTIONS_BEGIN
  private static int cost(int p, int d) {
    if (d-p>m) { // check page size
      return inf;
    }
    else {
      return distance[p][d];
    }
  }
GENERAL_FUNCTIONS_END

STATE_TYPE: (int p);

DECISION_VARIABLE: int d;

DECISION_SPACE: decisionSet(p)={p+1,..,N1};

GOAL:  f(0);

DPFE_BASE_CONDITIONS:
  f(p)=0.0 WHEN (p==N1);

DPFE: f(p) = MIN_{d IN decisionSet} { r(p,d) + f(t(p,d))};

REWARD_FUNCTION: r(p,d) = cost(p,d);

TRANSFORMATION_FUNCTION: t(p,d) = (d);

END
```

4.42 gDPS source for SELECT

```
BEGIN
```

```
  NAME select;  // optimal selection problem

  GENERAL_VARIABLES_BEGIN
    private static int N = 10; //size of data to select from
    private static int[] item= {3,6,8,10};
        //select k-th out of N {1<=k<=N}
    private static int M = item.length; //M=4 items to select
  GENERAL_VARIABLES_END

  STATE_TYPE: (int i, int j, int p, int q);
      //(i,j)=subset of items

  DECISION_VARIABLE: int k;
  DECISION_SPACE: decisionSet(i,j,p,q)={i,..,j};

  GOAL: f(1,M,1,N);

  DPFE_BASE_CONDITIONS: f(i,j,p,q)=0.0 WHEN (i>j);
  DPFE: f(i,j,p,q)=MIN_{k IN decisionSet}
            { r(i,j,p,q,k) + f(tLeft(i,j,p,q,k))
                           + f(tRight(i,j,p,q,k)) };

  REWARD_FUNCTION: r(i,j,p,q,k) = q-p+1;

  TRANSFORMATION_FUNCTION:
    //Note: subscript is k-1 for zero-indexing
    tLeft(i,j,p,q,k)  = (i,k - 1,p,item[k - 1] - 1);
    tRight(i,j,p,q,k) = (k+1, j, item[k - 1]+1, q);

END
```

4.43 gDPS source for SPA

The SPA problem instance from section 2.43 can be coded in gDPS as follows.

```
BEGIN
  NAME SPA; //ShortestPathAcyclic;

  GENERAL_VARIABLES_BEGIN
    private static final int infty=Integer.MAX_VALUE;
    //adjacency matrix for acyclic shortest path (Lew85, p.337)
    //The nodes (s,x,y,t) are coded as (0,1,2,3)
    private static int[][] distance =
      {
```

```
    {infty,     3,      5,  infty},
    {infty,  infty,     1,      8},
    {infty,  infty,  infty,     5},
    {infty,  infty,  infty,  infty}
   };
  //Note: infty entry represents the fact that there is no
  //      edge between nodes.
GENERAL_VARIABLES_END

GENERAL_FUNCTIONS_BEGIN
  private static NodeSet possibleNextNodes(int node) {
    NodeSet result = new NodeSet();
    for (int i=0; i<distance[node].length; i++) {
      if (distance[node][i]!=infty) {
        result.add(new Integer(i));
      }
    }
    return result;
  }
GENERAL_FUNCTIONS_END

STATE_TYPE: (int currentNode);

DECISION_VARIABLE: int d;
DECISION_SPACE: possibleSuccessors(currentNode)
                =possibleNextNodes(currentNode);

GOAL:
  f(0); //We start from node number zero, so
        //the goal is to compute f(0).

DPFE_BASE_CONDITIONS:
  f(currentNode)=0.0 WHEN (currentNode==3);

DPFE: f(currentNode)
      =MIN_{d IN possibleSuccessors}
        { cost(currentNode,d)
          +f(t(currentNode,d))};

REWARD_FUNCTION: cost(currentNode,d)
                 =distance[currentNode][d];

TRANSFORMATION_FUNCTION: t(currentNode,d)
                         =(d);
```

```
END
```

Note the flexibility offered by the optional `GENERAL_FUNCTIONS` section that allows one to efficiently define the current decision set via the function `possibleNextNodes()`. By eliminating nodes connected to the current node by infinitely weighted edges from the decision set we can reduce the complexity of the problem somewhat.

4.44 gDPS source for SPC

Two different gDPS sources are given. The first model is based on the DP functional equation (2.45) that keeps track of the set of nodes already visited.

```
BEGIN
  NAME SPCalt; //ShortestPathCyclicAlt;
  //single source shortest path, cycles allowed,
  //no negative weights allowed
  //DPFE chosen here: set approach similar to TSP problem

  GENERAL_VARIABLES_BEGIN
    private static final int infty=Integer.MAX_VALUE;
    //adjacency matrix for cyclic shortest path (Lew85, p.337)
    //The nodes (s,x,y,t) are coded as (0,1,2,3)
    private static int[][] distance =
      {
       {infty,      3,      5,  infty},
       {infty,  infty,      1,      8},
       {infty,      2,  infty,      5},
       {infty,  infty,  infty,  infty}
      };
    //Note: infty entry represents the fact that there is no
    //      edge between nodes
  GENERAL_VARIABLES_END

  SET_VARIABLES_BEGIN
    Set goalSet={0};
  SET_VARIABLES_END

  GENERAL_FUNCTIONS_BEGIN
    private static NodeSet possibleNextNodes(int node) {
      NodeSet result = new NodeSet();
      for (int i=0; i<distance[node].length; i++) {
        if (distance[node][i]!=infty) {
          result.add(new Integer(i));
        }
```

```
    }
    return result;
  }
GENERAL_FUNCTIONS_END

STATE_TYPE: (int currentNode, Set nodesVisited);

DECISION_VARIABLE: int alpha;
DECISION_SPACE: possibleSuccessors(currentNode, nodesVisited)
    = possibleNextNodes(currentNode) SETMINUS nodesVisited;

GOAL:
  f(0,goalSet); //that is: (0,{0});

DPFE_BASE_CONDITIONS:
  f(currentNode,nodesVisited)=0.0
    WHEN (currentNode==3);

DPFE: f(currentNode,nodesVisited)
      =MIN_{alpha IN possibleSuccessors}
        { cost(currentNode,nodesVisited,alpha)
          +f(t(currentNode,nodesVisited,alpha))};

REWARD_FUNCTION: cost(currentNode,nodesVisited,alpha)
                 =distance[currentNode][alpha];

TRANSFORMATION_FUNCTION: t(currentNode,nodesVisited,alpha)
    =(alpha, nodesVisited SETUNION {alpha});

END
```

The second model is based on the DP functional equation (2.46) that uses the "relaxation" approach, as discussed in Sect. 1.1.9.

```
BEGIN
  NAME SPC; //ShortestPathCyclic;
  //single source shortest path, cycles allowed,
  //no negative weights allowed

  GENERAL_VARIABLES_BEGIN
    private static final int infty=Integer.MAX_VALUE;
    //adjacency matrix for cyclic shortest path (Lew85, p.337)
    //The nodes (s,x,y,t) are coded as (0,1,2,3)
    private static int[][] distance =
      {
       {infty,      3,       5,  infty},
```

```
      {infty,  infty,      1,      8},
      {infty,      2,  infty,      5},
      {infty,  infty,  infty,  infty}
    };
  //Note: infty entry represents the fact that there is no
  //      edge between nodes
GENERAL_VARIABLES_END

GENERAL_FUNCTIONS_BEGIN
  private static NodeSet possibleNextNodes(int node) {
    NodeSet result = new NodeSet();
    for (int i=0; i<distance[node].length; i++) {
      if (distance[node][i]!=infty) {
        result.add(new Integer(i));
      }
    }
    return result;
  }
GENERAL_FUNCTIONS_END

STATE_TYPE: (int currentNode, int noOfEdgesToTarget);

DECISION_VARIABLE: int d;
DECISION_SPACE:
    possibleSuccessors(currentNode, noOfEdgesToTarget)
    = possibleNextNodes(currentNode);

GOAL: f(0,3);
    //We start from node number zero, and with a total of 4
    //nodes in the graph, we do not need more than 3 edges to
    //the target, otherwise there is a (nonegative cycle)
    //that could be eliminated and the path shortened.

DPFE_BASE_CONDITIONS:
  f(currentNode,noOfEdgesToTarget)=0.0
    WHEN (currentNode==3);
  f(currentNode,noOfEdgesToTarget)=infty
    WHEN ((noOfEdgesToTarget==0)&&(currentNode!=3));

DPFE: f(currentNode,noOfEdgesToTarget)
      =MIN_{d IN possibleSuccessors}
        { cost(currentNode,noOfEdgesToTarget,d)
          +f(t(currentNode,noOfEdgesToTarget,d))};

REWARD_FUNCTION: cost(currentNode,noOfEdgesToTarget,d)
```

```
                    =distance[currentNode][d];

  TRANSFORMATION_FUNCTION: t(currentNode,noOfEdgesToTarget,d)
                           =(d, noOfEdgesToTarget - 1);

END
```

4.45 gDPS source for SPT

```
BEGIN
  NAME SPT; //SPT Scheduling;

  GENERAL_VARIABLES_BEGIN
    private static int[] proctime = {3,5,2};
    //number of processes:
    private static int n = proctime.length;
    //total of proc times:
    private static int ttl = 10;
  GENERAL_VARIABLES_END

  SET_VARIABLES_BEGIN
    Set setOfAllItems={0,..,n - 1};
    Set emptySet={};
  SET_VARIABLES_END

  GENERAL_FUNCTIONS_BEGIN
    private static int size(Set items) {
      return items.size();
    }
  GENERAL_FUNCTIONS_END

  STATE_TYPE: (int k,Set items);

  DECISION_VARIABLE: int d;
  DECISION_SPACE: decisionSet(k,items)=items;

  GOAL: f(0,setOfAllItems);

  DPFE_BASE_CONDITIONS:
    f(k,items)=0.0 WHEN (items SETEQUALS emptySet);

  DPFE: f(k,items)=MIN_{d IN decisionSet}
          { cost(k,items,d) + f(t(k,items,d)) };
```

```
  REWARD_FUNCTION: cost(k,items,d) = k + proctime[d];

  TRANSFORMATION_FUNCTION: t(k,items,d)
      =(k + proctime[d],items SETMINUS {d});

END
```

4.46 gDPS source for TRANSPO

```
BEGIN
  NAME transpo; // TransportationFlow-Production Problem
                // Bronson97 Problem 19.22;

  GENERAL_VARIABLES_BEGIN
    private static final double infty=Integer.MAX_VALUE;

    //production per unit cost:
    private static double[][] C = {
                 {0.0,35.0,74.0,113.0,infty,infty,infty},
                 {0.0,43.0,86.0,133.0,180.0,infty,infty},
                 {0.0,40.0,80.0,120.0,165.0,210.0,infty}
                               };
    //max no. units produced:
    private static int m=6;
    //no. of time stages
    private static int n=3;
    //demand
    private static int[] D = {2,2,2};
    //inventory perunitcost:
    private static double Icost = 3.0;
  GENERAL_VARIABLES_END

  GENERAL_FUNCTIONS_BEGIN
    private static double I(int s) {
      if (s < 0) return infty;
      else return Icost*s;  //inventory cost
    }
    private static double costfct(int g, int s, int x) {
      //produce enough to meet demand?
      if (s + x < D[g]) return infty;
      else return C[g][x]+I(s);
    }
  GENERAL_FUNCTIONS_END
```

```
  // g=stage, s=inventory
  STATE_TYPE: (int g, int s);

  DECISION_VARIABLE: int x; // x=amt produced

  DECISION_SPACE: decisionSet(g,s) = {0,..,m};

  GOAL:  f(0,0);

  DPFE_BASE_CONDITIONS: f(g,s)=0.0 WHEN (g==n);

  DPFE: f(g,s)
       =MIN_{x IN decisionSet}
                              { f(t(g,s,x))
                               +c(g,s,x)
                              };

  REWARD_FUNCTION: c(g,s,x) = costfct(g,s,x);

  TRANSFORMATION_FUNCTION: t(g,s,x) = (g+1,s+x - D[g]);

END
```

4.47 gDPS source for TSP

Two different gDPS sources are given. The first model uses the DP functional equation (2.49) keeping the set of nodes visited as a part of the state.

```
BEGIN
  NAME TSP; //TravelingSalesmanProblem;

  GENERAL_VARIABLES_BEGIN
    //adjacency matrix for TSP.
    private static int[][] distance =
      {
       { 0,  1,  8,  9, 60},
       { 2,  0, 12,  3, 50},
       { 7, 11,  0,  6, 14},
       {10,  4,  5,  0, 15},
       {61, 51, 13, 16,  0}
      };
    private static int n = distance.length;
    //number of nodes n=5. Nodes are named starting
```

```
    //at index 0 through n-1.
  GENERAL_VARIABLES_END

  SET_VARIABLES_BEGIN
    Set setOfAllNodes={0,..,n - 1};
    Set goalSet={0};
  SET_VARIABLES_END

  STATE_TYPE: (int currentNode, Set nodesVisited);

  DECISION_VARIABLE: int alpha;
  DECISION_SPACE: nodesNotVisited(currentNode,nodesVisited)
                 =setOfAllNodes SETMINUS nodesVisited;

  GOAL:
    f(0,goalSet); //that is: (0,{0});

  DPFE_BASE_CONDITIONS:
    f(currentNode, nodesVisited)
     =distance[currentNode][0]
        WHEN (nodesVisited SETEQUALS setOfAllNodes);

  DPFE: f(currentNode,nodesVisited)
        =MIN_{alpha IN nodesNotVisited}
          { cost(currentNode,nodesVisited,alpha)
            +f(t(currentNode,nodesVisited,alpha))};

  REWARD_FUNCTION: cost(currentNode,nodesVisited,alpha)
                   =distance[currentNode][alpha];

  TRANSFORMATION_FUNCTION: t(currentNode,nodesVisited,alpha)
      =(alpha, nodesVisited SETUNION {alpha});

END
```

The second model is an alternative formulation using the DP functional equation (2.50) that keeps the set of nodes not yet visited as a part of the state.

```
BEGIN
  NAME TSPalt; //TravelingSalesmanProblem;

  GENERAL_VARIABLES_BEGIN
    //adjacency matrix for TSP
    private static int[][] distance =
      {
```

```
      { 0,  1,  8,  9, 60},
      { 2,  0, 12,  3, 50},
      { 7, 11,  0,  6, 14},
      {10,  4,  5,  0, 15},
      {61, 51, 13, 16,  0}
     };
   private static int n = distance.length;
   //number of nodes n=5. Nodes are named
   //starting at index 0 through n-1.
 GENERAL_VARIABLES_END

 SET_VARIABLES_BEGIN
   Set setOfOtherNodes={1,..,n - 1};
   Set emptySet={};
 SET_VARIABLES_END

 STATE_TYPE: (int currentNode, Set nodesToBeVisited);

 DECISION_VARIABLE: int alpha;
 DECISION_SPACE: nodesNotVisited(currentNode,nodesToBeVisited)
                 =nodesToBeVisited;

 GOAL:  f(0,setOfOtherNodes);
     //start from node 0 (any other node would do)

 DPFE_BASE_CONDITIONS:
   f(currentNode, nodesToBeVisited)
    =distance[currentNode][0]
       WHEN (nodesToBeVisited SETEQUALS emptySet);

 DPFE: f(currentNode,nodesToBeVisited)
       =MIN_{alpha IN nodesNotVisited}
         { cost(currentNode,nodesToBeVisited,alpha)
           +f(t(currentNode,nodesToBeVisited,alpha))};

 REWARD_FUNCTION: cost(currentNode,nodesToBeVisited,alpha)
                  =distance[currentNode][alpha];

 TRANSFORMATION_FUNCTION: t(currentNode,nodesToBeVisited,alpha)
     =(alpha, nodesToBeVisited SETMINUS {alpha});

END
```

Note the convenient use of variables and literals of the type `Set` in both formulations.

5

Bellman Nets: A Class of Petri Nets

This chapter reviews fundamental Petri net concepts. Prior work on relationships between DP and Petri nets has dealt with using DP to solve optimization problems arising in Petri net models. Among the few who used a Petri net model to solve an optimization problem are [36, 46, 52]. Lew [36] explored interconnections between DP, Petri nets, spreadsheets, and dataflow and other nondeterministic computer architectures. Mikolajczak and Rumbut [46] gave a Petri net model for the MCM problem. Richard [52] showed that an integer linear program (ILP) can be associated with a certain PN class.

We introduce Bellman nets as a very specialized class of Petri nets that can be associated with dynamic programming. Petri nets are generalizations of directed graphs. Briefly, a *directed graph* is a mathematical model consisting of a set of objects, called *nodes* (or *vertices*), together with another set of objects, called *branches* (or *arcs* or *edges*), that connect pairs of nodes. A directed graph is a useful model of *state transition* systems, where states are modeled by nodes and transitions between states and next-states are modeled by branches. However, an ordinary directed graph cannot be used to model more complex systems, such as where there may be parallel next-states or concurrent transitions. Petri nets, on the other hand, can be used for this purpose.

5.1 Petri Net Introduction

Petri Nets are named for Prof. Dr. Carl Adam Petri, University of Hamburg, Germany, who developed schemata that allowed the modeling of parallel and independent events in an illustrative manner.

Detailed introductions to PNs are [51] and [47]. For colored PNs, see [27].

5.1.1 Place/Transition Nets

The following formal definitions follow the notation used in [51].

A. Lew and H. Mauch: *Bellman Nets: A Class of Petri Nets*, Studies in Computational Intelligence (SCI) **38**, 205–220 (2007)
www.springerlink.com

The first definition describes the basic syntactic properties that a net, as used in the following, should have.

Definition 5.1 (Net). *A triple $N = (S, T, F)$ is called a finite* net *iff*

1. *S and T are disjoint finite sets and*
2. *$F \subseteq (S \times T) \cup (T \times S)$ is a binary relation called the* flow relation *of N. Elements of F are also called* arcs.

Mathematically speaking our notion of a net is a directed bipartite graph. The nodes are partitioned into two sets S and T, with arcs connecting only nodes in different sets. For convenience we use the following notation.

Definition 5.2 (Preset and Postset). *Let $N = (S, T, F)$ be a net. For $x \in S \cup T$*
$\bullet x = \{y|(y, x) \in F\}$ is called the preset *of x,*
$x\bullet = \{y|(x, y) \in F\}$ is called the postset *of x.*
This notation extends to sets. For $X \subset S \cup T$ define

$$\bullet X = \cup_{x \in X} \bullet x$$

and

$$X\bullet = \cup_{x \in X} x \bullet .$$

In this book we use a low-level PN class called place/transition nets to define low-level Bellman nets.

Definition 5.3 (Place/Transition Net). *Let $\mathbf{N}$ denote the set of non-negative integers, and let ω denote an infinite capacity. A 6-tuple $N = (S, T, F, K, M, W)$ is called a* place/transition net *iff*

1. *(S, F, T) is a finite net,*
2. *$K : S \to \mathbf{N} \cup \{\omega\}$ gives a (possibly unlimited) capacity for each place,*
3. *$W : F \to \mathbf{N} - \{0\}$ attaches a weight to each arc of the net, and*
4. *$M : S \to \mathbf{N} \cup \{\omega\}$ is the initial marking where all the capacities are respected, i.e. $M(s) \leq K(s)$ for all $s \in S$.*

Elements of S are called places, *elements of T are called* transitions.

Graphically, places will be represented by circles; transitions will be represented by rectangles. The *marking* of a place corresponds to the number of *tokens* in that place. If $p \in S$ is a place, then the fact that p contains k tokens is represented by writing the number k inside p's circle. The weighted arcs of the flow relation are represented by weighted directed arcs in the bipartite graph.

The dynamic behavior of a place/transition net is described in the following definitions.

Definition 5.4. *Let $N = (S, T, F, K, M, W)$ be a place/transition net. A transition $t \in T$ is* activated *(or* enabled*) if*

1. *for every place p in the preset $\bullet t$ the number of tokens at p is greater than or equal to the weight $W(p,t)$ of the arc from p to t and*
2. *for every place p in the postset $t\bullet$ the number of tokens at p plus the weight $W(t,p)$ of the arc from t to p does not exceed the capacity $K(p)$ of p.*

Definition 5.5. *Let $N = (S,T,F,K,M,W)$ be a place/transition net. An activated transition $t \in T$ may* fire *by*

1. *decreasing the number of tokens for all $p \in \bullet t$ by $W(p,t)$ and*
2. *increasing the number of tokens for all $p \in t\bullet$ by $W(t,p)$.*

5.1.2 High-level Petri Nets

The place/transition nets introduced in Sect. 5.1.1 are considered "low-level" PNs. In this section we describe the basic properties of high-level PNs. The distinction is similar to that between low-level programming languages such as assembler languages and high-level programming languages such as Java. High-level PNs facilitate modeling on a different level of abstraction. Arcs and nodes are annotated by a special inscription language. Tokens are more complex data items, and nodes are inscribed with more complex symbolic expressions than low-level PNs. The most prominent examples of high-level PNs are *colored PNs* ([25, 26, 27], see Sect. 5.1.3 for a brief formal overview) and predicate/transition nets [14, 15].

In this section we will informally describe the basic common properties of high-level PNs. We use the terminology that describes the *Petri net markup language* (PNML) [6, 62], the emerging XML-based interchange format for PNs. PNML is flexible enough to integrate different types of high-level PNs and is open for future extensions. For our purpose it is sufficient to introduce the meta model of *basic PNML* — the features belonging to *structured PNML* are not needed.

In its most general definition a high-level PN is considered as a labeled graph. All additional information can be stored in labels that can be attached to the net itself, to the nodes of the net or to its arcs [6]. A PN type definition determines the legal labels for a particular high-level PN type.

Definition 5.6. *A high-level PN $\mathcal{N} = (O, L_{\mathcal{N}})$, when described in basic PNML, consists of a set O of objects and a set of labels $L_{\mathcal{N}}$ attached to the net. Every object $o \in O$ is associated with a set of labels L_o. The set $O = N \cup A$ is partitioned into the nodes N and the arcs A. The set $N = P \cup T$ is partitioned into the set of places P and the set of transitions T. An arc connects a source node to a target node.*

The PNML meta model defines high-level PNs in a *very* general way; for example it does not even forbid arcs between nodes of the same kind, which are apparently allowed in some more exotic PN types.

There are two types of labels, namely *annotations* and *attributes*. Annotations are typically displayed as text near the corresponding object. Object names, markings of places by tokens, arc inscriptions, and transition guards are examples of annotations. Attributes specify graphical properties of an object.

The label set $L_{\mathcal{N}}$ that is associated with the net $\mathcal{N}$ itself could for example consist of the declarations of functions and variables that are used in the arc inscriptions.

5.1.3 Colored Petri Nets

In 1981 Jensen introduced colored Petri nets (CPN) in [25]. At the same time Genrich and Lautenbach introduced predicate/transition Nets (PrTN) in [15]. The common idea is to allow distinguishable typed individuals as tokens. In CPNs this is achieved by attaching a "color" (commonly a numerical value) to each token in the net. Lautenbach and Pagnoni [34] showed in 1985 the duality of CPNs and PrTNs. The following definition follows [27].

Definition 5.7 (Colored Petri Net). *A 9-tuple* $N = (\Sigma, P, T, F, Q, C, G, E, I)$ *is called a* colored Petri net *(CPN) iff*

1. Σ *is a finite set of non-empty types, called* color sets, *which determine the types, operations and functions that can be used in the net inscriptions,*
2. (P, T, F) *is a finite net,*
3. $Q : F \to P \times T \cup T \times P$ *is the node function, which is similar to the flow relation* F*, but allows multiple arcs between the same ordered pair of nodes,*
4. $C : P \to \Sigma$ *is the color function, which means that each token on a place* p *must have a color token that belongs to the type* $C(p)$*,*
5. G *is a guard function that maps each transition* t *to a suitable boolean predicate, also known as the inscription of the transition (note that the types of the variables in the predicate must be a subset of* Σ*, i.e. the variables must have legal types),*
6. E *is an arc expression function that maps each arc into an expression of appropriate type, also known as the inscription of the arc (note that types must match), and*
7. I *is the initialization function, which maps each place* $p \in P$ *into a suitable closed expression (i.e. an expression without variables) which must be a multi-set over the type* $C(p)$*. The initial marking* M_0 *is obtained by evaluating the initialization expressions [27, p.74].*

The dynamic behavior of a CPN is described in the following definitions.

Definition 5.8. *Let* $N = (\Sigma, P, T, F, Q, C, G, E, I)$ *be a CPN. A transition* $t \in T$ *is* activated *(or* enabled*) if*

1. *for every place* p *in the preset* $\bullet t$ *there is a suitable binding of a token in* p *to the arc expression* $E(p, t)$ *and*

2. *for every place* p *in the postset* $t\bullet$ *there is a suitable binding to the arc expression* $E(t,p)$ *and*
3. *its guard expression* $G(t)$ *evaluates to true under the chosen binding.*

Definition 5.9. *Let* $N = (\Sigma, P, T, F, Q, C, G, E, I)$ *be a CPN. An activated transition* $t \in T$ *may* fire *(or* occur*) by*

1. *removing tokens, which are determined by the arc expressions, evaluated for the occuring bindings, from the input places and*
2. *adding tokens, which are determined by the arc expressions, evaluated for the occuring bindings, to the output places.*

5.1.4 Petri Net Properties

The following definitions are based on [47].

Definition 5.10. *A marking* M *is said to be* reachable *from an initial marking* M_0 *if there exists a sequence of firings that transforms* M_0 *to* M.

Definition 5.11. *The* submarking reachability problem *for a Petri net is the question of whether or not a subset of the places can have a certain marking, reachable from the initial marking.*

Definition 5.12. *A Petri net is* k-bounded *if the number of tokens in each place does not exceed a finite number* $k \in \mathbf{N}$ *for any marking reachable from its initial marking. A Petri net is* safe *if it is 1-bounded.*

Definition 5.13. *A transition* t *in a Petri net is* L1-live (potentially firable) *if* t *can be fired at least once in some firing sequence from the initial marking. A Petri net is* L1-live *if every transition in the net is L1-live. (Stronger definitions of liveness specify that a transition can be fired more than once.)*

Definition 5.14. *A transition* t *in a Petri net is* dead *if* t *can never be fired. A Petri net is* dead *if every transition in the net is dead.*

We say that a Petri net is *deadlocked* if no transition is activated; i.e. according to the previous definition if it is in a state (to be interpreted as the initial marking) where it is dead. Note that a Petri net might not be dead in its initial marking, but it might become deadlocked at some later point in time.

Definition 5.15. *A Petri net is* persistent *if, for any two enabled transitions, the firing of one transition will not disable the other one.*

Definition 5.16. *A Petri net is* acyclic *if it has no directed circuit.*

Note that the nodes of an acyclic Petri net can be topologically sorted.

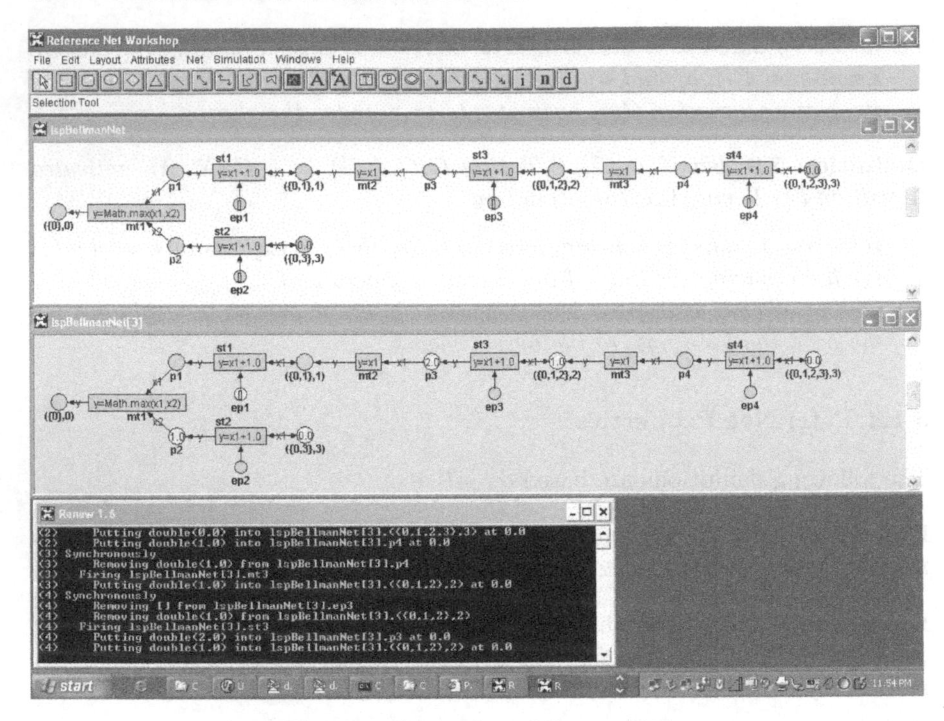

Fig. 5.1. Screenshot of Renew Tool

5.1.5 Petri Net Software

Petri Nets are supported by numerous software tools, which allow the creation of PNs with graphical editors, simulate the dynamic behavior of PNs, animate the "token game" to visually display movements of tokens, and perform an analysis of PNs for net theoretic properties. In 1998 a study [60] evaluated 91 existing Petri Net tools. The website at `http://www.daimi.au.dk/PetriNets/tools/` contains a more recent searchable database of currently available tools.

Throughout this book the Renew tool [32, 33], which is freely available and open source, was used to import PNs in the XML-based PNML standard, to modify or create PN figures and to simulate the execution of Petri net models of dynamic programming problems. Renew is based on a high-level PN model called "reference nets" [31] which is compatible with both the LLBNs and HLBNs introduced in Sect. 5.2. Figure 5.1 shows a screenshot of Renew (version 1.6).

5.2 Petri Net Models of Dynamic Programming

A basic introduction to both low-level and high-level PNs is given in Sect. 5.1. The following sections discuss two special purpose PN models, which are suitable to represent and solve DP problem instances. We refer to PNs that are capable of representing and solving DP problems as *Bellman nets*, in honor of Richard E. Bellman, who established the fundamentals of DP. The details of two different Bellman net models are discussed. First, a low-level Bellman net (LLBN) model based on place/transition nets, as introduced in [43], is described. Second, high-level Bellman Nets (HLBN) [37, 40, 41] based on the high-level PN concepts of colored Petri nets (CPN) [25, 26, 27] or predicate/transition nets (PrTN) [14, 15] are introduced. For the purpose of defining HLBNs it makes no difference whether we use CPNs or PrTNs as the underlying concept. We only use features that appear in both CPNs and PrTNs, and the subtle differences between CPNs and PrTNs do not come into play. We actually go one step further and describe HLBNs based on the emerging PN standard PNML, which is so general that it allows for future extensions of the HLBN model without having to worry about conforming to either CPNs or PrTNs. There is software support for both LLBNs and HLBNs (see Sect. 5.1.5).

Why use a PN model at all? For problems where a decision d leads from one state s to $K > 1$ successor states $s^{(1)}, \ldots, s^{(K)}$ (e.g. MCM, BST have $K = 2$ successor states) a weighted directed graph representation is not sufficient. These problems have solutions which are represented as K-ary trees (rather than paths) in the PN, since after each decision exactly K successor states need to be evaluated.

The DP state graph representation where there is one arc for each decision-(next-state) pair $(d, s^{(k)})$ is not suitable, because attaching a weight to an arc according to the reward of the decision involved would model multiple identical rewards for a single decision, which is incorrect.

So why not use a tree model? First, we have the same dilemma as before that one state is associated with K successor states per decision, but is also associated with exactly one arc weight representing the reward of the decision taken (see Fig. 5.2). Second, we want to take advantage of overlapping subproblems, which should only be solved once instead of multiple times. A tree model, e.g. a parse tree model with functions like addition and minimization as interior nodes and terminals as leaves, would not allow for such a topology — nodes that hold the result of overlapping subproblems would appear multiple times and not just once, which results in an inefficient model.

A PN model overcomes all these problems in an elegant and natural way.

While it is easier to transform HLBNs to executable code, the LLBN model from Sect. 5.3 is easier to examine with respect to consistency, and other net theoretic issues. For certain DP problem classes equivalence between the two models can be proven to exist and the proof is a constructive one. A practical transformation algorithm between the two models is possible [44], so we can take advantage of the benefits of both models.

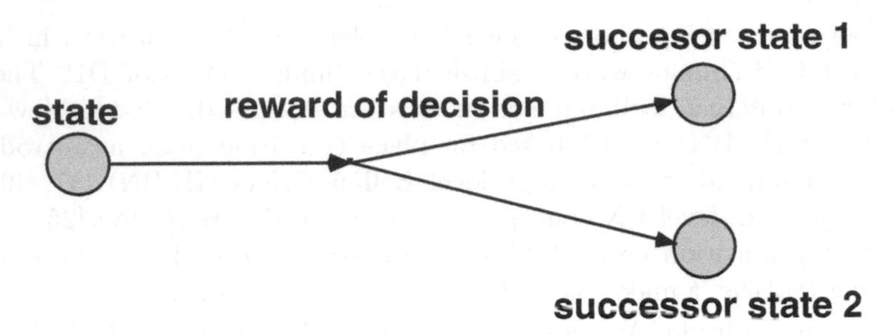

Fig. 5.2. Dilemma when a state has two successor states

5.3 The Low-Level Bellman Net Model

The details of the suggested LLBN model (see [43]) for DP problems are discussed in this section.

Definition 5.17. *A* low-level Bellman net (LLBN) *is a place/transition net that has the special structure described in Sects. 5.3.1 through 5.3.4.*

Note that a LLBN is indeed a place/transition net with places and transitions that adhere to the low-level PN conventions. When we introduce several types of places and transitions in the following, it is important to keep in mind that these types only help us with the *interpretation* of their functionality. Their semantics is exactly the semantics described for place/transition nets in Sect. 5.1.1.

5.3.1 Construction of the Low-Level Bellman Net Model

Suppose a discrete optimization problem can be solved by an integer dynamic programming equation of the form

$$f(s) = \min_{d \in D(s)} \{r(s,d) + \sum_{k=1}^{K} f(s^{(k)})\}, \tag{5.1}$$

with nonnegative integral base cases values $f(s_0)$ given.

Here s denotes a state, f is the recursion functional, d is the decision variable that chooses from alternatives feasible in state s, r is the reward function with nonnegative integral values, $s^{(k)}$ are the K next-states, and s_0 are base case (initial condition) states. The LLBN model corresponding to this DP problem has

1. a state place p_s for each state s, that has, except for the base case state places, a unique minimization transition in its preset,

2. a min-in place p_m for each decision d, that has a minimization transition in its postset, and whose initial marking equals $r(s, d) \in \mathbf{N}$,
3. a minimization transition associated with each state place p_s, except for the base case state places, that has in its preset the min-in places for each decision a,
4. a copy transition that transfers tokens from the state place associated with $s^{(k)}$ to those min-in places whose decision d involves the summand $f(s^{(k)})$.

5.3.2 The Role of Transitions in the Low-Level Bellman Net Model

Transitions serve two different purposes in the model.

- *Processing Transitions*: Transitions can be considered as elements processing the arithmetic functions required for the evaluation of the DP functional equation. So far, only *minimization transitions* are used. They have the advantage that they conform to the standard place/transition net semantics described in Sect. 5.1.1.
  Since place/transition nets, extended with inhibitor arcs, constitute a Turing-complete model of computation [47] there is the potential to design processing subnets for more complex operations such as maximization, multiplication, etc. (See e.g. [24] for the concept of a "weak PN computer".) In other words, it would be possible to have meta-transitions (subnets) representing arbitrary functions f. More formally, let t be a meta-transition representing the function f. Then its preset $\bullet t = \{p_1, \ldots, p_n\}$ consists of places representing the function variables $\{x_1, \ldots, x_n\}$. The postset $t\bullet$ consists of exactly one place representing the function value $f(x_1, \ldots, x_n)$. In this book we only consider LLBNs with minimization transitions. The summation operation can be trivially achieved, since it is the default semantics of place/transition nets to accumulate (black) tokens in a place. More complex meta-transitions are possible, but will not be discussed here.
- *Copy Transitions*: Another purpose of transitions is to make the results of solved subproblems available to the superproblems requesting these results. Each of these special "copy transitions" distributes a result from one "state place" to multiple "min-in places".

5.3.3 The Role of Places in the Low-Level Bellman Net Model

There are two types of places in the LLBN model.

- *State Places*: Such a place represents a state encountered during the solution process of a DP problem. Except for the base case state places, state places are the output places for minimization transitions.
- *Min-in Places*: These are the input places for minimization transitions. They are also the output places for copy transitions.

5.3.4 The Role of Markings in the Low-Level Bellman Net Model

- *Marking of State Places*: Let p_s be a state place. Immediately after all minimization transition in its preset $\bullet p_s$ have fired, but just before firing any of the copy transitions in its postset $p_s\bullet$, the marking of a state place p_s representing state s contains $f(s)$ tokens, where $f(s)$ equals the optimal value of the subproblem associated with state s. Base case state places are initially marked with an appropriate number of tokens as specified in the base cases of the DP functional equation. That is, a base case state place associated with a base case state s_0 is initially marked with $f(s_0)$ tokens. All other state places are initially marked with no tokens.
- *Marking of Min-in Places*: Let p_m be a min-in place. After all transitions in its preset $\bullet p_m$ have fired, but just before firing any transitions in its postset $p_m\bullet$, the marking of these places corresponds to the values to be minimized by a minimization transition. Min-in places are initially marked with $r(d)$ tokens, so the reward of a decision gets added implicitly.

5.3.5 Advantages of the Low-Level Bellman Net Model

The main features of the suggested LLBN model are:

- The standard place/transition net semantics are used.
- There are no conflicts in the net. The order in which transitions fire is not relevant for the value of the final result (i.e. the final marking of the goal state place). However, the marking of an intermediate state place during execution is only meaningful, if a "controlled" firing order is used.
- The optimal policy can easily be reconstructed from the final marking. We may just trace the tree from the final state place back to the base case state places following empty min-in places.

5.4 Low-Level Bellman Net Properties

In this section we describe net theoretic properties of LLBNs. We assume here that these LLBNs are "legal" or "proper" in the sense that they represent correctly specified DP problems. Of course, if a DP formulation is not correct, e.g. if a base condition is omitted, neither is its associated LLBN.

- LLBNs are pure (there are no double arcs).
- LLBNs are ordinary (all arc weights are 1).
- LLBNs are not state machines, since typically minimization transitions have more than one incoming arc, and copy transitions more than one outgoing arc.
- LLBNs are not marked graphs, since min-in places might have more than one incoming arc.
- Every LLBN is a free-choice net, since $|p\bullet| \leq 1$ for all places p.

- LLBNs are conflict-free. Every LLBN is a forward-conflict-free net, since each place has at most one outgoing arc.
- LLBNs allow the concurrent firing of transitions.
- LLBNs represent a dataflow model of computation [47, p.545].
- The reachability graph of a LLBN is acyclic and has exactly one sink node, referred to as the final marking (in which the net is dead). The final marking can be reached from any marking in the reachability graph.
- Let $M(p)$ denote the marking of a place p. For every non-base state place p_s in a LLBN $M(p_s) \leq f(s)$, i.e. p_s is $f(s)$-bounded. Every LLBN is structurally bounded, since if in its initial marking M_0 it holds that $M_0(p) < \infty$ for all places p then the LLBN is bounded.
- LLBNs are persistent — for any two enabled transitions the firing of one transition will not disable the other.
- LLBNs are acyclic.
- LLBNs do not have isolated places or transitions.
- Every LLBN has exactly one sink place, the goal state place.
- Every LLBN has one or more source places, the base state places.

5.5 The High-Level Bellman Net Model

Lew introduces in [37] another PN model, which is suitable as an intermediate DP problem representation: high-level Bellman nets (HLBN). HLBNs can model DP problems with DP functional equations of the general form as described in Chap. 1. In contrast to the LLBN model described earlier, these are high-level PNs with numerically-colored tokens. For all places the color set from which tokens are picked is the set of real numbers. Special types of transitions are introduced — M-transitions for minimization/maximization and E-transitions for expression evaluation. In the original definition input tokens are not removed when firing a transition. To allow for standard PN semantics [40] modifies the HLBN model, such that input tokens are removed unless they are replaced using double arcs, as is the case for E-transitions. In [41] two additional types of transitions (to allow for comparisons and multiplexing) are introduced for HLBNs.

HLBNs are the underlying logical structure used in DP2PN2Solver. They are described in detail in [37, 40, 41, 45]. The notation used here is based on [41].

Definition 5.18 (High-level Bellman net). *A* high-level Bellman net (HLBN) *is a high-level Petri net as defined in Sect. 5.1.2, with a set of place nodes P, a set of transition nodes T, and a set of arcs A. The place nodes P can be partitioned $P = S \cup I \cup E$, where S denote* state places, *I denote* intermediate places *and E denote* enabling places. *Furthermore, there is a special* goal place *$g \in S$ that describes the computational goal of the DPFE, and there is a set of* base state places *$B \subseteq S$. Tokens are of the type* *"real number" (for state places and intermediate places) or "black" (for enabling places). The base state places B are initially marked with a single real-valued token (according to the initial conditions in the DPFE); all enabling places E are initially marked with a single black token; all other places do not have any token in their initial marking. The transition nodes of HLBNs are* B-transitions, *which are governed by the following rules.*

1. *Each B-transition has one unique* designated *output place* $p \in S \cup I$.
2. *Every B-transition is labeled with an annotation describing a function f as defined below. In addition to the function type (e.g. "min", "max", "+", "*") the annotation can optionally contain a constant value $c \in \mathbf{R}$ or other information that is essential for computing the function f. When firing, the value of its output token is the function value using the value of its input tokens as arguments.*
3. *Every B-transition t can fire only once. To be consistent with standard Petri net semantics this is ensured by having an enabling place p for every B-transition t along with a connecting arc from p to t. Sometimes we will not show the enabling place and the connecting arc explicitly, but rather implicitly assume their presence to simplify the drawing of HLBNs.*
4. *When a B-transition fires, the input tokens are returned unchanged back to their respective input places (exception: enabling places), which is modeled by adding* reverse arcs. *All its input places (its preset) with the exception of enabling places are connected to the B-transition with* double arcs, *i.e., a symmetric pair of arcs (from a place to a transition and a reverse one from the transition to the place, also called a* self-loop *in Murata [47]). While, formally, these input places serve also as output places, in essence f only has an effect on the designated output place.*

The arcs A connect place nodes with B-transitions and vice versa. In HLBNs, arcs are not labeled. Having said this, to make the graphical representation of HLBNs more readable, we inscribe arcs for illustrative purposes only; labels "$x_1, \ldots, x_n$" denote arcs delivering the input values to a B-transition, whereas an arc labeled "y" originates from a B-transition and ends at the designated output place.

We now define two "basic" types of B-transitions: E-transitions (for arithmetic expression evaluation) and M-transitions (for minimization or maximization). What distinguishes these B-transitions is the nature of the function f.

Definition 5.19. *An* E-transition *outputs a token to its designated output place having a numerical value determined by evaluating a simple arithmetic $(+, *)$ function of the values of its input tokens and of its associated constant value c. Specifically, they can compute functions $f : \mathbf{R}^n \rightarrow \mathbf{R}$ of the form*

$$f(x_1, \ldots, x_n) = c + \sum_{i=1}^{n} x_i$$

or

$$f(x_1, \ldots, x_n) = c \prod_{i=1}^{n} x_i$$

It is possible to define E-transitions in a more general way, e.g. by allowing other arithmetic operators. However, for all DP examples considered, the above definition proved to be sufficiently general.

Definition 5.20. *An* M-transition *compares the numerical values of its input tokens and outputs a token with value equal to the minimum (or maximum). That is, they can compute functions* $f : \mathbf{R}^n \rightarrow \mathbf{R}$ *of the form*

$$f(x_1, \ldots, x_n) = \min(x_1, \ldots, x_n)$$

or

$$f(x_1, \ldots, x_n) = \max(x_1, \ldots, x_n)$$

For M-transitions the reverse arcs and the enabling places will be omitted in our drawings of HLBNs to simplify the figures; also, when firing, the tokens in the preset of M-transitions will be consumed rather than be preserved.

The two types of B-transitions may be regarded as special cases of transitions with the function f defined appropriately. More specific transition types can be defined. An A-transition is the special case of an E-transition where the function is addition. M-transitions must have at least one input; the minimum (or maximum) of a set of size 1 is the value in the set. Rather than generalizing the types of B-transitions that a HLBN can have, we will limit ourselves to those transition types that are sufficient to handle all of the example DP problems solved in this book, which are M-transitions and E-transitions.

To summarize, a HLBN is a special high-level colored Petri net with the following properties.

1. The color type is numerical in nature, tokens are real numbers. In addition, single black tokens, depicted `[]` in the figures of Sect. 6.2 are used to initialize *enabling places*, which are technicalities that prevent transitions from firing more than once.
2. The postset of a transition contains exactly one *designated output place*, which contains the result of the computational operation performed by the transition. (In the example HLBNs for the MCM problem shown in Fig. 6.25 and 6.26 designated output places are labeled `(1,4)`, `(1,3)`, `(2,4)`, `(1,2)`, `(2,3)`, `(3,4)` and `p1` through `p10`; base state places are labeled `(1,1)`, `(2,2)`, `(3,3)` and `(4,4)`, and enabling places are labeled starting with the prefix `ep`. In all examples of Sect. 6.2 base state places are labeled with the base state they represent.)
3. A place contains at most one token at any given time. This follows since each transition can only contribute a new token to its only designated output place once.

4. There are two different types of transitions, *M-transitions* and *E-transitions*. An M-transition performs a minimization or maximization operation using the tokens of the places in its preset as operands and puts the result into its designated output place. An E-transition evaluates a basic arithmetic expression (involving operators like addition or multiplication) using fixed constants and tokens of the places in its preset as operands and puts the result into its designated output place. (In the example HLBNs for the MCM problem shown in Fig. 6.25 and 6.26 M-transitions are labeled `mt1` through `mt6`, E-transitions are labeled `st1` through `st10`. The other HLBN figures of Sect. 6.2 are labeled similarly.)
5. There are double arcs between an E-transition and all places (exception: enabling places) in its preset. Their purpose is to conserve operands serving as input for more than one E-transition.

A numerical token as the marking of a place in a HLBN can be interpreted as an intermediate value that is computed in the course of calculating the solution of a corresponding DP problem instance. To clarify the concept of a HLBN, numerous examples are given in Sect. 6.2.

The following two theorems put HLBNs in context with the two most prevalent high-level PN models.

Theorem 5.21. *Every HLBN is a colored Petri net (as defined in [27].)*

Proof. A constructive sketch of a proof is given. Given a HLBN an equivalent CPN is constructed.

The only variables used in arc expressions and transition guards are the real-valued variables $x_1, x_2, \ldots, x_n, y$.

The places, transitions and arcs of a HLBN are mapped canonically to places, transitions and arcs of a CPN, which automatically defines the node function of the CPN.

What is left to show is that the labeling of objects and the special functionality of a HLBN can be achieved with a CPN.

The color sets used in the CPN are $\Sigma = \{\mathbf{R}, \{\text{black}\}\}$.

The color function C is defined as follows. The color set of the enabling places E is the singleton $\{\text{black}\}$ (i.e. $C(p) = \{\text{black}\}$ for all $p \in E$); the color set of the other places $S \cup I$ equals $\mathbf{R}$ (i.e. $C(p) = \mathbf{R}$ for all $p \in S \cup I$).

The guard function G is defined as follows. If a B-transition in a HLBN computes the function $f(x_1, x_2, \ldots, x_n)$ then the corresponding transition t in the CPN has the guard $y = f(x_1, x_2, \ldots, x_n)$ inscribed. This has the effect that t only fires after the guard evaluates to true, which is the case after the variable y is bound to the real value $f(x_1, x_2, \ldots, x_n)$.

The arc expression function is defined as follows. Let $A_{t,in}$ be the set of $n = |A_{t,in}|$ arcs entering a transition t. Inscribe an arc $a_i \in A_{t,in}$ with the variable expression x_i for all $i \in \{1, \ldots, n\}$. An arc leaving t is inscribed with the variable expression y. Arcs from enabling places to transitions are inscribed with the constant expression "black".

The initialization function is defined as follows. For the CPN's initial marking, every enabling place gets a single black token and every base case state place gets the same constant real-valued token as in the HLBN. All other places do not have an initial marking. □

Note that the CPN constructed in the previous proof was built such that it resembles the HLBN in the most natural way. In particular the transition labels of the HLBN reappear as guard inscriptions in the CPN. It should be pointed out that there is an alternative way of constructing a CPN from a HLBN that avoids transcription inscriptions completely and uses more complex arc inscriptions instead. An arc entering a transition would still have the same variable expression x_i inscribed, but the arc leaving the transition would be inscribed with $f(x_1, x_2, \ldots, x_n)$ and the variable y would no longer be used.

Theorem 5.22. *Every HLBN is a predicate/transition net (as defined in [14, p.216].)*

Proof. Due to the similarity of CPNs and PrTNs the proof is almost analogous to that of theorem 5.21. Given a HLBN an equivalent strict PrTN is constructed. Strict means that multiple occurences of tuples (i.e. tokens) on places are not allowed.

The places, transitions and arcs of a HLBN are mapped canonically to the directed net underlying the PrTN definition.

The variables used in the first-order language are the real-valued variables $x_1, x_2, \ldots, x_n, y$. Terms are built from the variables and of n-ary operators f, which correspond to the functions associated with B-transitions in the HLBN. Atomic formulas of the form $y = f(x_1, x_2, \ldots, x_n)$ can be built from terms. These are the formulas annotating the transitions in the PrTN. The predicates annotating the places are unary and real-valued (exception: the enabling places are treated as 0-ary predicates). The arcs are annotated by unary predicates (exception: arcs from enabling places are annotated by a zero-tuple indicating a no-argument predicate). The initial marking of the PrTN consists of exactly one real-valued constant for each base state place and of exactly one zero-tuple per enabling place. □

5.6 High-Level Bellman Net Properties

The data model of computation [47, p.545] is used.

HLBNs are not state machines, since transitions can have more than one incoming or outgoing arcs.

HLBNs are not marked graphs, since places usually have more than one incoming or outgoing arcs.

Certain transitions/events in a HLBN are concurrent, but there are no true conflicts.

While, strictly speaking, HLBNs do contain conflicts, these are mere technicalities. A state place node can have two or more output transitions; however these are connected with double arcs, so there is no true representation of decisions. The nondeterminism merely expresses the order in which transitions fires. But the order in which transitions fire is irrelevant in a HLBN; the final marking will always be the same.

Synchronization is implicit in HLBNs. An M-transition has to wait until all its operands are computed and available before it can fire.

Theorem 5.23. *Every HLBN is safe.*

Proof. In its initial marking a HLBN N contains by definition exactly one token in each of its base state places B and enabling places E; all other places do not contain any token. Enabling places have an empty preset ($\bullet e = \emptyset$ for all $e \in E$), so they are safe. Base state places B have E-transitions in their preset, however connected via double arcs (implying no net gain of tokens), so they are safe. The other state places $S - B$ also have E-transitions in their preset, also connected via double arcs (implying no net gain of tokens) and in addition they have a single M-transition in their preset; but since each M-transition fires at most once, they are safe. Intermediate places have a single E-transition in their preset, and since each E-transition may fire at most once, they are safe. □

Theorem 5.24. *Every HLBN is persistent.*

Proof. In a HLBN, any two enabled transitions either have disjoint presets (all transitions in a HLBN have disjoint postsets), or the places which lie in the intersection of the presets are connected to the transition with double arcs. So in no case will the firing of one enabled transition disable the other one. □

6

Bellman Net Representations of DP Problems

In Chap. 1, we showed how the following spreadsheet can be generated from and essentially represent the DPFE for SPA.

```
A1: =min(A5,A6)
A2: =min(A7,A8)
A3: =min(A9)
A4: 0.0
A5: =A2+3.0
A6: =A3+5.0
A7: =A3+1.0
A8: =A4+8.0
A9: =A4+5.0
```

This spreadsheet model may also be interpreted as a tabular representation of a Petri net where the formula cells in the spreadsheet correspond (1) to transition nodes in the Petri net, to be evaluated, and (2) to place nodes in the Petri net that hold the computed values of the transition-formulas. That is, cell A1 corresponds to a transition that computes the minimum of the values in cells A5 and A6; cell A5 corresponds to a transition that computes the sum of the value in cell A2 plus the constant 3.0. In the Petri net, when these transitions fire, their computed values are placed in their respective place nodes, for possible use by other transition-formulas; e.g., the computed value of A5 is used to compute the value of A1. In this chapter, we show how all of the DP problems solved in this book (in Chap. 2), or to be more precise, how the DPFEs that solve these DP problems, can be modeled and solved using Petri nets. Specifically, we show use of both the special classes of Petri nets, called Bellman nets, which were designed expressly for DP problems and defined in Chap. 5.

A. Lew and H. Mauch: *Bellman Net Representations of DP Problems*, Studies in Computational Intelligence (SCI) **38**, 221–244 (2007)
www.springerlink.com

6.1 Graphical Representation of Low-Level Bellman Net Examples

In this section, several example DP problems are discussed. They were chosen to illustrate the basic ideas, and are not meant to reflect the class of problems that can be handled. Some problems (e.g. PERM) are easier than other problems (e.g. MCM) in the sense that an optimal solution is represented as a path in the PN, since after each decision exactly one successor state needs to be evaluated. For these kinds of problems a weighted directed graph representation, where states are nodes and branch weights represent the cost of a decision, would be sufficient.

Other problems (e.g. MCM, BST) have solutions which are represented as (binary) trees in the PN, since after each decision exactly two successor states need to be evaluated. For these problems a weighted directed graph representation as mentioned before is obviously no longer sufficient, since we have the dilemma that one state is associated with two successor states, but is also associated with exactly one branch weight (see Fig. 5.2.)

Problems with a more complex state space, such as the TSP, can be handled as well with the LLBN model.

6.1.1 Low-Level Bellman Net for BST

The LLBN model of the BST instance (Sect. 2.6), in its initial marking, is depicted in Fig. 6.1 — however this figure is simplified since the base case state place $\emptyset$, which never contains any tokens, is left out, as well as the copy transition belonging to it. Furthermore the trivial minimalization transitions in the presets of the places denoted $\{A\}$, $\{B\}$, $\{C\}$, $\{D\}$ and $\{E\}$ are left out because they minimize over a single input. Instead these places are initialized with an appropriate number of tokens. In order to get integral values for tokens, probability values are scaled, i.e. multiplied by 100. E.g. the initial marking of state $\{A\}$ is 25 instead of 0.25. These simplifications are equivalent to using

$$f(S) = p(x) \text{ if } S = \{x\}$$

as the base case in DPFE (2.9).

To obtain the final marking of the PN we fire transitions until the net is dead. The PN, in its final marking, is depicted in Fig. 6.2. We can see that the final state place $\{A, B, C, D, E\}$ contains $f(\{A, B, C, D, E\}) = 190$ tokens now, corresponding to the optimal unscaled value of 1.9 using probability values.

6.1.2 Low-Level Bellman Net for LINSRC

The LLBN model of the LINSRCS version of the LINSRC instance (Sect. 2.24), in its initial marking, is depicted in Fig. 6.3. In order to get integral values

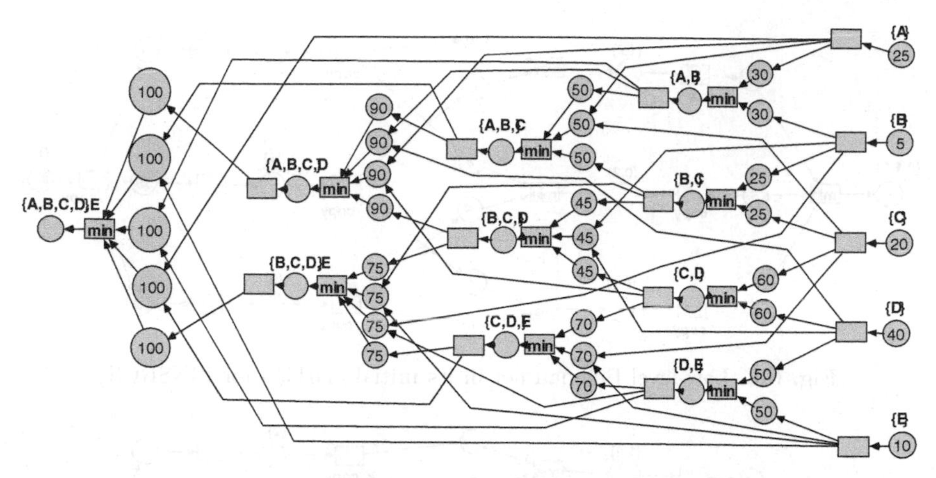

Fig. 6.1. Low-level Bellman net in its initial marking for BST

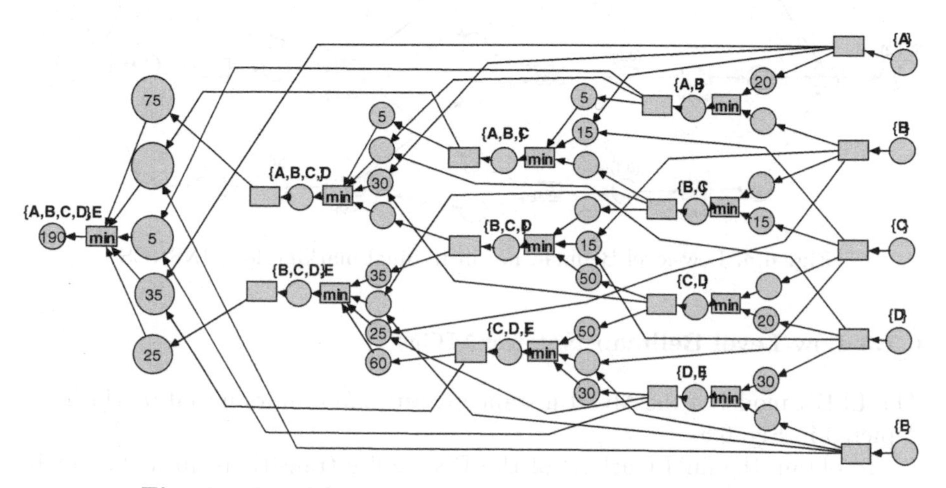

Fig. 6.2. Low-level Bellman net in its final marking for BST

for tokens, fractions are scaled, i.e. multiplied by 10 for this instance. E.g. a value of 1.5 is represented by 15 black tokens.

To obtain the final marking of the PN fire transitions until the net is dead. The PN, in its final marking, is depicted in Fig. 6.4. We can see that the final state place $\{0, 1, 2\}$ contains 17 tokens now, corresponding to the optimal unscaled value of $f(\{0, 1, 2\}) = 1.7$. The actual solution can easily be constructed from the final marking. Just trace the path from the final state place $\{0, 1, 2\}$ back to the base case state place $\emptyset$ following empty min-in places.

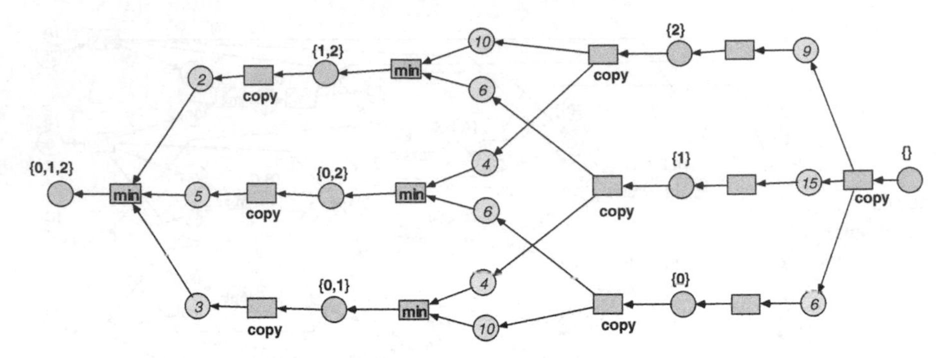

Fig. 6.3. Low-level Bellman net in its initial marking for LINSRCS

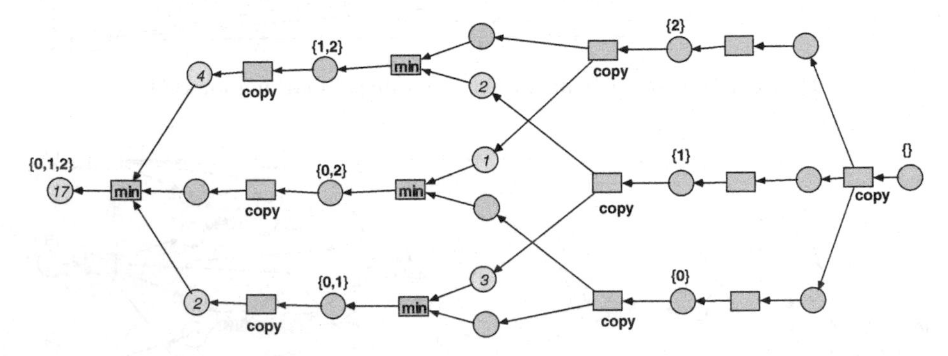

Fig. 6.4. Low-level Bellman net in its final marking for LINSRCS

6.1.3 Low-Level Bellman Net for MCM

The LLBN model of the MCM instance (Sect. 2.27), in its initial marking, is depicted in Fig. 6.5.

To obtain the final marking of the PN we fire transitions until the net is dead. The PN in its final marking is depicted in Fig. 6.6. We can see that the final state place $(1,4)$ contains $f(1,4) = 76$ tokens now, the correct value indeed.

Note that the order in which transitions fire is not relevant for the value of the final result $f(1,4)$. However values of intermediate states, e.g. $f(1,3)$, are only meaningful, if a "controlled" firing order is used.

The optimal policy can easily be reconstructed from the final marking. Just trace the tree from the goal state place $(1,4)$ back to the base case state places following empty min-in places.

6.1.4 Low-Level Bellman Net for ODP

The LLBN model of the ODP instance (Sect. 2.31), in its initial marking, is depicted in Fig. 6.7. In order to avoid too many details in this figure we

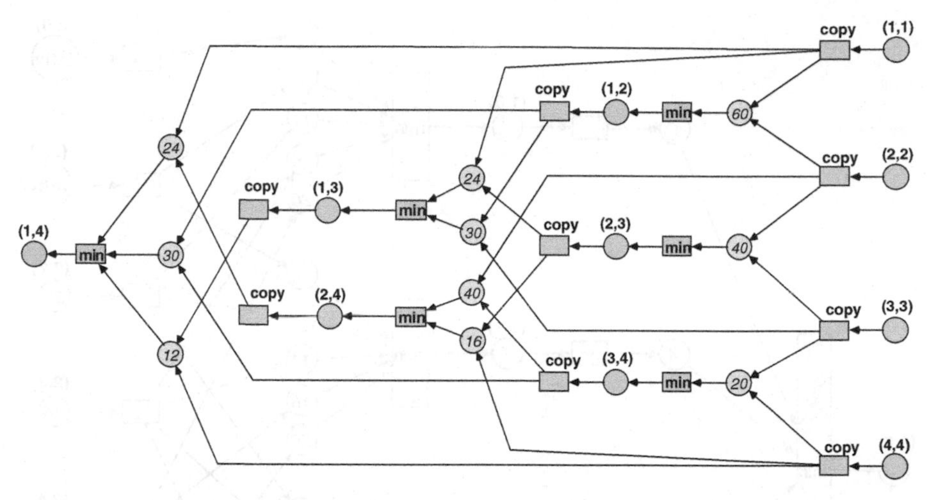

Fig. 6.5. Low-level Bellman net in its initial marking for MCM

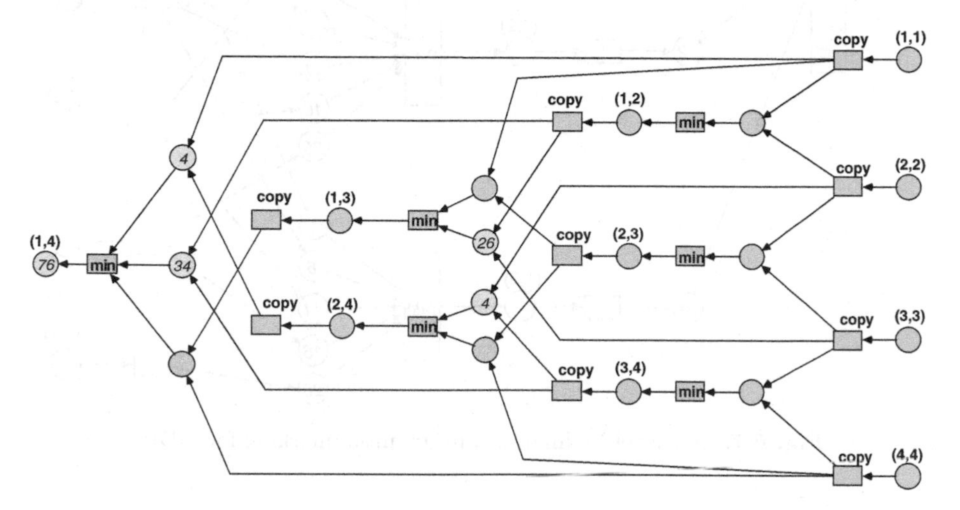

Fig. 6.6. Low-level Bellman net in its final marking for MCM

simplified it in the following way. Base case state places at stage 3, which contain either 0 or infinitely many tokens, are left out, as well as the copy transitions which belong to these places. Furthermore the minimization transitions in the presets of the stage 2 places are left out. Instead the stage 2 states are initialized with an appropriate number of tokens. That is,

$$f(2,x) = \begin{cases} \infty & \text{if } 0 \leq x \leq 1 \\ 22 & \text{if } x = 2 \\ 16 & \text{if } 3 \leq x \leq 5 \\ 0 & \text{if } 6 \leq x \leq 7. \end{cases}$$

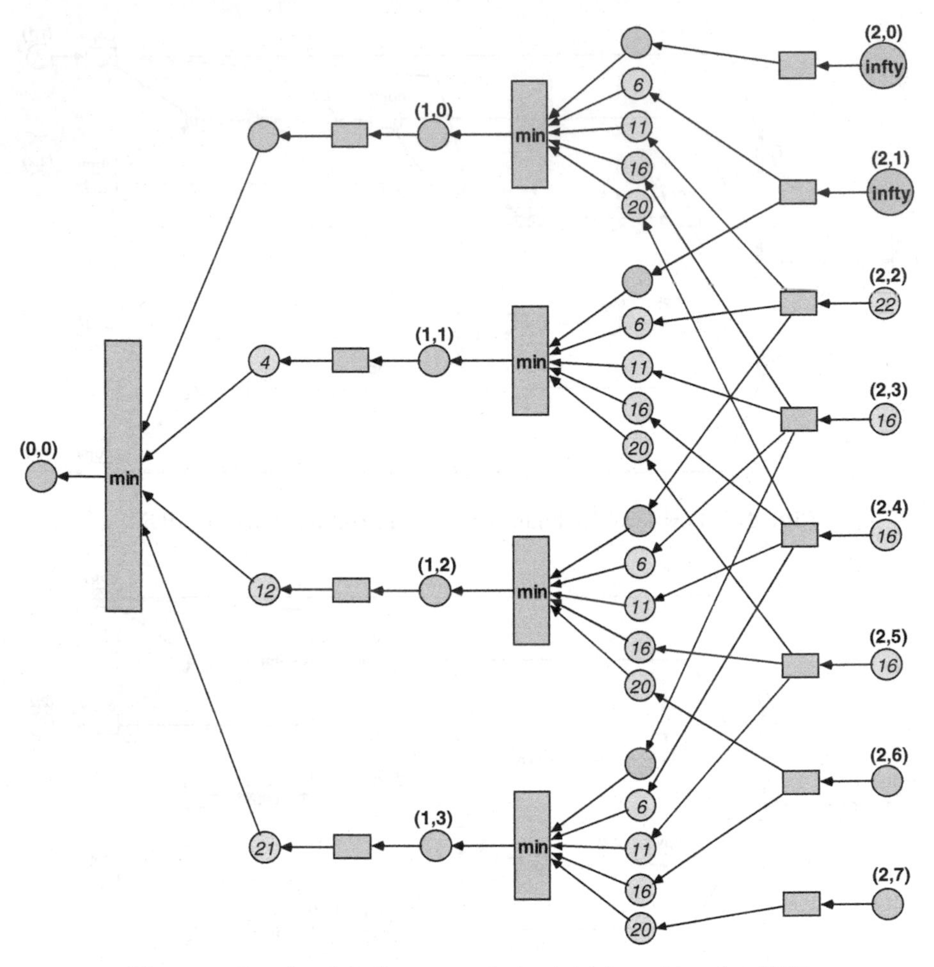

Fig. 6.7. Low-level Bellman net in its initial marking for ODP

To obtain a final marking of the PN we fire transitions until the net is dead (i.e. no transition can fire any more), or if there are places with infinitely many tokens (which is the case here) until only the copy transitions in the postset of these places are activated, and no other transitions are activated.

The PN, in its final marking, is depicted in Fig. 6.8. We can see that at stage 0 the goal state place associated with $x = 0$ contains $f(0, 0) = 31$ tokens now, the correct value indeed. The actual solution, i.e. the decision to make at each stage can easily be constructed from the final marking. Just trace the path from the goal state place back to the base case state place following empty min-in places.

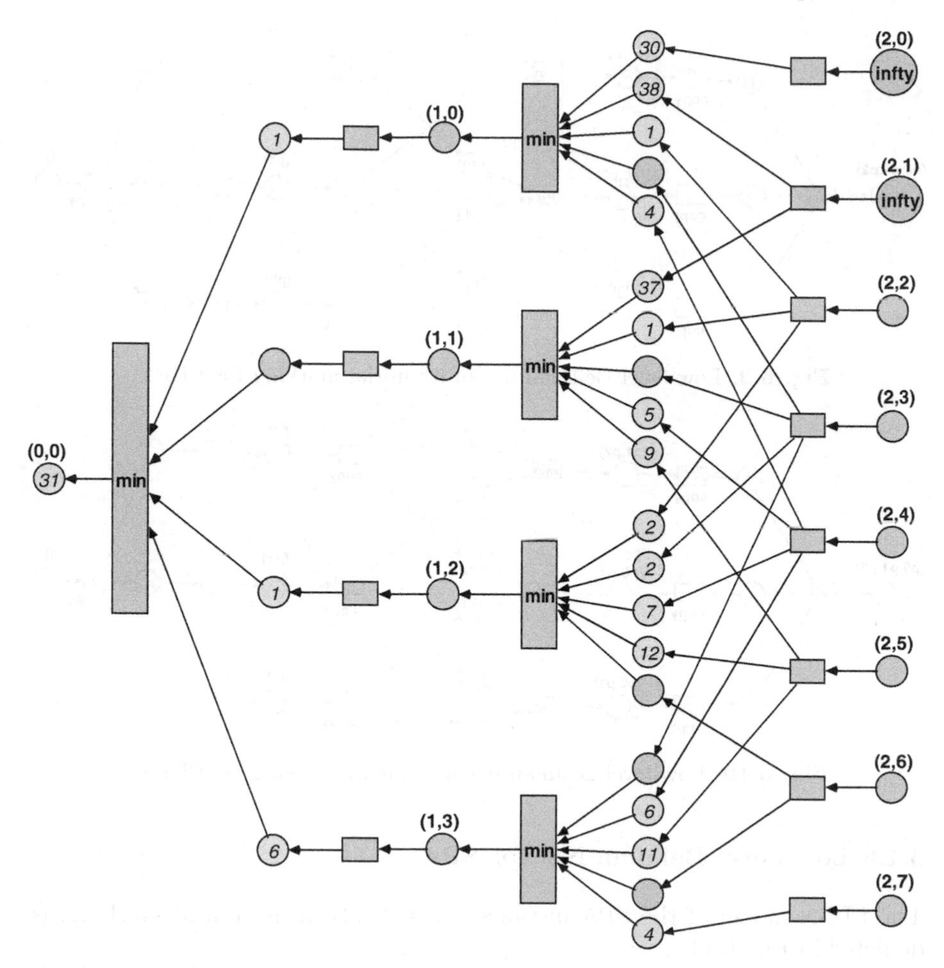

Fig. 6.8. Low-level Bellman net in its final marking for ODP

6.1.5 Low-Level Bellman Net for PERM

The LLBN model of the PERM instance (Sect. 2.32), in its initial marking, is depicted in Fig. 6.9.

To obtain the final marking of the PN fire transitions until the net is dead. The PN, in its final marking, is depicted in Fig. 6.10. We can see that the final state place $\{p_0, p_1, p_2\}$ contains $f(\{p_0, p_1, p_2\}) = 17$ tokens now, the correct value indeed. The actual solution, i.e. the order of the programs on the tape can easily be constructed from the final marking. Just trace the path from the final state place $\{p_0, p_1, p_2\}$ back to the base case state place $\emptyset$ following empty min-in places.

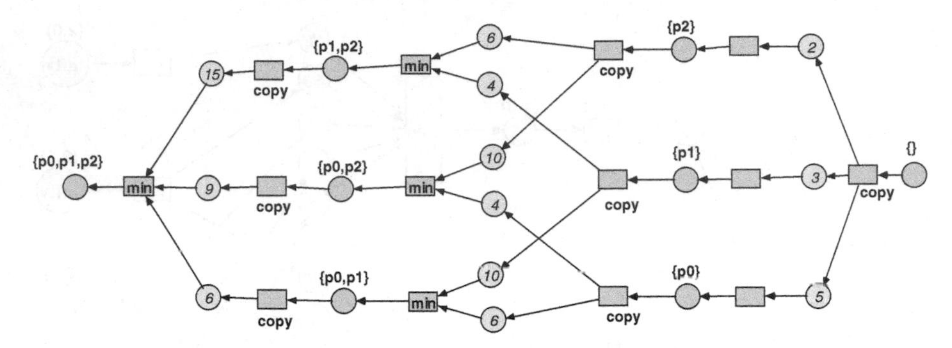

Fig. 6.9. Low-level Bellman net in its initial marking for PERM

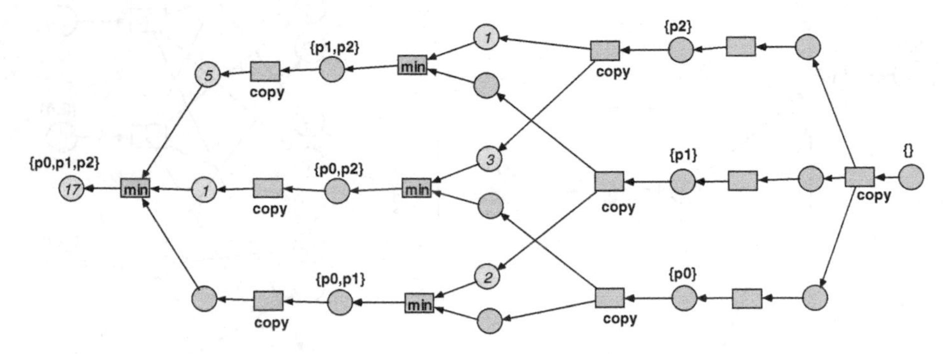

Fig. 6.10. Low-level Bellman net in its final marking for PERM

6.1.6 Low-Level Bellman Net for SPA

The LLBN model of the SPA instance (Sect. 2.43), in its initial marking, is depicted in Fig. 6.11.

To obtain the final marking of the PN fire transitions until the net is dead. The PN in its final marking is depicted in Fig. 6.12. We can see that the final state place (0) contains $f(0) = 9$ tokens now, indeed the correct length of the shortest path from node 0 to node 3. The actual solution, i.e. shortest path itself, can easily be constructed from the final marking. Just trace back the path from the final state place (0) to the base case state place (3) following empty min-in places, which yields the shortest path $(0, 1, 2, 3)$.

6.2 Graphical Representation of High-Level Bellman Net Examples

In the following examples, note how the intermediate HLBN representation gives an illustrative graphical representation of the problem instance.

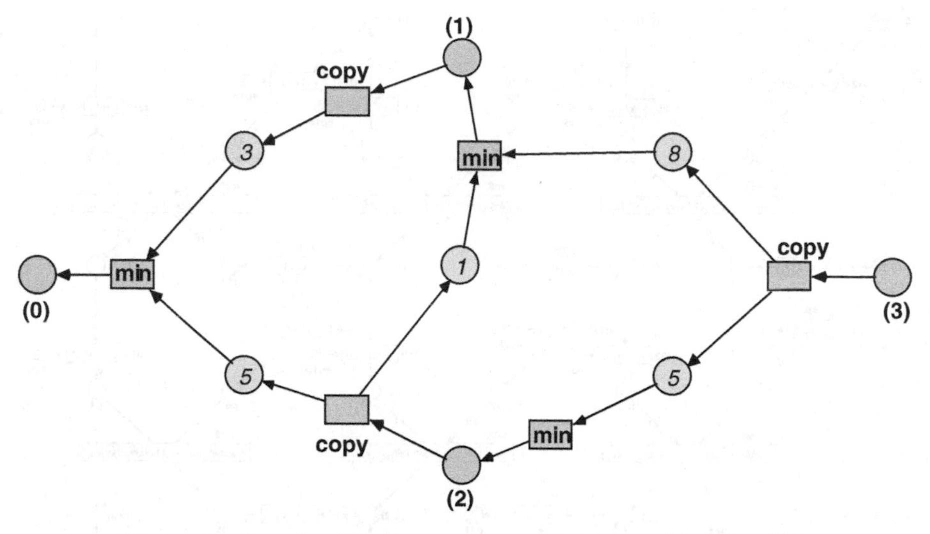

Fig. 6.11. Low-level Bellman net in its initial marking for SPA

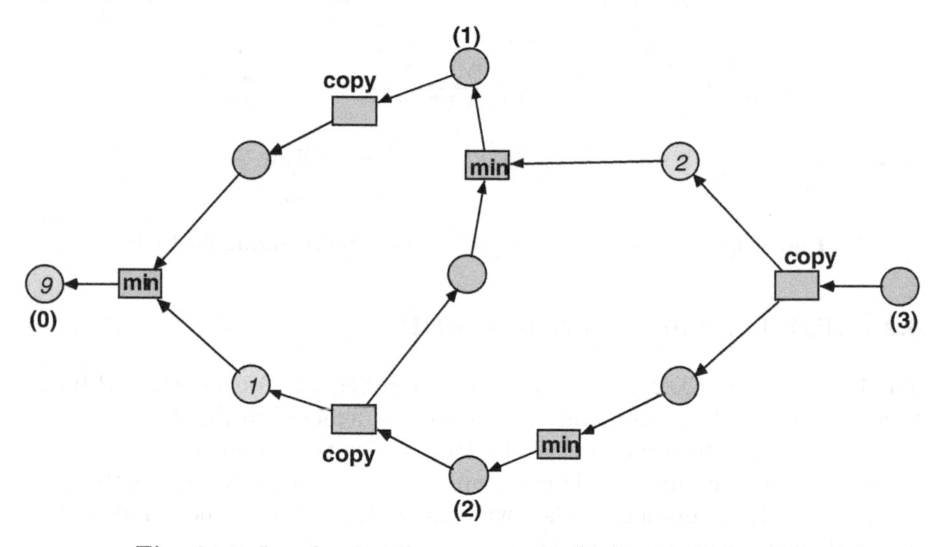

Fig. 6.12. Low-level Bellman net in its final marking for SPA

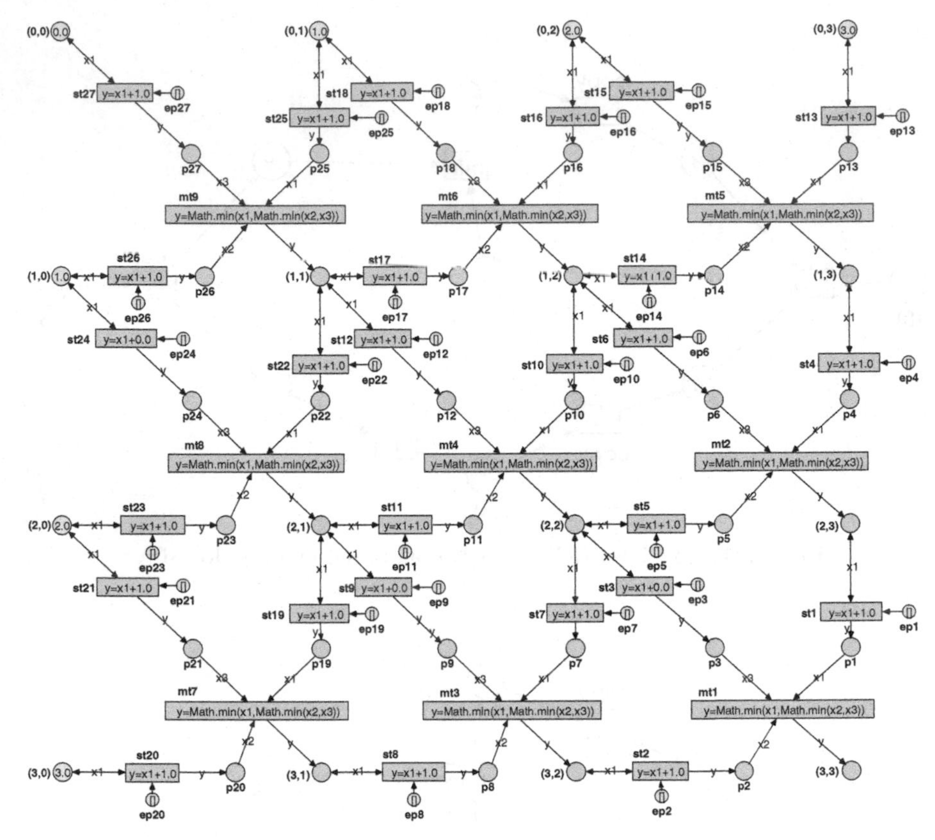

Fig. 6.13. High-level Bellman net in its initial marking for EDP

6.2.1 High-Level Bellman Net for EDP

The HLBN of the EDP problem instance from Sect. 2.10 using the DP functional equation (2.12), in its initial marking, is depicted in Fig. 6.13.

To obtain the final marking of the HLBN, fire transitions until the net is dead. The net in its final marking is depicted in Fig. 6.14. Note that the goal state place `(3,3)` contains a token with value $f(X_3, Y_3) = 2$ now, indeed the minimal cost for an edit sequence.

6.2.2 High-Level Bellman Net for ILP

The HLBN of the ILP problem instance from Sect. 2.14 in its initial marking, is depicted in Fig. 6.15.

To obtain the final marking of the HLBN, fire transitions until the net is dead. The net in its final marking is depicted in Fig. 6.16. Note that the goal state place `(0,4,12,18)` contains a token with value $f(0,(4,12,18)) = 36$ now, indeed the maximum value of the ILP problem instance.

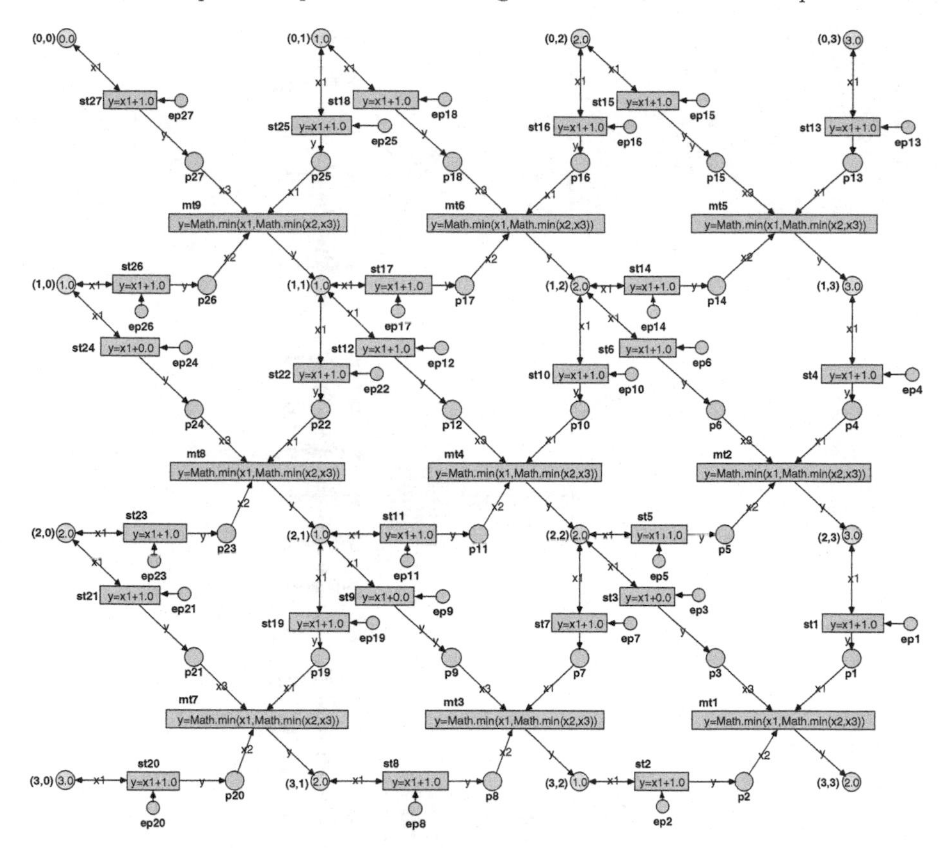

Fig. 6.14. High-level Bellman net in its final marking for EDP

6.2.3 High-Level Bellman Net for KS01

The HLBN of the KS01 problem instance from Sect. 2.20 in its initial marking, is depicted in Fig. 6.17.

To obtain the final marking of the HLBN, fire transitions until the net is dead. The net in its final marking is depicted in Fig. 6.18. Note that the goal state place (2,22) contains a token with value $f(2, 22) = 25$ now, indeed the maximum value of a knapsack.

6.2.4 High-Level Bellman Net for LCS

The HLBN of the LCS problem instance from Sect. 2.23 using the DP functional equation (2.29), in its initial marking, is depicted in Fig. 6.19.

To obtain the final marking of the HLBN, fire transitions until the net is dead. The net in its final marking is depicted in Fig. 6.20. Note that the goal state place (7,6) contains a token with value $f(X_7, Y_6) = 4$ now, indeed the correct length of an LCS.

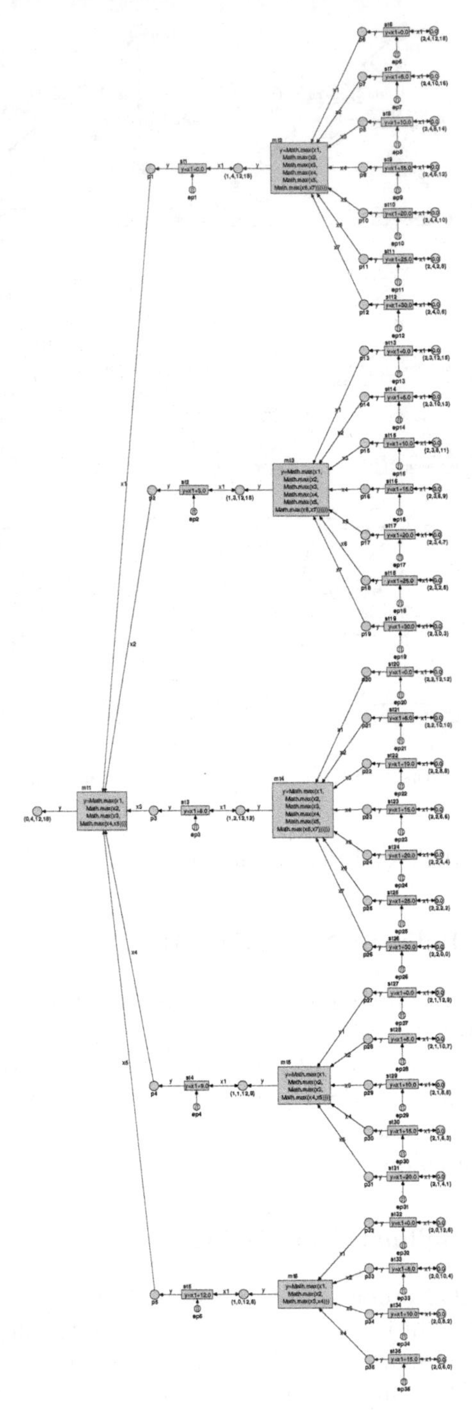

Fig. 6.15. High-level Bellman net in its initial marking for ILP

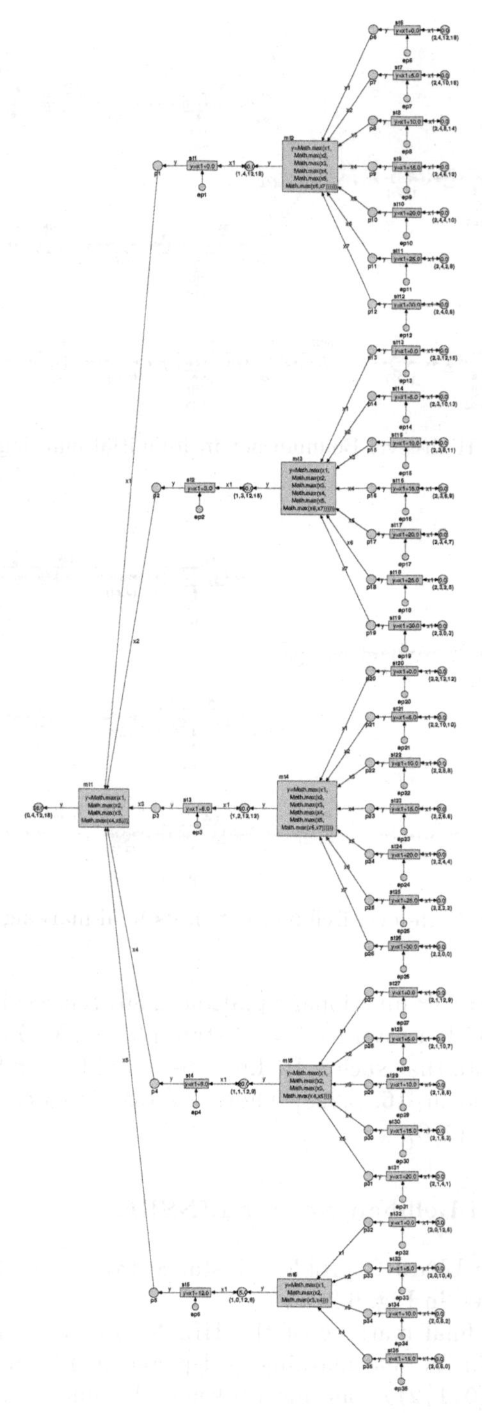

Fig. 6.16. High-level Bellman net in its final marking for ILP

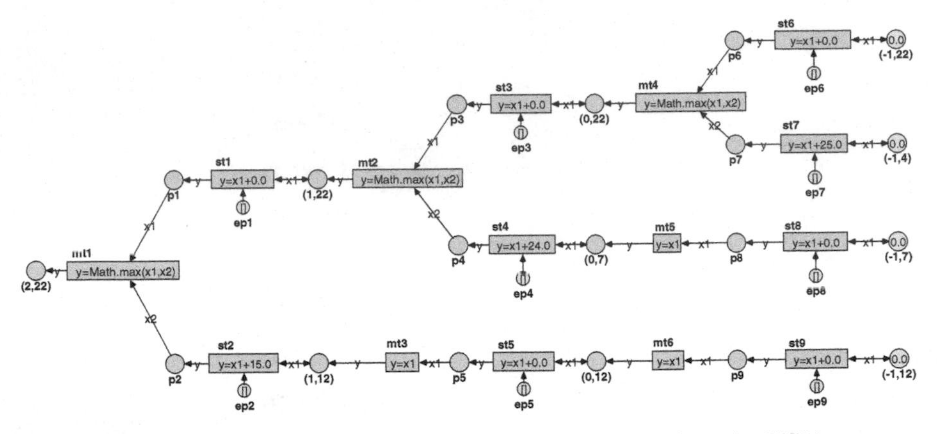

Fig. 6.17. High-level Bellman net in its initial marking for KS01

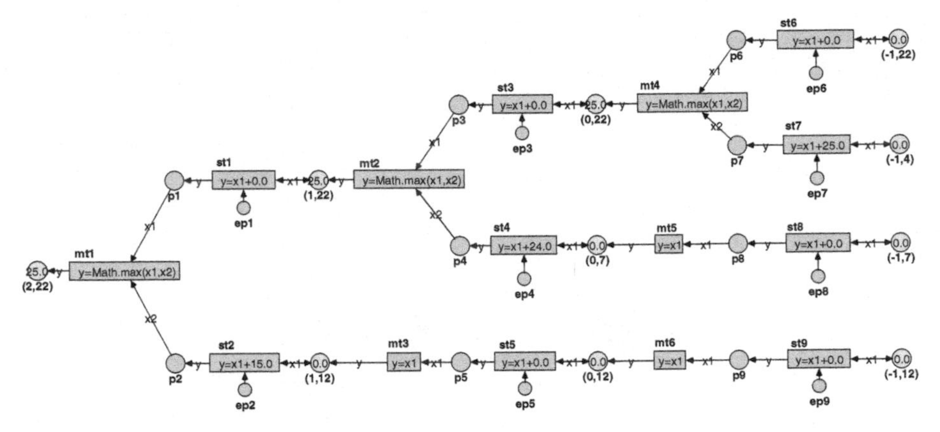

Fig. 6.18. High-level Bellman net in its final marking for KS01

Had we used the DP functional equation (2.30) the resulting HLBN would have been larger with a total $8 \cdot 7 = 56$ state places (X_i, Y_j) where $0 \leq i \leq 7$ and $0 \leq j \leq 6$. Note that such a Bellman net would resemble an "alignment graph" as described in [16, p.228] where the task is to compute the longest path from $(0, 0)$ to $(7, 6)$.

6.2.5 High-Level Bellman Net for LINSRC

The HLBN of the LINSRC problem instance from Sect. 2.24, in its initial marking, is depicted in Fig. 6.21.

To obtain the final marking of the HLBN, fire transitions until the net is dead. The net in its final marking is depicted in Fig. 6.22. Note that the goal state place `({0,1,2})` contains a token with value $f(\{0, 1, 2\}) = 1.7$ now, indeed the correct value.

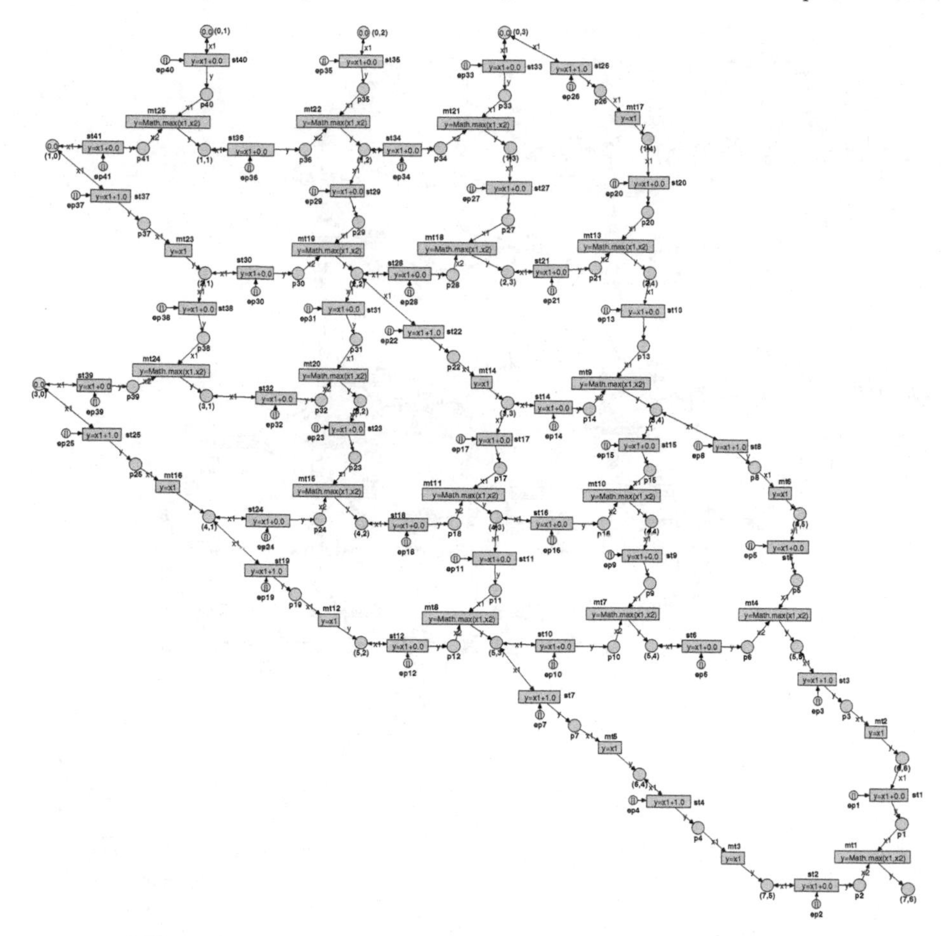

Fig. 6.19. High-level Bellman net in its initial marking for LCS

6.2.6 High-Level Bellman Net for LSP

The HLBN of the LSP problem instance from Sect. 2.26, in its initial marking, is depicted in Fig. 6.23.

To obtain the final marking of the HLBN, fire transitions until the net is dead. The net in its final marking is depicted in Fig. 6.24. Note that the goal state place `({0},0)` contains a token with value $f(\{0\},0) = 3$ now, indeed the maximum length of a simple path from node 0 to node 3. Also, the marking of an intermediate state place represents the optimal value for the associated subproblem. For example, the value of the intermediate state place `({0,1,2},2)` is $f(\{0,1,2\},2) = 1$ which is the maximum length of a path from node 2 to node 3 *given that nodes 0,1,2 have already been visited.* Without this precondition the longest path from node 2 to node 3 would be $(2,1,0,3)$ and have length 3.

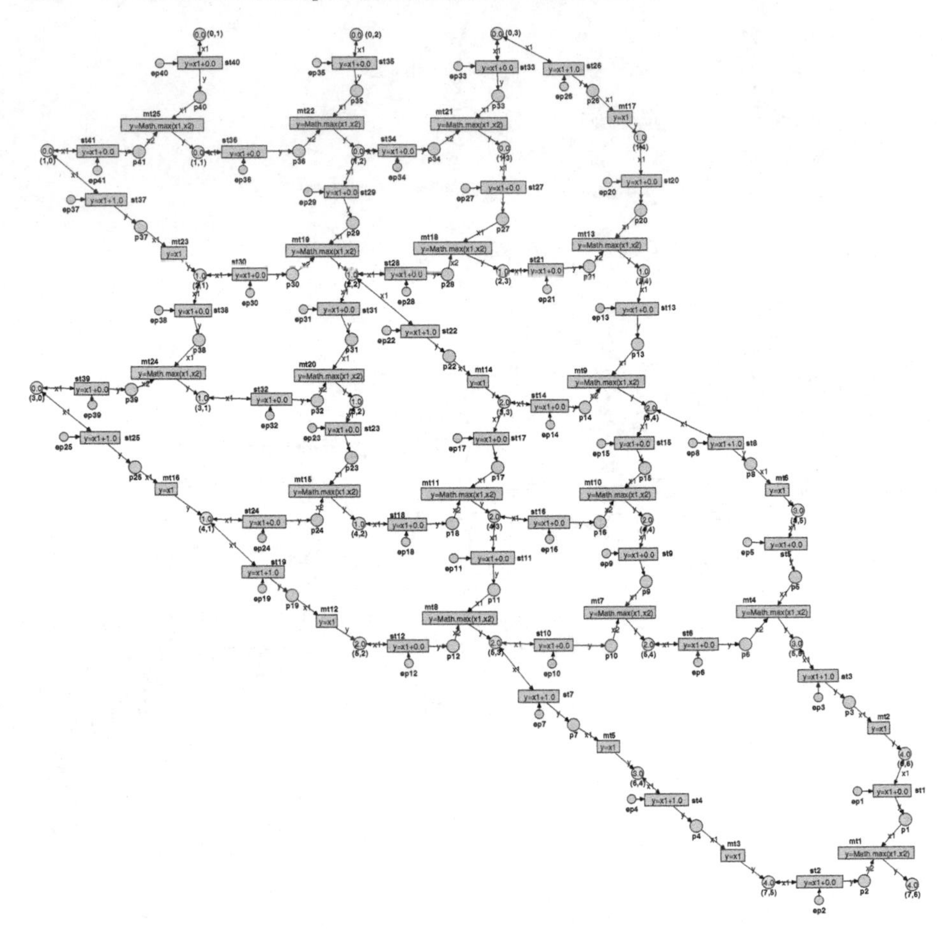

Fig. 6.20. High-level Bellman net in its final marking for LCS

6.2.7 High-Level Bellman Net for MCM

The HLBN of the MCM problem instance from Sect. 2.27, in its initial marking, is depicted in Fig. 6.25.

To obtain the final marking of the HLBN, fire transitions until the net is dead. The net in its final marking is depicted in Fig. 6.26. Note that the goal state place (1,4) contains a token with value $f(1,4) = 76$ now, the correct value indeed. Also, the marking of every state place represents the optimal value for the associated subproblem. For example, the value of the state place (2,4) is $f(2,4) = 56$ which is the number of componentwise multiplications for the optimally parenthesized matrix product $A_2 \cdots A_4$.

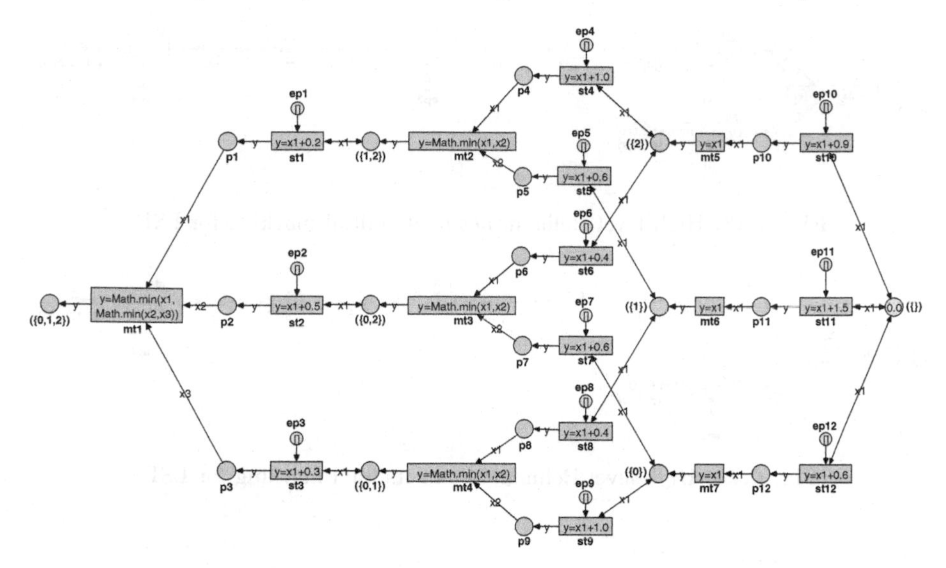

Fig. 6.21. High-level Bellman net in its initial marking for LINSRCS

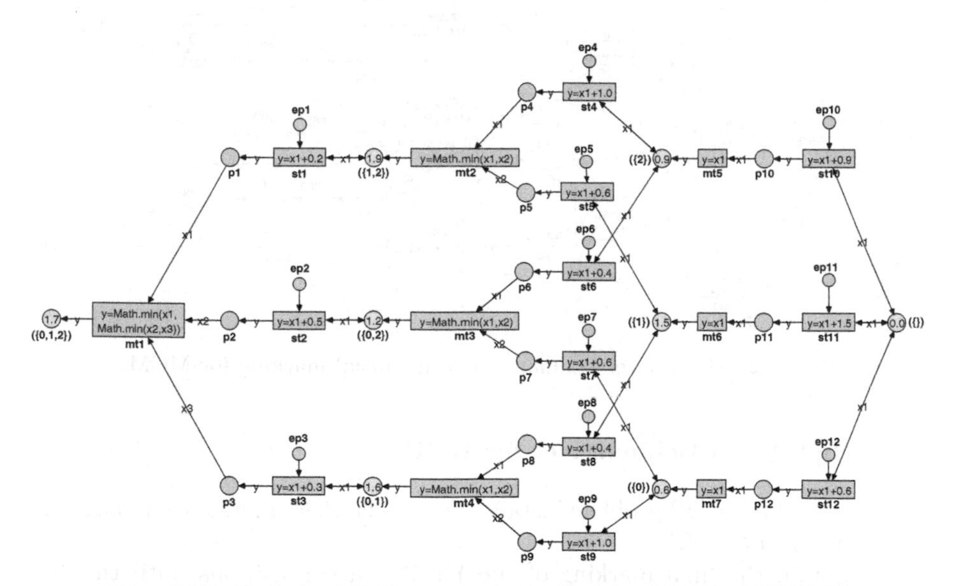

Fig. 6.22. High-level Bellman net in its final marking for LINSRCS

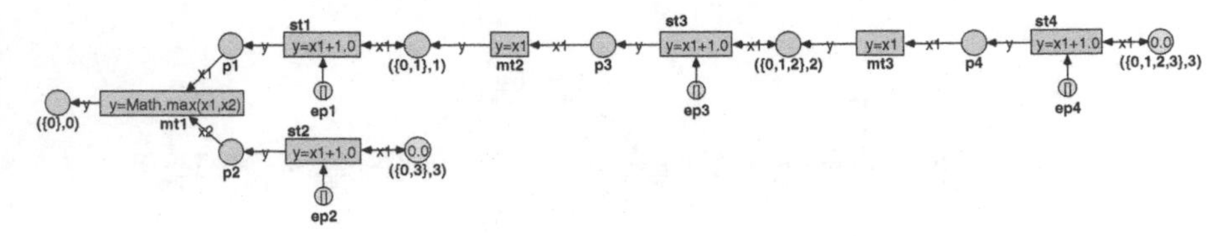

Fig. 6.23. High-level Bellman net in its initial marking for LSP

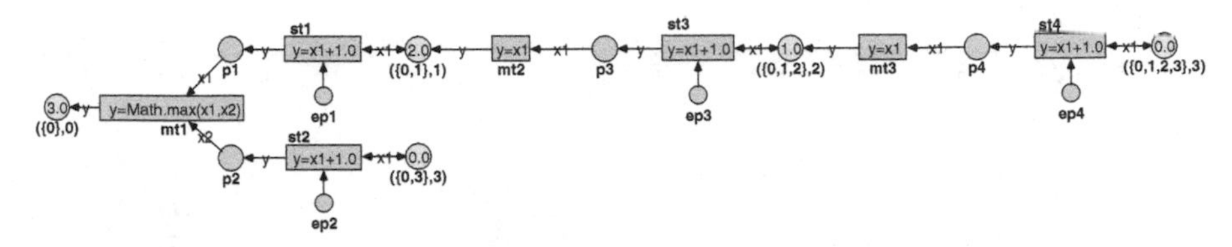

Fig. 6.24. High-level Bellman net in its final marking for LSP

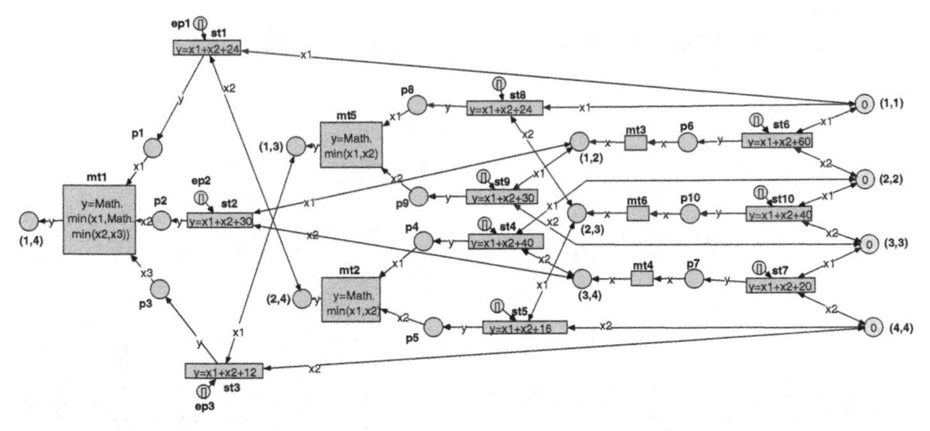

Fig. 6.25. High-level Bellman net in its initial marking for MCM

6.2.8 High-Level Bellman Net for RDP

The HLBN of the RDP problem instance from Sect. 2.36, in its initial marking, is depicted in Fig. 6.27.

To obtain the final marking of the HLBN, fire transitions until the net is dead. The net in its final marking is depicted in Fig. 6.28. Note that the goal state place (2,105) contains a token with value $f(2, 105) = 0.648$ now, indeed the correct value for a system with maximal reliability.

6.2.9 High-Level Bellman Net for SCP

The HLBN of the SCP problem instance from Sect. 2.38, in its initial marking, is depicted in Fig. 6.29.

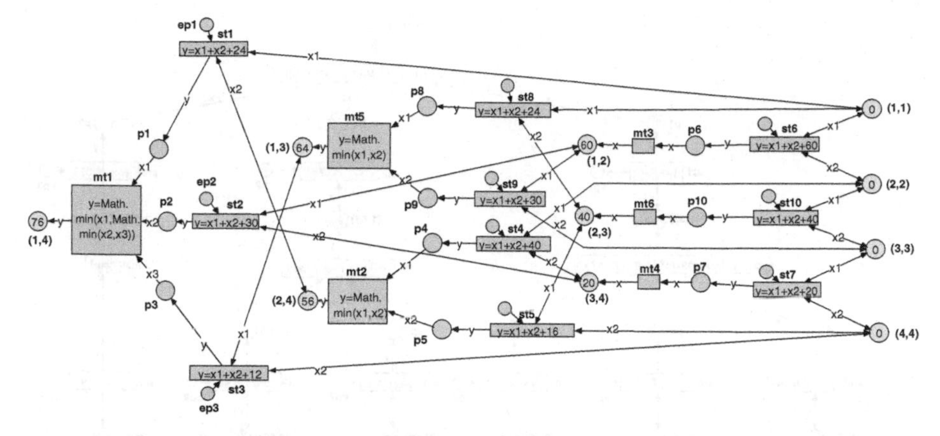

Fig. 6.26. High-level Bellman net in its final marking for MCM

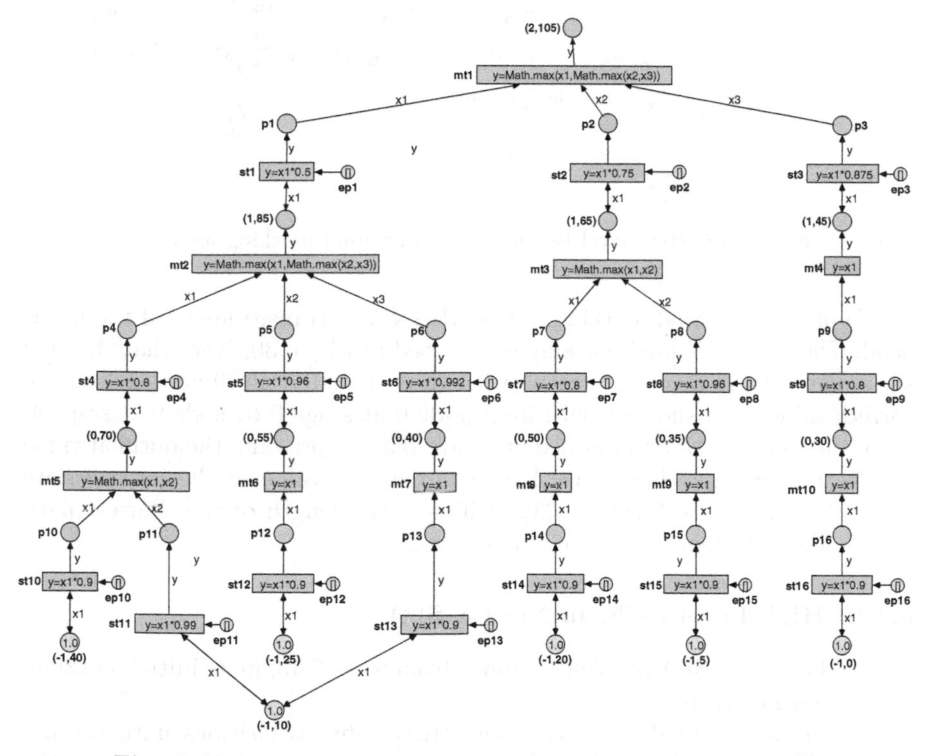

Fig. 6.27. High-level Bellman net in its initial marking for RDP

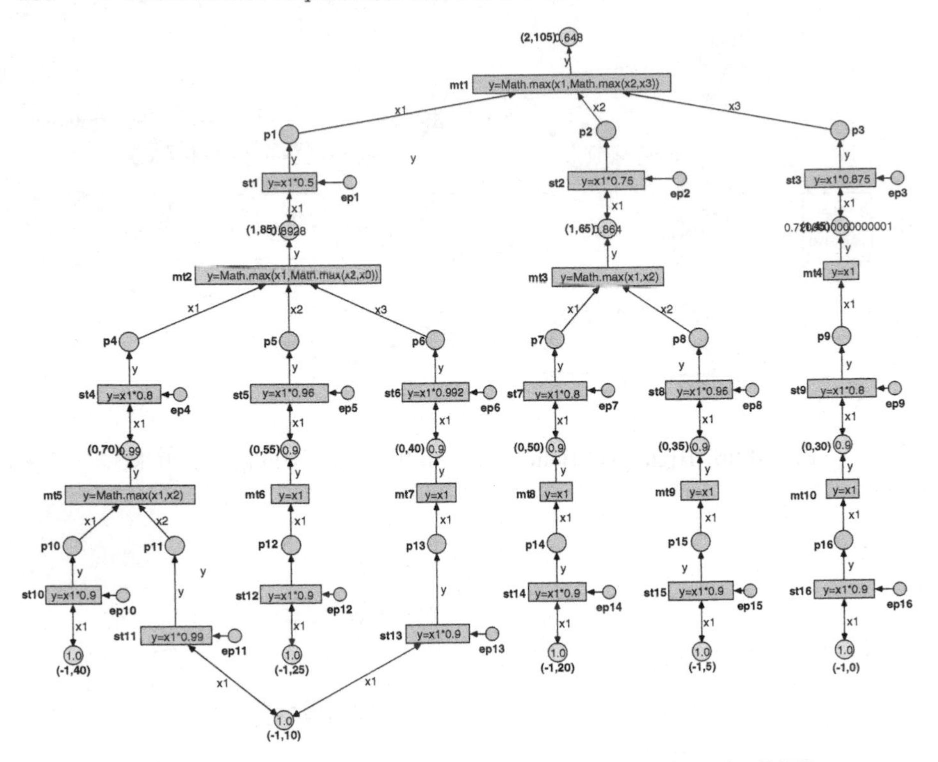

Fig. 6.28. High-level Bellman net in its final marking for RDP

To obtain the final marking of the HLBN, fire transitions until the net is dead. The net in its final marking is depicted in Fig. 6.30. Note that the goal state place (0,0) contains a token with value $f(0,0) = 2870$ now, indeed the correct value for a shortest path from node 0 at stage 0 to node 9 at stage 4. Also, the marking of an intermediate state place represents the optimal value for the associated subproblem. For example, the value of the intermediate state place (1,1) is $f(1,1) = 2320$ which is the length of the shortest path from node 1 at stage 1 to node 9 at stage 4.

6.2.10 High-Level Bellman Net for SPA

The HLBN of the SPA problem instance from Sect. 2.43, in its initial marking, is depicted in Fig. 6.31.

To obtain the final marking of the HLBN, fire transitions until the net is dead. The net in its final marking is depicted in Fig. 6.32. Note that the goal state place (0) contains a token with value $f(0) = 9$ now, indeed the correct value for a shortest path from node 0 to node 3. Also, the marking of an intermediate state place represents the optimal value for the associated subproblem. For example, the value of the intermediate state place (1) is $f(1) = 6$ which is the length of the shortest path from node 1 to node 3.

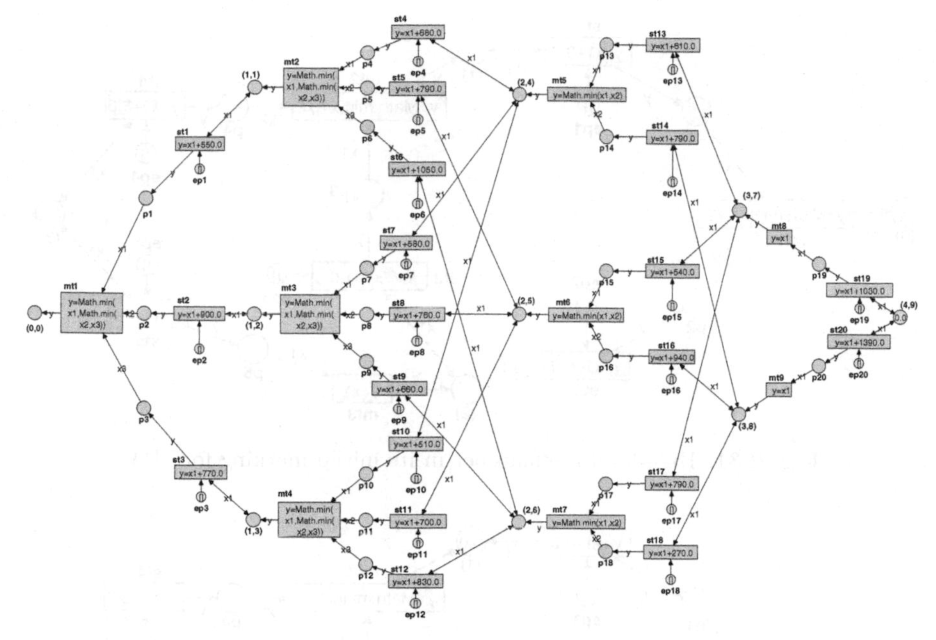

Fig. 6.29. High-level Bellman net in its initial marking for SCP

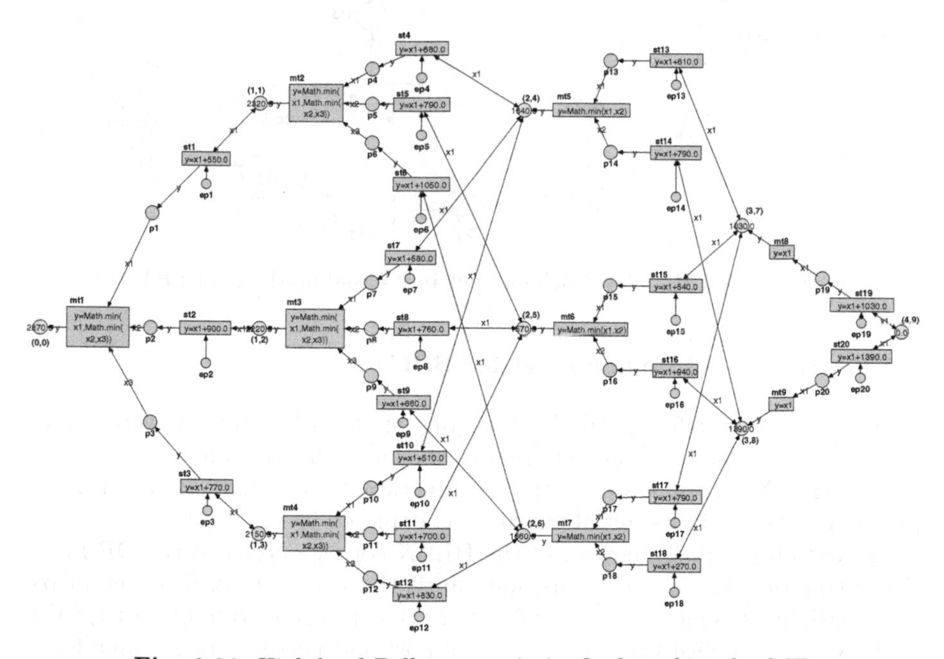

Fig. 6.30. High-level Bellman net in its final marking for SCP

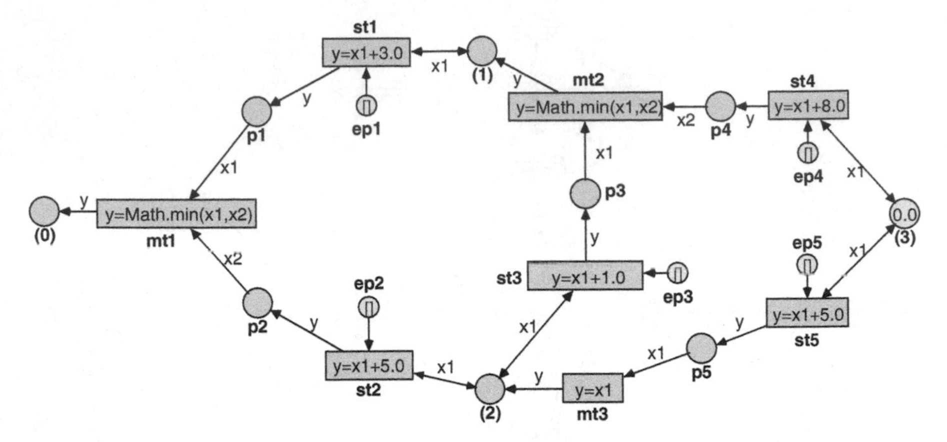

Fig. 6.31. High-level Bellman net in its initial marking for SPA

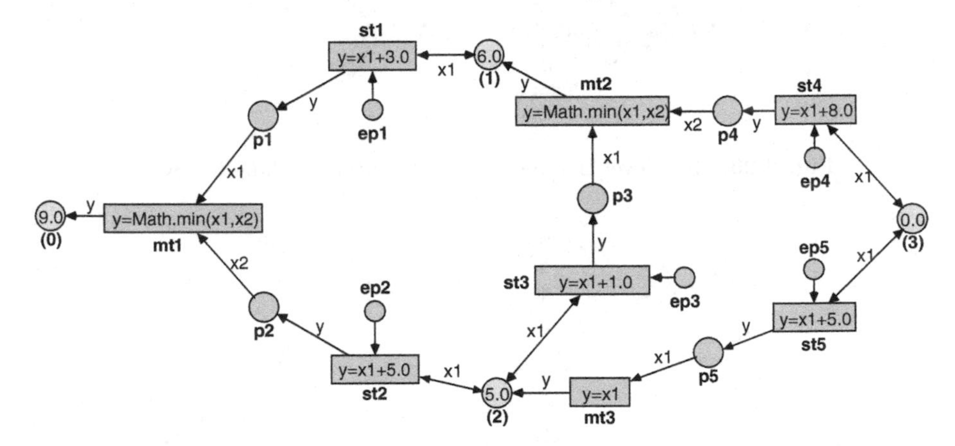

Fig. 6.32. High-level Bellman net in its final marking for SPA

6.2.11 High-Level Bellman Net for SPC

In Sect. 2.44 two different DP formulations of the SPC problem were given. These result in the two different HLBNs shown in this section.

The HLBN corresponding to the DP functional equation (2.45) of the SPC problem instance, in its initial marking, is depicted in Fig. 6.33.

To obtain the final marking of the HLBN corresponding to the DP functional equation (2.45), fire transitions until the net is dead. The net in its final marking is depicted in Fig. 6.34. Note that the goal state place `(0,{0})` contains a token with value $f(0, \{0\}) = 9$ now, indeed the correct value for a shortest path from node 0 to node 3.

The HLBN corresponding to the DP functional equation (2.46) of the SPC problem instance, in its initial marking, is depicted in Fig. 6.35.

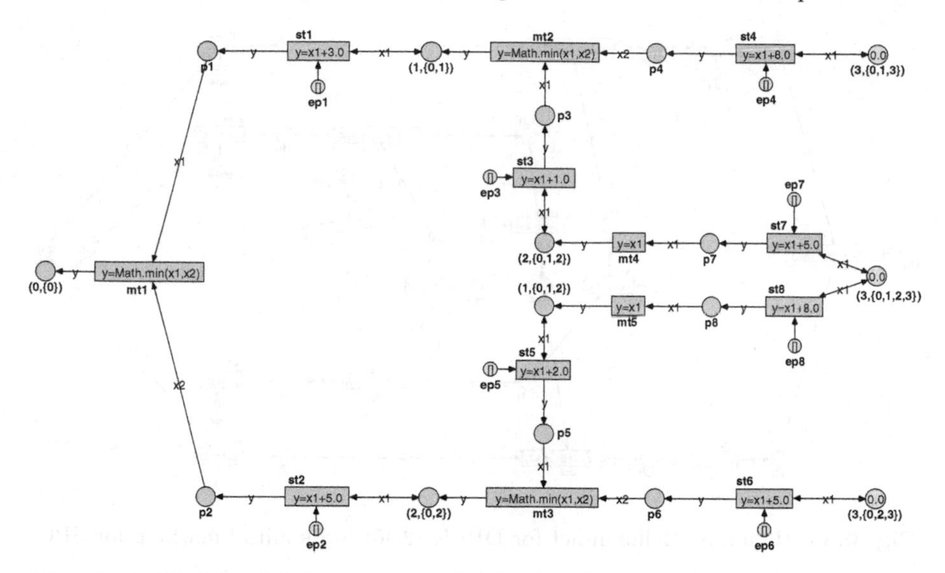

Fig. 6.33. High-level Bellman net for DPFE (2.45) in its initial marking for SPC

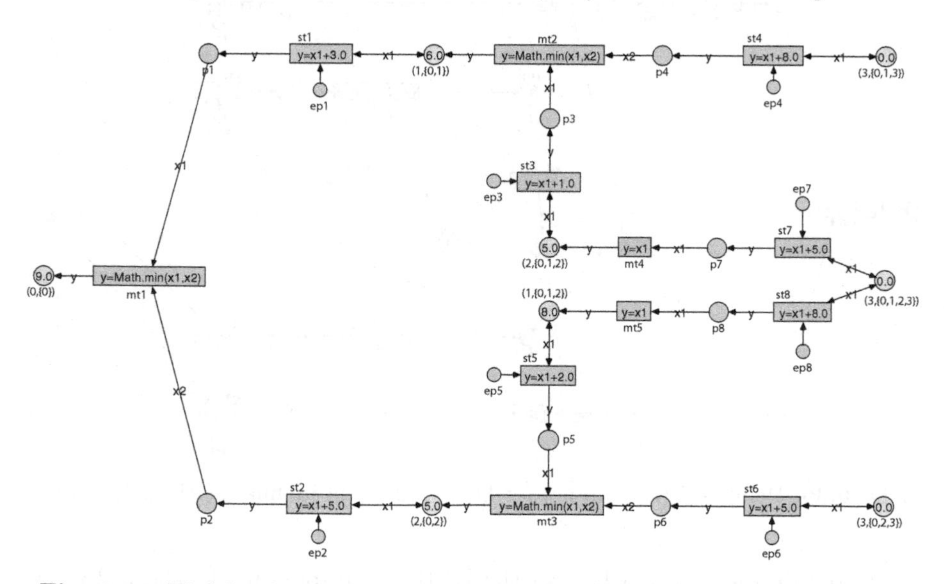

Fig. 6.34. High-level Bellman net for DPFE (2.45) in its final marking for SPC

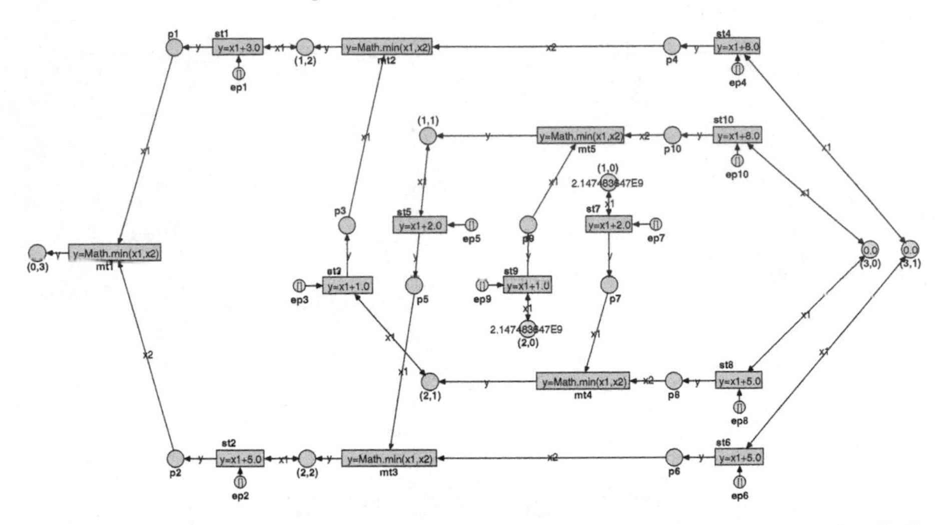

Fig. 6.35. High-level Bellman net for DPFE (2.46) in its initial marking for SPC

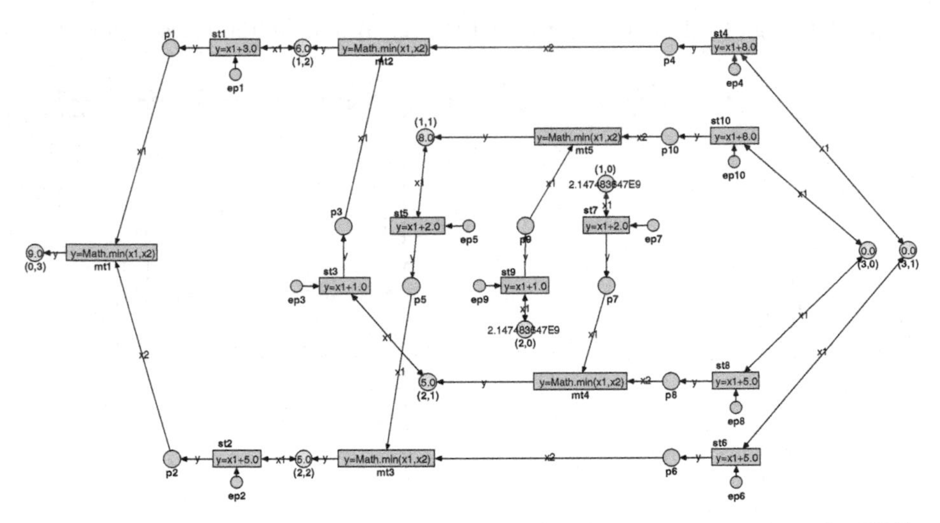

Fig. 6.36. High-level Bellman net for DPFE (2.46) in its final marking for SPC

To obtain the final marking of the HLBN corresponding to the DP functional equation (2.46), fire transitions until the net is dead. The net in its final marking is depicted in Fig. 6.36. Note that the goal state place `(0,3)` contains a token with value $f(0,3) = 9$ now, indeed the correct value for a shortest path from node 0 to node 3.

Part III

Design and Implementation of DP Tool

7

DP2PN2Solver Tool

This chapter gives an overview of the DP2PN2Solver Software. After describing the general architecture in Sect. 7.1 we focus on the file representation of Bellman nets (Sect. 7.2). Section 7.3 shows how DP2PN2Solver compiles and executes DP programs. For integer linear programming (ILP) problems, a gDPS source can be generated automatically (Sect. 7.4).

7.1 Overview

The DP2PN2Solver software system we have developed [44, 45] and describe in this book is a general, flexible, and expandable software tool that solves DP problems. It consists of modules that may be grouped into two phases. Phase One modules (gDPS2PN, bDPS2PN, etc.) take the specification of a discrete DP problem instance as input and produce an intermediate Petri net representation, the high-level *Bellman net* as defined in Chap. 5, as output. This Bellman net concisely captures all the essential elements of a DP problem instance in a standardized and mathematically precise fashion. Phase Two modules (PN2Java, PN2XML, PN2Spreadsheet, etc.) take the intermediate Bellman net representation as input and produce as output "executable" solver code, that is, any kind of code from which the numerical solution can be obtained by executing a "solver" system. For example, a Java system could compile and interpret solver code in the form of a Java source program, a spreadsheet application could open a spreadsheet file and calculate all cells, or a Petri net tool could simulate a Bellman net provided in a standard XML-based Petri net file interchange format.

Software tools that allow a user to conveniently solve arbitrary DP problems using the terminology and techniques that have been established in the DP field are not well developed. Early tools (such as DYNACODE [18]) were characterized as too problem specific by Sniedovich [57, p.193], who concluded that "no general-purpose dynamic progamming computer codes seem to be available commercially." Over the years this situation has not changed much,

A. Lew and H. Mauch: *DP2PN2Solver Tool*, Studies in Computational Intelligence (SCI) **38**, 247–257 (2007)
www.springerlink.com

arguably because others did not focus on DP alone. Currently available software systems can only solve a narrow class of DP problems having relatively simple DP functional equations. The solver software DP2PN2Solver presented in this book is intended to fill this gap.

One of the difficulties in designing a DP solver system is to come up with a specification language that is on one hand general enough to capture the majority of DP problems arising in practice, and that is on the other hand simple enough to be user-friendly and to be parsed efficiently. For linear programming (LP) problems, it is very easy to achieve both of these goals (using the *canonical*, *standard*, or *general form* of LP [49, p.27]); for DP problems, however, it is much harder to satisfy these conflicting goals. The general DP specification language gDPS (see Chap. 3) was designed with this in mind. It allows users to *program* the solution of DP problems by simply specifying its DPFE in a text-based source language. Users would not need to be aware of the internal Bellman net model that was adopted as an internal computer representation of the DP functional equations specified in their gDPS source programs. They need only perform the DP modeling to construct the DP functional equation required by DP2PN2Solver.

DP2PN2Solver's architecture is expandable and open to use of alternative source DP specification languages (see Fig. 7.1). For example, if the user desires to solve only very basic DP problems that involve only integer types as states or decisions, so that declarations (or general functions and set variables) are not necessary, then a much simpler language, say, a *basic* DP specification (bDPS) language, is sufficient. By adding a compiler to DP2PN2Solver that translates bDPS into the Bellman net representation, it becomes possible to use the other parts of the DP2PN2Solver system without any further changes. Even compilers for established mathematical programming languages like AMPL, GAMS, OPL, or other languages can be envisioned. Alternative compilers may translate into gDPS source instead of translating directly into internal Bellman net representations. Such compilers could easily be integrated into DP2PN2Solver (even though their development might require a substantial initial implementation effort), so that users can write the DP specification in their favorite language.

The software tool we have developed is only applicable to "discrete" optimization problems for which a DP functional equation with a finite number of states and decisions can be obtained. Therefore, continuous DP problems cannot be solved with DP2PN2Solver. (Some continuous DP problems may be solved approximately by discretization.) When the number of states is excessive, it is common to resort to approximation or search techniques to solve DP problems. While sometimes useful, in general such techniques cannot guarantee that an exact optimal solution will be found, hence will not be considered further in this book. We have surveyed currently existing software supporting the exact optimal solution of DP problems, and have found no system that can apply DP to the wide variety of problems treated in this book. Other optimization software systems can only solve narrower classes of DP problems with simpler DP functional equations. Thus, the value of the DP2PN2Solver system for DP problems is apparent.

The design of DP2PN2Solver is illustrated by typical DP problems. DP2PN2Solver can be used to solve problems which are more complicated than finding, say, the shortest path in a multistage graph. The matrix chain multiplication (MCM) problem, for example, is nonserial, i.e., there is not only one, but two successor states in its DP functional equation that have to be evaluated. The traveling salesman problem (TSP) can be formulated with a DP functional equation that requires the use of set theoretic operators. Solving integer linear programming (ILP) problems with DP2PN2Solver illustrates that this important class of problems, which appears frequently in real-world applications, can also be handled. Other examples show that DP2PN2Solver can successfully solve probabilistic DP problems (PROD, INVEST) and discounted DP problems (DPP), and nonoptimization problems (POUR, HANOI) as well.

Figure 7.1 gives an overview of the architecture of the DP2PN2Solver software tool (in the form of a Petri net). The static components such as input files, intermediate representations, and output files are depicted as places. The dynamic components such as compilers and code generators are shown as transitions. Solid arcs connect fully working components while dotted arcs connect components that have not been implemented but should be available in future versions of DP2PN2Solver.

To alleviate the task of having to specify problems in gDPS for which a standardized format exists, as is the case for ILP problems which can be described in a tableau format, a preprocessor module can save a great amount of work. Currently we have implemented a preprocessor module "ILP2gDPS", described later (see Sect. 7.4), that takes an ILP tableau as input and outputs the gDPS source specification.

To summarize, the following are the steps that need to be performed to solve a DP problem instance; only step 1 is performed by the human DP modeler whereas all other steps are automatically performed by the DP2PN2Solver system.

1. Model the real-world problem by formulating a DPFE that solves it, and for this formulation create the DP specification file represented, for example, as a textfile in the gDPS language (see Chap. 3). A graphical input mask, or some other "intelligent" front-end could be useful to simplify this task. For problems that have a standard specification such as ILP, a preprocessor module can automatically derive the DP model.
2. Use the gDPS2PN compiler (see Chap. 8) to produce the intermediate Bellman net representation (see Sect. 5.5) as a `.csv` file (see Sect. 7.2).
3. Use one of the PN2Solver modules (see Chap. 9) to produce runnable Java code, or a spreadsheet, or another form of executable solver code, which is capable of solving the problem instance.

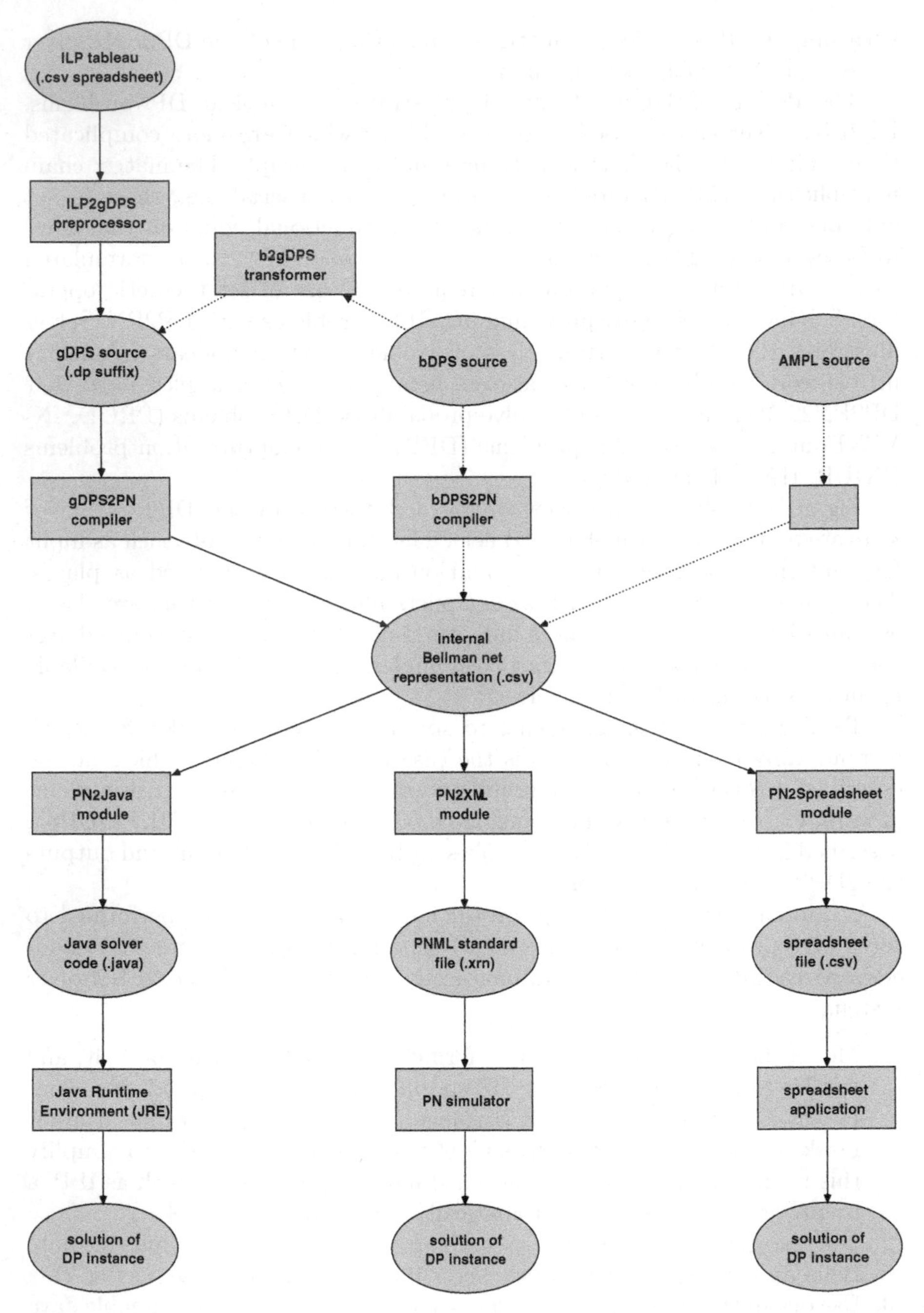

Fig. 7.1. Architecture of the DP2PN2Solver Tool

4. Run the resulting executable solver code and output the solution of the problem instance (see Chap. 10 for the output of the Java solver).

One distinguishing characteristic of our design is that the gDPS2PN compiler is generated in part by a compiler-compiler rather than being directly coded. This provides the system with added flexibility.

7.2 Internal Representation of Bellman Nets

Section 6.2 displayed HLBNs in their graphical representation. The graphical representation is most readable and understandable for humans. However, when a computer reads or writes a Bellman net it relies on a file representation. This section describes the file representation used in the DP2PN2Solver software.

Our design of the file format is based on a matrix representation of Bellman nets. Since a Bellman net is a class of directed graphs, it can be represented in adjacency or *incidence* matrix form. Since the graph is bipartite, we let the rows of the matrix correspond to place nodes of the Bellman net, and let the columns of the matrix correspond to transition nodes. Columns are labeled with the transition names (which should be quoted) and rows are labeled with the place names (which also should be quoted). An arc from a transition j to a place i is represented by a matrix entry $a_{i,j} = 1$, whereas an arc from place i to transition j is represented by a matrix entry $a_{i,j} = -1$. If there is no arc in either direction between i and j, then $a_{i,j} = 0$.

There are two columns with additional information about the places. One column is labeled `PLACETYPES` and takes entries from the set {`i`,`s`} designating each place as either an *intermediate* place or as a *state* place, respectively. Another column labeled `INIT/DEC` serves two purposes, depending on the type of the place. It stores the initial marking, if any, as a floating-point number for a state place. (There is an initial marking if and only if the place is a base state place.) It also stores (as a quoted string) the decision, if any, that is associated with an intermediate place.

There are two rows with additional information about the transitions. The row labeled `TRANSTYPE` takes entries from the set {`min`,`max`,`+`,`*`} designating a transition as a minimization, maximization, addition, or multiplication transition, respectively. The former two operations declare the transition to be a M-transition, the latter two make it an E-transition. Another row labeled `ETRANSCONST` contains floating-point number entries that are the constant summands or factors associated with E-transitions; for M-transitions these entries are left blank.

Note that reverse arcs cannot be coded with an incidence matrix, and in the case of HLBNs they do not have to — since every arc originating from a state place is a double arc, and since all other arcs are ordinary arcs, we can implicitly deduce this information from the type of the place from which the arc originates.

Enabling places and the arcs that connect them to E-transitions are also pieces of information not explicitly represented in the tabular file format; they are simply left out to keep the table small. The rule is that for every E-transition, one should implicitly assume the presence of an enabling place, initially marked with a single black token. An enabling place is connected to its E-transition by means of an ordinary arc. Formally, double arcs and enabling places are associated with M-transitions also, but for simplicity of presentation we omit them in our examples.

The resulting incidence matrix can be represented as a spreadsheet and saved as an unformatted text file in `.csv` form, where each row of the matrix has comma separated values listing the values in the columns. The HLBN file for the SPA problem instance from Sect. 2.43, named `SPABN.csv`, is given below.

```
PLACETYPES,PNAMES\TNAMES,"mt1","st1","st2","mt2","st3","st4","mt3","st5",INIT/DEC
s,"(0)",1,0,0,0,0,0,0,0,
i,"p1",-1,1,0,0,0,0,0,0,"d=1"
s,"(1)",0,-1,0,1,0,0,0,0,
i,"p2",-1,0,1,0,0,0,0,0,"d=2"
s,"(2)",0,0,-1,0,-1,0,1,0,
i,"p3",0,0,0,-1,1,0,0,0,"d=2"
i,"p4",0,0,0,-1,0,1,0,0,"d=3"
s,"(3)",0,0,0,0,0,-1,0,-1,0.0
i,"p5",0,0,0,0,0,0,-1,1,"d=3"
,TRANSTYPE,min,+,+,min,+,+,min,+,
,ETRANSCONST,,3.0,5.0,,1.0,8.0,,5.0,
```

7.3 Compiling and Executing DP Programs

In this section, we summarize the process of solving (compiling and executing) a DP problem that has a given DPFE formulation expressed as a gDPS program. Additional details are provided in subsequent chapters. For specificity, we trace the process for the linear search (LINSRCS) example.

The tool starts by invoking the compiler, which reads the linear search specification file `linsrcs.dp` as input, and produces a (High-Level) Bellman net representation `LINSRCSBN.csv` as output. It does this indirectly by first generating the program `LINSRCSMain.java` based upon the elements of the linear search DPFE as extracted from `linsrcs.dp`; when `LINSRCSMain.class`, the compiled `LINSRCSMain.java`, is executed, the SDRT table is generated and then from this table the Bellman net is constructed. A trace of this process is provided in the output file `buildBNlog.txt`, as shown below.

```
Starting...
Goal State:
({0,1,2})
Base States with values:
({}) 0.0


Operator associated with transitions:
+
Direction of optimization:
min

StateDecisionRewardTransformationTable:
({0,1,2}) [d=0] 0.2 (({1,2})) ()
```

```
({0,1,2}) [d=1] 0.5 (({0,2})) ()
({0,1,2}) [d=2] 0.3 (({0,1})) ()
({1,2}) [d=1] 1.0 (({2})) ()
({1,2}) [d=2] 0.6 (({1})) ()
({0,2}) [d=0] 0.4 (({2})) ()
({0,2}) [d=2] 0.6 (({0})) ()
({0,1}) [d=0] 0.4 (({1})) ()
({0,1}) [d=1] 1.0 (({0})) ()
({2}) [d=2] 0.8999999999999999 (({})) ()
({1}) [d=1] 1.5 (({})) ()
({0}) [d=0] 0.6000000000000001 (({})) ()

Make a place for the goal state ({0,1,2})
Make a min transition mt1
make an arc from mt1 to state({0,1,2})
Make an intermediate place p1 (for decision d=0)
make an arc from p1 to mt1
make a + transition st1 with value 0.2
make an arc from st1 to p1
make an enabling place ep1 containing 1 black token
make an an arc from ep1to st1
Make a place for the state ({1,2})
Make an arc from state ({1,2}) to + transition st1 and a return arc.
Make an intermediate place p2 (for decision d=1)
make an arc from p2 to mt1
make a + transition st2 with value 0.5
make an arc from st2 to p2
make an enabling place ep2 containing 1 black token
make an an arc from ep2to st2
Make a place for the state ({0,2})
Make an arc from state ({0,2}) to + transition st2 and a return arc.
Make an intermediate place p3 (for decision d=2)
make an arc from p3 to mt1
make a + transition st3 with value 0.3
make an arc from st3 to p3
make an enabling place ep3 containing 1 black token
make an an arc from ep3to st3
Make a place for the state ({0,1})
Make an arc from state ({0,1}) to + transition st3 and a return arc.
Make a min transition mt2
make an arc from mt2 to state({1,2})
Make an intermediate place p4 (for decision d=1)
make an arc from p4 to mt2
make a + transition st4 with value 1.0
make an arc from st4 to p4
make an enabling place ep4 containing 1 black token
make an an arc from ep4to st4
Make a place for the state ({2})
Make an arc from state ({2}) to + transition st4 and a return arc.
Make an intermediate place p5 (for decision d=2)
make an arc from p5 to mt2
make a + transition st5 with value 0.6
make an arc from st5 to p5
make an enabling place ep5 containing 1 black token
make an an arc from ep5to st5
Make a place for the state ({1})
Make an arc from state ({1}) to + transition st5 and a return arc.
Make a min transition mt3
make an arc from mt3 to state({0,2})
Make an intermediate place p6 (for decision d=0)
make an arc from p6 to mt3
make a + transition st6 with value 0.4
make an arc from st6 to p6
make an enabling place ep6 containing 1 black token
make an an arc from ep6to st6
Make an arc from state ({2}) to + transition st6 and a return arc.
Make an intermediate place p7 (for decision d=2)
make an arc from p7 to mt3
make a + transition st7 with value 0.6
make an arc from st7 to p7
make an enabling place ep7 containing 1 black token
make an an arc from ep7to st7
Make a place for the state ({0})
Make an arc from state ({0}) to + transition st7 and a return arc.
Make a min transition mt4
make an arc from mt4 to state({0,1})
Make an intermediate place p8 (for decision d=0)
make an arc from p8 to mt4
make a + transition st8 with value 0.4
make an arc from st8 to p8
make an enabling place ep8 containing 1 black token
make an an arc from ep8to st8
Make an arc from state ({1}) to + transition st8 and a return arc.
Make an intermediate place p9 (for decision d=1)
make an arc from p9 to mt4
make a + transition st9 with value 1.0
make an arc from st9 to p9
make an enabling place ep9 containing 1 black token
make an an arc from ep9to st9
```

```
Make an arc from state ({0}) to + transition st9 and a return arc.
Make a min transition mt5
make an arc from mt5 to state({2})
Make an intermediate place p10 (for decision d=2)
make an arc from p10 to mt5
make a + transition st10 with value 0.8999999999999999
make an arc from st10 to p10
make an enabling place ep10 containing 1 black token
make an an arc from ep10to st10
Make a place for the state ({})
Make an arc from state ({}) to + transition st10 and a return arc.
Make a min transition mt6
make an arc from mt6 to state({1})
Make an intermediate place p11 (for decision d=1)
make an arc from p11 to mt6
make a + transition st11 with value 1.5
make an arc from st11 to p11
make an enabling place ep11 containing 1 black token
make an an arc from ep11to st11
Make an arc from state ({}) to + transition st11 and a return arc.
Make a min transition mt7
make an arc from mt7 to state({0})
Make an intermediate place p12 (for decision d=0)
make an arc from p12 to mt7
make a + transition st12 with value 0.6000000000000001
make an arc from st12 to p12
make an enabling place ep12 containing 1 black token
make an an arc from ep12to st12
Make an arc from state ({}) to + transition st12 and a return arc.
Add a token with value 0.0 into state ({})
End.
```

The constructed incidence matrix representation of the Bellman net for the linear search example can be represented as a spreadsheet (in .CSV form) and is saved as `LINSRCSBN.csv` as follows.

```
PLACETYPES,PNAMES\TNAMES,"mt1","st1","st2","st3","mt2","st4","st5","mt3","st6","st7","mt4","st8","st9","mt5","st10","mt6",
   "st11","mt7","st12",INIT/DEC
s,"({0,1,2})",1,0,0,0,0,0,0,0,0,0,0,0,0,0,0,0,0,0,0,
i,"p1",-1,1,0,0,0,0,0,0,0,0,0,0,0,0,0,0,0,0,0,"d=0"
s,"({1,2})",0,-1,0,0,1,0,0,0,0,0,0,0,0,0,0,0,0,0,0,
i,"p2",-1,0,1,0,0,0,0,0,0,0,0,0,0,0,0,0,0,0,0,"d=1"
s,"({0,2})",0,0,-1,0,0,0,0,1,0,0,0,0,0,0,0,0,0,0,0,
i,"p3",-1,0,0,1,0,0,0,0,0,0,0,0,0,0,0,0,0,0,0,"d=2"
s,"({0,1})",0,0,0,-1,0,0,0,0,0,0,1,0,0,0,0,0,0,0,0,
i,"p4",0,0,0,0,-1,1,0,0,0,0,0,0,0,0,0,0,0,0,0,"d=1"
s,"({2})",0,0,0,0,0,-1,0,0,-1,0,0,0,0,1,0,0,0,0,0,
i,"p5",0,0,0,0,-1,0,1,0,0,0,0,0,0,0,0,0,0,0,0,"d=2"
s,"({1})",0,0,0,0,0,0,-1,0,0,0,0,-1,0,0,0,1,0,0,0,
i,"p6",0,0,0,0,0,0,0,-1,1,0,0,0,0,0,0,0,0,0,0,"d=0"
i,"p7",0,0,0,0,0,0,0,-1,0,1,0,0,0,0,0,0,0,0,0,"d=2"
s,"({0})",0,0,0,0,0,0,0,0,0,-1,0,0,-1,0,0,0,0,1,0,
i,"p8",0,0,0,0,0,0,0,0,0,0,-1,1,0,0,0,0,0,0,0,"d=0"
i,"p9",0,0,0,0,0,0,0,0,0,0,-1,0,1,0,0,0,0,0,0,"d=1"
i,"p10",0,0,0,0,0,0,0,0,0,0,0,0,0,-1,1,0,0,0,0,"d=2"
s,"({})",0,0,0,0,0,0,0,0,0,0,0,0,0,0,-1,0,-1,0,-1,0.0
i,"p11",0,0,0,0,0,0,0,0,0,0,0,0,0,0,0,-1,1,0,0,"d=1"
i,"p12",0,0,0,0,0,0,0,0,0,0,0,0,0,0,0,0,0,-1,1,"d=0"
,TRANSTYPE,min,+,+,+,min,+,+,min,+,+,min,+,+,min,+,min,+,min,+,
,ETRANSCONST,,0.2,0.5,0.3,,1.0,0.6,,0.4,0.6,,0.4,1.0,,0.8999999999999999,,1.5,,0.6000000000000001,
```

The `LINSRCSBN.csv` file given above is the internal representation of the HLBN for the linear search DPFE. We emphasize that this spreadsheet is a description of the DPFE rather than an executable program that computes its solution. The remaining problem is to obtain the numerical solution of the DPFE from this Bellman net specification. This can be done in at least three ways. One option is to use a Petri net simulator. Our Petri net solver code module converts our internal HLBN representation to that required by a standard Petri net tool; we discuss this in Chap. 11. A second option is to produce an executable spreadsheet that performs the operations of the M-transition and E-transition nodes. Our spreadsheet solver module generates such a spreadsheet; we also discuss this in Chap. 11.

A third option is to generate solver code in a conventional procedural programming language, such as Java. For the linear search example, the following excerpt from the Java solver code, `LINSRCSJavaSolver.java`, would be generated. This is discussed in greater detail in Chap. 9. Here, we simply note that the firing E- or M-transitions is accomplished by means of "CalculationObjects", one for each transition.

```
public class linsrcsJavaSolver {
  public static void main(String[] args) throws IOException {
    final String subDirName="linsrcsSolverCode";
    String currentWorkingDir=System.getProperty("user.dir");
    if(!currentWorkingDir.endsWith(subDirName)) {
         currentWorkingDir=currentWorkingDir+"/"+subDirName;
    }
    Out.pw=new PrintWriter(new FileWriter(
        new File(currentWorkingDir+"/"+"linsrcsSolutionTree.txt")));
    List argumentList; //reusable temporary variable used for min/max
    List operandList; //reusable temporary variable for eval. (Etransitions)
    //code to initialize base state place variables
    CalculationObject I35 = new CalculationObject("({})",0.0);

    //code that simulates the transitions firing
    operandList=new ArrayList();
    operandList.add(I35);
    ECalculationObject I33 = new ECalculationObject(
        "d=2",
        I35.getValue()+0.8999999999999999,
        operandList);
     ...

    argumentList=new ArrayList();
    argumentList.add(I3);
    argumentList.add(I7);
    argumentList.add(I11);
    MCalculationObject I1 = new MCalculationObject("({0,1,2})");
    I1.performMinimization(argumentList);

    //code that gives us the final answer
    Out.put("The optimal value is: ");
    Out.putln(I1.getValue());
    Out.putln("The solution tree is: ");
    Out.putln(I1.predecessorSubtree(0));

    Out.pw.close(); //close the output stream
  } //end of main()
} //end of class
```

When the program `LINSRCSJavaSolver.class` (obtained by compiling `LINSRCSJavaSolver.java`) is executed, the numerical solution is placed in the output file `LINSRCSSolutionTree.txt`, which is as follows.

```
The optimal value is: 1.7000000000000002
The solution tree is:
State ({0,1,2}) has optimal value: 1.7000000000000002
 Decision d=1
  State ({0,2}) has optimal value: 1.2000000000000002
   Decision d=2
    State ({0}) has optimal value: 0.6000000000000001
     Decision d=0
      Base state ({}) has initial value: 0.0
```

The optimal solution is 1.7 obtained by making the sequence of decisions 1, 2, and 0, with costs 0.5, 0.6, and 0.6, respectively.

7.4 The ILP2gDPS Preprocessor Module

Our DP software tool allows additional modules to be added that permits users to provide inputs in other than the gDPS source language, and to produce solver code output in other than Java. We give an example of the former in this section.

Integer linear programs can be defined in a standard fashion using a tableau. This tableau is identical to the tableau specification of LP problems for use in LP solver software. For example, in order to use a simplex LP solver, the user would specify a "simplex tableau".

It is possible to support the DP modeler who wishes to solve integer linear programs satisfying the nonnegativity assumptions of the ILP problem (r, b and A have nonnegative entries, as defined in section 2.14) by providing a preprocessor tool that automatically generates the gDPS source from the tableau. Such a preprocessor has been implemented and has been named "ILP2gDPS". The gDPS source code produced follows the DP functional equation (2.18) described in section 2.14. This illustrates the generality and flexibility of our approach, not necessarily its practicality. In the following we describe the details of ILP2gDPS.

The user of DP2PN2Solver might possess a spreadsheet file containing the tableau for an ILP problem instance. The tableau in `.csv` format (commas separate the values) serves as the input for the ILP2gDPS preprocessor. The input file for the example instance from section 2.14 would read as follows.

```
3,5,
1,0,4
0,2,12
3,2,18
```

The first line must contain the coefficients of the objective function (i.e. representing the r vector). Each following line represents a constraint. For constraint i it is given by the constraint coefficients of the left-hand side (i.e. row i of the matrix A), followed by the right-hand side value (i.e. the b_i entry of the b vector). Values must be separated by commas and the end of a line is indicated by carriage return and/or line feed. Empty lines are not permitted since they would be interpreted as constraints with an insufficient number of values.

The ILP2gDPS preprocessor is called by
`java ILPpreprocessor input.csv output.dp`
or simply by
`java ILPpreprocessor file.csv`
in which case the gDPS output file is named `file.dp` automatically.

The gDPS output for the sample ILP instance is given next.

```
//This gDPS code was generated by the ILPpreprocessor
BEGIN
  NAME ilpAuto;
  GENERAL_VARIABLES_BEGIN
    //Must assume a,b,c have all pos. int. entries.
    //objective function coefficients:
    private static int[] c = {3, 5};
    //right hand side of constraints vector:
    private static int[] b = {4, 12, 18};
    //constraint matrix:
    private static int[][] a=
      {
```

```
        {1, 0},
        {0, 2},
        {3, 2}
      };
    private static int n = c.length;
    private static int m = b.length;
    private static final int infty=Integer.MAX_VALUE;
  GENERAL_VARIABLES_END

  GENERAL_FUNCTIONS_BEGIN
    private static NodeSet calculateDecisionSet(int stage,
                                                int y1,
                                                int y2,
                                                int y3) {
      NodeSet result = new NodeSet();
      int maxpc=infty; //max. possible choice
      if(a[0][stage]!=0){
        maxpc=Math.min(maxpc, y1/a[0][stage]);
      }
      if(a[1][stage]!=0){
        maxpc=Math.min(maxpc, y2/a[1][stage]);
      }
      if(a[2][stage]!=0){
        maxpc=Math.min(maxpc, y3/a[2][stage]);
      }
      for (int i=0; i<=maxpc; i++) {
        result.add(new Integer(i));
      }
      return result;
    }
  GENERAL_FUNCTIONS_END

  STATE_TYPE: (int stage,
               int y1,
               int y2,
               int y3);
  DECISION_VARIABLE: int d;
  DECISION_SPACE: decisionSet(stage,y1,y2,y3)
                 =calculateDecisionSet(stage,y1,y2,y3);
  GOAL: f(0,b[0],b[1],b[2]);
  DPFE_BASE_CONDITIONS:
    f(stage,y1,y2,y3)=0.0 WHEN (stage == n);
  DPFE: f(stage,y1,y2,y3)
        =MAX_{d IN decisionSet}
                             { f(t(stage,y1,y2,y3,d))
                              +r(stage,y1,y2,y3,d)
                             };
  REWARD_FUNCTION: r(stage,y1,y2,y3,d)=c[stage]*d;
  TRANSFORMATION_FUNCTION: t(stage,y1,y2,y3,d)
                           =(stage+1,
                             y1-a[0][stage]*d,
                             y2-a[1][stage]*d,
                             y3-a[2][stage]*d);

END
```

Now this gDPS code can be used as the input for the DP2PN2Solver software in the same way a gDPS source hand-crafted by the modeler would be used.

8

DP2PN Parser and Builder

As introduced in Chap. 7.1 the gDPS2PN compiler performs the tasks of parsing the DP specification file and of building an internal Bellman net representation. Figure 8.1 shows the design overview of the gDPS2PN compiler in form of a Petri net. The compiler consists of a parser module and a builder module. The following sections describe the design, implementation, and test of these modules in more detail.

8.1 Design of the DP2PN modules

The complete grammar of the gDPS language in Backus-Naur form (BNF) is given in Sect. 3.4 and is incorporated into the parser specification file `DPspecificationParser.jj`. From this input, JavaCC 3.2 [tm], a compiler-compiler that facilitates both the lexical analysis and the parsing of the grammar (i.e. it has the same functionality as the C-language based "Lex" and "Yacc" combined, but can be integrated seamlessly into a Java environment), is used to obtain a DP2PN parser that is written completely in Java.

Execution of the DP2PN parser module results in the generation of the main part of the DP2PN builder module. The DP2PN parser module remains fixed for a chosen source language (in this case the source language is gDPS) and does not need to be recompiled unless changes to the source language are desired. The DP2PN builder module needs to be recompiled and then executed for every new gDPS input. All steps can be automated in a single batch file, so these intermediate steps are hidden from the user. From the user's perspective, if a gDPS source file is provided as input, then the gDPS2PN compiler produces a Bellman net as output (and also a log file documenting errors encountered during this process).

A. Lew and H. Mauch: *DP2PN Parser and Builder*, Studies in Computational Intelligence (SCI) **38**, 259–269 (2007)
www.springerlink.com

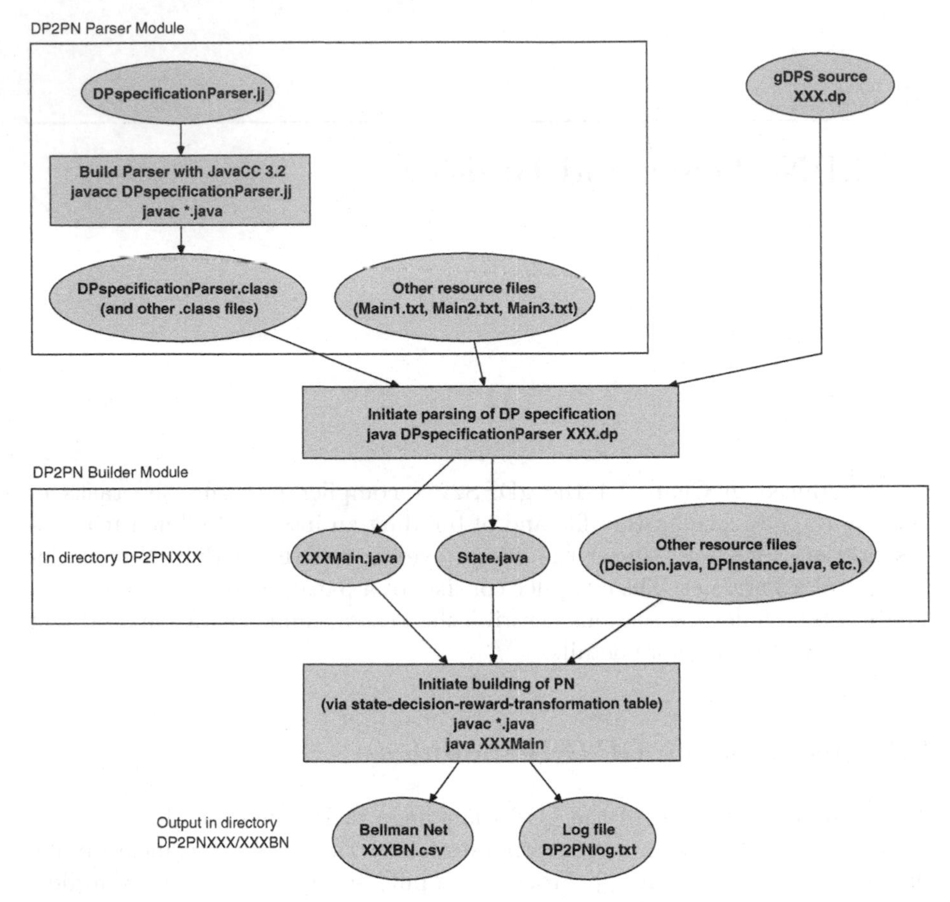

Fig. 8.1. Overview of the gDPS2PN Compiler

8.2 Implementation of the DP2PN modules

After parsing the gDPS source, the DP2PN parser module generates two problem dependent Java classes. The class `State` in `State.java` is the data structure for the heterogeneous state vector with the appropriate number of components as discussed in Chapter 1. The class `XXXMain` in `XXXMain.java` contains the main procedure that builds the Bellman net (i.e. the desired output of the DP2PN builder module) where `XXX` is to be read as the problem name under consideration. Internal details of this procedure are described next. For a more comprehensive listing of the pseudocode of DP2PN see Appendix A.1.

The idea is to construct a state-decision-reward-transformation table (SDRT table, see Sect. 1.2.2) that captures almost all of the essential characteristics of a DP instance. It only lacks the direction of optimization (minimization or maximization) and the information about base states, goal state, and the mathematical operation that is to be applied to "connect" the reward value and the next-state functionals. This additional information is stored in appropriate data structures and functions in an obvious way. The SDRT table is built in a top-down manner by exploring newly found states in a breadth-first fashion. Starting from the goal state, as defined in the GOAL section of the gDPS source, all reachable states are generated; for every decision in the decision set, a set of successor states unfolds. Newly discovered states are put into a queue that determines which states are created next. In order to avoid duplication of states in the queue over time, some bookkeeping is necessary in the form of maintaining a set of the states that have already been seen.

The following algorithm generates the SDRT table for a DP problem instance. This algorithm is incorporated within the main method of the PN builder module, e.g. XXXMain.java, as discussed further in the next section. (Note: The step numbers given here are used in our later discussion.)

```
         // SDRT generation algorithm:
[STEP 1] initialize the set statesSeen to contain the goal state
[STEP 2] initialize the queue stateQueue to contain the goal state
[STEP 3] while stateQueue not empty
[STEP 4]   currentState:=dequeue state from stateQueue
[STEP 5]   if currentState is not a base state
[STEP 6]     loop over all possible decisions
[STEP 7]       generate table entry (a row for the
                 state-decision-reward-transformation-table)
[STEP 8]       enqueue those next states which are not in statesSeen
                 into the stateQueue and also add them to statesSeen
```

For the linear search example, whose gDPS source is LINSRCS.dp, the SDRT table that is built is as follows:

```
StateDecisionRewardTransformationTable:
({0,1,2}) [d=0] 0.2 (({1,2})) ()
({0,1,2}) [d=1] 0.5 (({0,2})) ()
({0,1,2}) [d=2] 0.3 (({0,1})) ()
({1,2}) [d=1] 1.0 (({2})) ()
({1,2}) [d=2] 0.6 (({1})) ()
({0,2}) [d=0] 0.4 (({2})) ()
({0,2}) [d=2] 0.6 (({0})) ()
({0,1}) [d=0] 0.4 (({1})) ()
({0,1}) [d=1] 1.0 (({0})) ()
({2}) [d=2] 0.8999999999999999 (({})) ()
({1}) [d=1] 1.5 (({})) ()
({0}) [d=0] 0.6000000000000001 (({})) ()
```

As a second example, where there are multiple next states, specifically for the MCM problem instance of Sect. 2.27, the SDRT table is shown in Table 8.1.

Table 8.1. State-Decision-Reward-Transformation-Table for MCM instance

state	decision	reward	next states
(1,4)	k=1	24.0	((1,1), (2,4))
(1,4)	k=2	30.0	((1,2), (3,4))
(1,4)	k=3	12.0	((1,3), (4,4))
(2,4)	k=2	40.0	((2,2), (3,4))
(2,4)	k=3	16.0	((2,3), (4,4))
(1,3)	k=1	24.0	((1,1), (2,3))
(1,3)	k=2	30.0	((1,2), (3,3))
(1,2)	k=1	60.0	((1,1), (2,2))
(2,3)	k=2	40.0	((2,2), (3,3))
(3,4)	k=3	20.0	((3,3), (4,4))

Probabilistic and discounted DP problem instances have an additional "weight" column which accounts for the list of probability or discount weights associated with each successor state. For the PRODRAP problem (Sect. 2.35) the result is shown in Table 8.2.

Table 8.2. State-Decision-Reward-Transformation-Table for PRODRAP instance

state	decision	reward	next states	weights
(1)	xn=0	0.0	((2))	(1.0)
(1)	xn=1	4.0	((2))	(0.5)
(1)	xn=2	5.0	((2))	(0.25)
(1)	xn=3	6.0	((2))	(0.125)
(1)	xn=4	7.0	((2))	(0.0625)
(1)	xn=5	8.0	((2))	(0.03125)
(2)	xn=0	0.0	((3))	(1.0)
(2)	xn=1	4.0	((3))	(0.5)
(2)	xn=2	5.0	((3))	(0.25)
(2)	xn=3	6.0	((3))	(0.125)
(2)	xn=4	7.0	((3))	(0.0625)
(2)	xn=5	8.0	((3))	(0.03125)
(3)	xn=0	0.0	((4))	(1.0)
(3)	xn=1	4.0	((4))	(0.5)
(3)	xn=2	5.0	((4))	(0.25)
(3)	xn=3	6.0	((4))	(0.125)
(3)	xn=4	7.0	((4))	(0.0625)
(3)	xn=5	8.0	((4))	(0.03125)

Following the building of the state-decision-reward-transformation-table for the linear search example, the Bellman net representation (as described in Sect. 7.2) is then constructed. A trace of this construction process was given in Sect. 7.3. We emphasize that the building of the SDRT table and the construction of the Bellman net are done indirectly. They are generated not by the DP2PN compiler itself, but by a generated Java program, LINSRCSMain.java, which when executed as LINSRCSMain.class produces the desired results. In the next section, we indicate how LINSRCSMain.java can be produced from LINSRCS.dp.

8.3 The Module LINSRCSMain

A pseudocode listing of LINSRCSMain.java is given in App. A.1.1. In this section, we relate the statements in LINSRCS.dp to certain generated statements in LINSRCSMain.java to better explain the design of our system. This is meant to give interested readers the information needed to decipher the details provided in our appended listings.

- The name declaration in the LINSRCS.dp source program is used to generate the name that is inserted into the header portion of the LINSRCSMain.java object program. Specifically, the source statement

```
NAME LINSRCS; //OptimalLinearSearch-S;
```

  yields the following object statements.

```
//This is the file LINSRCSMain.java which has been
//automatically generated by the DPspecificationParser
//and its helper CodeCreator
import java.io.*;
import java.util.*;
import bellman_net.*;

public class LINSRCSMain {
  public static String problemName="LINSRCS";
```

- The source statements in the GENERAL_VARIABLES section, namely,

```
private static double[] prob= {0.2,0.5,0.3};
private static int N = prob.length;
```

  are inserted directly, without any modifications, into the object program.
- The source statements in the SET_VARIABLES section, namely,

```
Set setOfAllItems={0,..,N - 1};
Set emptySet={};
```

  are parsed, and the set variables, setOfAllItems and emptySet, are declared and initialized accordingly in the object program:

```
private static NodeSet setOfAllItems = _make_setOfAllItems();
private static NodeSet _make_setOfAllItems(){
  NodeSet _setInDDNO=new NodeSet();
  { //extra block so _i reusable
  for (int _i=0;_i<=N-1;_i++) {
```

```
      _setInDDNO.add(new Integer(_i));
      }
    } //close extra block
    return _setInDDNO;
  }

  private static NodeSet emptySet = _make_emptySet();
  private static NodeSet _make_emptySet(){
    NodeSet _setExplicit1=new NodeSet();
    return _setExplicit1;
  }
```

- The source statements in the GENERAL_FUNCTIONS section, namely,

```
  private static int size(Set items)
    return items.size();
  }
```

  are inserted directly, without any modifications, into the object program.
- The object program then includes statements that prepare the output files. These statements are copied from the system file named Main1.txt. A complete listing of Main1.txt is given in Appendix A.2.

```
//////////////////////////////////////////////////////
//beginning of the fixed code Main1.txt for Main.java
public static void main(String[] args) throws IOException{
  String outputFileName="buildBNlog.txt";
  final String subDirName="DP2PN"+problemName;
  String currentWorkingDir=System.getProperty("user.dir");
  if(!currentWorkingDir.endsWith(subDirName)) {
    currentWorkingDir=currentWorkingDir+"/"+subDirName;
  }
  //create an output directory for the Bellman net
  File newDir=new File(currentWorkingDir+"/"+problemName+"BN");
  newDir.mkdir(); //make the directory
  Out.pw=new PrintWriter(new FileWriter(
             new File(currentWorkingDir+"/"+problemName+"BN"+"/"
                 +outputFileName)));
  Out5.pw=new PrintWriter(new FileWriter(
              new File(currentWorkingDir+"/"+problemName+"BN"+"/"
                  +problemName+"BN.csv")));
  Out.putln("Starting...");
//end of the fixed code Main1.txt for Main.java
/////////////////////////////////////////////////////
```

- The GOAL section of the source program

```
GOAL: f(setOfAllItems);
```

  is then parsed, resulting in the placement of a statement in the object program that creates the goalState object and initializes it to the setOfAllItems. (Note: the State class is discussed at the end of this section.)

```
State goalState = new State(setOfAllItems);
```

- The DPFE_BASE section of the source program

```
DPFE_BASE: f(emptySet)=0.0 ;
```

is then parsed, resulting in the placement of statements in the object program that define a mapping of base states to their base state values. In this example, there is only one entry in the baseStatesWithValues map — the base-state ($\emptyset$) maps to the initial value 0.0.

```
StateValueMap baseStatesWithValues = new StateValueMap();
baseStatesWithValues.put(new State(emptySet),new Double(0.0));
```

- Now, the DPInstance object that will hold all the information of a DP instance is created:

```
DPInstance theInstance
  = new DPInstance(goalState,baseStatesWithValues);
```

- The DPFE section of the source program

```
DPFE: f(items)=MIN_{d IN decisionSet}
    { cost(items,d)+f(t(items,d)) };
```

is then parsed. The minimization and addition operators are extracted, resulting in the initialization of the appropriate instance variables of the DPInstance object theInstance.

```
theInstance.setTransitionOperator("+");
theInstance.setMinOrMax("min");
```

This information is used later.

- The code for the algorithm for building the SDRT table is then inserted. The first portion is copied from the system file Main2.txt. A complete listing of Main2.txt is given in Appendix A.2. The step numbers added here are from the above pseudocode.

```
         ///////////////////////////////////////////////////////
         //beginning of the fixed code Main2.txt for Main.java
           // make table entries automatically
           StateSet statesSeen =new StateSet();
           //add goal state to statesSeen
[Step 1]   statesSeen.add(goalState);
           LinkedList stateQueue = new LinkedList();
           //initialize queue with the goal state
[Step 2]   stateQueue.addLast(goalState);
           //loop as long as there is something in the queue
[Step 3]   while (stateQueue.size()>0) {
           //dequeue the first state and make it the current state
[Step 4]   State currentState=(State) stateQueue.removeFirst();
           //check whether the current state is a base state as
           //defined in DPFE_BASE_CONDITIONS section; if so, add it
           //and its value to baseStatesWithValues
```

```
[Step 5] if (determineWhetherBaseState(currentState)==true) {
           baseStatesWithValues.put(currentState,
             new Double(determineBaseStateValue(currentState)));
         }
         //if current state is not a base state...
         if (!baseStatesWithValues.containsKey(currentState)) {
         //now loop over all possible decisions
        //end of the fixed code Main2.txt for Main.java
        ///////////////////////////////////////////////////
```

At this point, information about the decision set and decision variable is needed.

- The decision set is obtained from the DECISION_SPACE section.

```
DECISION_SPACE: decisionSet(items)=items;
```

For this source statement, the following statements are inserted into the object program.

```
         //create the decision set
         NodeSet items=((NodeSet) ((NodeSet)
             currentState.getCoordinate(0)).clone());
         NodeSet decisionSet=items;
         //loop over all decisions now
[Step 6] for (Iterator it=decisionSet.iterator();it.hasNext();) {
           Integer currentDecisionCandidate = (Integer) it.next();
           Decision d
               = new Decision(currentDecisionCandidate.intValue());
```

- The decision variable is obtained from the DECISION_VARIABLE section.

```
DECISION_VARIABLE: int d;
```

For this source statement, the following statement is inserted into the object program. Its purpose is to provide labels for the decisions by setting an instance variable of the Decision object d. These labels are attached to decision arcs in the resulting Bellman net, and in the final SolverCode output.

```
d.setDecisionPrefix("d=");
```

- The remainder of the SDRT generation algorithm is from Main3.txt (see Appendix A.2).

```
///////////////////////////////////////////////////
//beginning of the fixed code Main3.txt for Main.java
        //determine the successor states of current state
        StateList successorStates
            =calculateTransformation(currentState,d);
        //determine the transition weights given the current state
        ArrayList transitionWeights
            =calculateTransitionWeights(currentState,d);
```

```
          //generate table entry
[Step 7]  DPInstanceTableEntry entry = new DPInstanceTableEntry(
                                        currentState,
                                        d,
                                        calculateReward(currentState,d),
                                        successorStates,
                                        transitionWeights);
          theInstance.addTableEntry(entry);

          //enqueue unseen destination states by iterating
          //over successorStates
[Step 8]  for(Iterator it2=successorStates.iterator();
              it2.hasNext();) {
            State currentSuccessor = (State) it2.next();
            if(!statesSeen.contains(currentSuccessor)) {
              stateQueue.addLast(currentSuccessor);
              statesSeen.add(currentSuccessor);//mark state as seen
            }
          }
        } //end of for loop over the decisions
      } //end of if
      else { //base state
        //do nothing
      }
    } //end of loop once queue is empty
[...]
    //build the Bellman PN from the instance
[...]
      BellmanNet bn=theInstance.buildBellmanPNTopDown(problemName);
      //write the BellmanNet as incidence matrix
      Out5.putln(bn.toIncidenceMatrix());
[...]
  //end of the fixed code Main3.txt for Main.java
  //////////////////////////////////////////////////
```

- From the REWARD_FUNCTION section,

```
REWARD_FUNCTION: cost(items,d) = (N+1-size(items)) * prob[d];
```

the function calculateReward() is generated in the object program, as follows.

```
public static double calculateReward(State s, Decision d) {
  double result;
  result=(N+1
      -size(((NodeSet) ((NodeSet) s.getCoordinate(0)).clone())))
      *prob[d.getDecision()];
  return result;
}
```

- From the TRANSFORMATION_FUNCTION section,

```
TRANSFORMATION_FUNCTION: t(items,d)=(items SETMINUS {d});
```

the function calculateTransformation() is generated in the object program, as follows.

```
private static StateList calculateTransformation(State s,
    Decision d) {
  StateList result=new StateList();
  {
  NodeSet items=((NodeSet) ((NodeSet) s.getCoordinate(0)).clone());
  NodeSet _setExplicit2=new NodeSet();
  _setExplicit2.add(new Integer(d.getDecision()));
  items.removeAll(_setExplicit2);
  result.add(new State(items));
  }
  return result;
}
```

This function returns a *list* of states, since for some DP problems such as MCM we have multiple successor states.

- Source code for the methods defined in earlier steps (determineWhether BaseState, determineBaseStateValue, and calculateTransitionWeights if used) are inserted at the end of LINSRCSMain.java.
- The state type declaration in the gDPS source is handled is a special way. For design reasons this declarative information is incorporated into a separate State class in the Java file State.java rather than in LINSRCSMain.java. For example, in LINSRCS.dp, the source statement

```
STATE_TYPE: (Set items);
```

results in a specialized, problem-specific constructor for State objects in State.java as follows:

```
public State(NodeSet items) {
  theState=new ArrayList();
  theState.add(items);
} //end of constructor
```

8.4 Error Detection in DP2PN

The parser module of DP2PN using gDPS as the source language has been tested and is capable of detecting the following syntactical errors in the DP specification.

- No goal state provided – the modeler did not specify what the goal of the computation is for the given instance of the DP problem.
- No or illegal DP functional equation provided.
- No base case for the DP functional equation provided.
- No decision variable for the DP functional equation provided.
- No decision set for the DP functional equation provided.

- No reward function provided. (If the DP functional equation does not contain a reward function, it is still necessary to define a reward function in gDPS, e.g. the constant function $r(s, d) = 0.0$ for additive DP functional equations, or $r(s, d) = 1.0$ for multiplicative DP functional equations.)
- No transformation function provided.

If a "syntax" error is detected the DP2PN module will report an error message and will not attempt to produce the intermediate PN representation. No attempt of error recovery is made. All syntax errors lead to the termination of DP2PN.

"Semantical" errors cannot be detected by the DP2PN module. Some might be detectable subsequently by checking certain net theoretic properties (such as cyclicness and deadlock) of the resulting PN. For all gDPS examples tested so far, the following general observations have been made.

Note 8.1. Given a syntactically correct gDPS source representing a proper DP problem formulation, the DP2PN module produces a PN that is correct in the sense that it produces the optimal objective function value as output.

Note 8.2. Given a syntactically correct but semantically improper gDPS source, the DP2PN module produces a PN that is incorrect in the sense that it does not necessarily produce the optimal objective function value as output.

Some examples of specific errors detectable by an analysis of net properties are the existence of circular definitions and incorrect or omitted base conditions. Certain transformation function errors are also detectable since they lead to cyclicness or deadlock.

9

The PN2Solver Modules

The general terminology "PN2Solver" is used to describe a software component that produces a solver code from the internal Bellman net representation. In Sect. 9.2 we describe PN2Java, one of three components that have been implemented to produce solver code. PN2Java produces solver code in the general programming language Java. Later we describe PN2Spreadsheet (Sect. 9.3), which produces a spreadsheet to be used by spreadsheet applications, and PN2XML (Sect. 9.4), which produces a file in the standardized Petri net exchange format PNML.

As an alternative to the presented PN2Solver modules one could envision a module that transforms the Bellman net directly into machine code (for applications that require the utmost performance at the expense of machine independence). Computations that can take place concurrently are identified by transitions that are enabled concurrently; these computations could be assigned to parallel processors for a further gain in performance.

9.1 The Solver Code Generation Process

In Chap. 8, we showed how a Bellman net representation of a dynamic programming problem can be obtained. In this chapter, we show how we can automatically generate code that numerically solves the problem. We will assume here that we are given the Bellman net representation of a dynamic programming problem in the form of the incidence matrix descibed in Sect. 7.2. For example, for the linear search example, this matrix representation was given as the following `LINSRCSBN.csv` file, shown here aligned in columns.

We emphasize that this matrix is a description of the DPFE rather than an executable code that computes its solution. The remaining problem is to numerically solve the DPFE given this Bellman net specification.

```
             TNAMES
PT PNAMES    mt1 st1 st2 st3 mt2 st4 st5 mt3 st6 st7 mt4 st8 st9 mt5 st10mt6 st11mt7 st12INIT/DEC
s  ({0,1,2}) 1   0   0   0   0   0   0   0   0   0   0   0   0   0   0   0   0   0   0
i  p1        -1  1   0   0   0   0   0   0   0   0   0   0   0   0   0   0   0   0   0   d=0
```

A. Lew and H. Mauch: *The PN2Solver Modules*, Studies in Computational Intelligence (SCI) **38**, 271–289 (2007)
www.springerlink.com

```
s  ({1,2})    0  -1  0   0   1   0   0   0   0   0   0   0   0   0   0   0   0   0   0
i  p2        -1   0  1   0   0   0   0   0   0   0   0   0   0   0   0   0   0   0   0   d=1
s  ({0,2})    0   0 -1   0   0   0   0   1   0   0   0   0   0   0   0   0   0   0   0
i  p3        -1   0  0   1   0   0   0   0   0   0   0   0   0   0   0   0   0   0   0   d=2
s  ({0,1})    0   0  0  -1   0   0   0   0   0   0   1   0   0   0   0   0   0   0   0
i  p4         0   0  0   0  -1   1   0   0   0   0   0   0   0   0   0   0   0   0   0   d=1
s  ({2})      0   0  0   0   0  -1   0   0  -1   0   0   0   0   1   0   0   0   0   0
i  p5         0   0  0   0  -1   0   1   0   0   0   0   0   0   0   0   0   0   0   0   d=2
s  ({1})      0   0  0   0   0   0  -1   0   0   0   0  -1   0   0   0   1   0   0   0
i  p6         0   0  0   0   0   0   0  -1   1   0   0   0   0   0   0   0   0   0   0   d=0
i  p7         0   0  0   0   0   0   0  -1   0   1   0   0   0   0   0   0   0   0   0   d=2
s  ({0})      0   0  0   0   0   0   0   0   0  -1   0   0  -1   0   0   0   0   1   0
i  p8         0   0  0   0   0   0   0   0   0   0  -1   1   0   0   0   0   0   0   0   d=0
i  p9         0   0  0   0   0   0   0   0   0   0  -1   0   1   0   0   0   0   0   0   d=1
i  p10        0   0  0   0   0   0   0   0   0   0   0   0   0  -1   1   0   0   0   0   d=2
s  ({})       0   0  0   0   0   0   0   0   0   0   0   0   0   0  -1   0  -1   0  -1   0
i  p11        0   0  0   0   0   0   0   0   0   0   0   0   0   0   0  -1   1   0   0   d=1
i  p12        0   0  0   0   0   0   0   0   0   0   0   0   0   0   0   0   0  -1   1   d=0
TRANSTYPE    min  +  +   +  min  +   +  min  +   +  min  +   +  min  +  min  +  min  +
ETRANSCONST      0.2 0.5 0.3     1  0.6     0.4 0.6     0.4  1      0.9     1.5     0.6
```

We start by first topologically sorting this Bellman net, resulting in the following list of states.

`(base-state,st10,mt5,st4,st6,st11,mt6,st5,st8,mt2,st1,st12, mt7,st7,st9,mt3,st2,mt4,st3,mt1).`

Note that the base state is first and the goal state is last. Permuting the rows and columns of the incidence matrix in this topological order, we obtain the following equivalent matrix.

```
                     1   2   3   4   5   6   7   8   9   10  11  12  13  14  15  16  17  18  19  20
    PT  PN \ TN  INIT/DEC        st10        mt5 st4 st6 st11    mt6 st5 st8 mt2 st1 st12    mt7 st7 st9 mt3 st2 mt4 st3 mt1
1   s   ({})         0   -1  0   0   0   -1  0   0   0   0   0   -1  0   0   0   0   0   0   0   0
2   i   p10 d=2  1   -1  0   0   0   0   0   0   0   0   0   0   0   0   0   0   0   0   0
3   s   ({2})            0   1   -1  -1  0   0   0   0   0   0   0   0   0   0   0   0   0   0   0
4   i   p4  d=1  0   0   1   0   0   0   0   0   -1  0   0   0   0   0   0   0   0   0   0
5   i   p6  d=0  0   0   0   1   0   0   0   0   0   0   0   0   0   0   -1  0   0   0   0
6   i   p11 d=1  0   0   0   0   1   -1  0   0   0   0   0   0   0   0   0   0   0   0   0
7   s   ({1})            0   0   0   0   0   1   -1  -1  0   0   0   0   0   0   0   0   0   0   0
8   i   p5  d=2  0   0   0   0   0   0   1   0   -1  0   0   0   0   0   0   0   0   0   0
9   i   p8  d=0  0   0   0   0   0   0   0   1   0   0   0   0   0   0   0   0   -1  0   0
10  s   ({1,2})          0   0   0   0   0   0   0   0   1   -1  0   0   0   0   0   0   0   0   0
11  i   p1  d=0  0   0   0   0   0   0   0   0   0   1   0   0   0   0   0   0   0   0   -1
12  i   p12 d=0  0   0   0   0   0   0   0   0   0   0   1   -1  0   0   0   0   0   0   0
13  s   ({0})            0   0   0   0   0   0   0   0   0   0   0   0   1   -1  -1  0   0   0   0   0
14  i   p7  d=2  0   0   0   0   0   0   0   0   0   0   0   0   1   0   -1  0   0   0   0
15  i   p9  d=1  0   0   0   0   0   0   0   0   0   0   0   0   0   1   0   0   -1  0   0
16  s   ({0,2})          0   0   0   0   0   0   0   0   0   0   0   0   0   0   0   1   -1  0   0   0
17  i   p2  d=1  0   0   0   0   0   0   0   0   0   0   0   0   0   0   0   1   0   0   -1
18  s   ({0,1})          0   0   0   0   0   0   0   0   0   0   0   0   0   0   0   0   0   1   -1  0
19  i   p3  d=2  0   0   0   0   0   0   0   0   0   0   0   0   0   0   0   0   0   1   -1
20  s   ({0,1,2})            0   0   0   0   0   0   0   0   0   0   0   0   0   0   0   0   0   0   0   1
        TRANSTYPE            +   min +   +   +   min +   +   min +   +   min +   +   min +   min +   min
        ETRANSCONST          0.9     1   0.4 1.5     0.6 0.4     0.2 0.6     0.6 1       0.5     0.3
```

Note that the connectivity portion of the matrix has an upper triangular form. Solver code can be generated for each transition by processing the matrix columnwise from left to right, generating either a minimization operation if the transition type is a min (i.e. an M-transition) or an addition operation if the transition type is a + (i.e. an E-transition), as specified in the next-to-last row. The input operands of this operation are indicated by a -1 entry in the column, whereas the output result operand is indicated by a +1 entry; an addition operation also has an input operand that is found in the last row of the matrix. Each transition operation can have several inputs but only one output.

Following the above procedure, we can generate solver code in the form of the following sequence of assignment statements. (The reason for naming the temporary variables "Bi" will be made clear later.) Each statement evaluates a transition (mininimization or expression) operation, or initializes a base-state.

```
B1  = 0
B2  = B1+0.9
B3  = min(B2)
B4  = B3+1.0
B5  = B3+0.4
B6  = B1+1.5
B7  = min(B6)
B8  = B7+0.6
B9  = B7+0.4
B10 = min(B4,B8)
B11 = B10+0.2
B12 = B1+0.6
B13 = min(B12)
B14 = B13+0.6
B15 = B13+1.0
B16 = min(B5,B14)
B17 = B16+0.5
B18 = min(B9,B15)
B19 = B18+0.3
B20 = min(B11,B17,B19)
```

It should be emphasized that the upper triangular form of the Bellman net matrix (based on the topological ordering) ensures that the generated Java code evaluates the expressions in a correct order, so that each statement only involves variables of prior statements. We note also that, as a consequence, the resulting code computes overlapping subproblems exactly once. Execution of this solver code using a conventional programming language, such as Java, will result in the final variable B20 being set equal to the desired goal (of 1.7). However, since we are generally interested in intermediate results, including knowledge of what the optimal decisions are, we will generate solver code in a different form. Details of this are discussed in the next section.

9.2 The PN2Java Module

At the end of Sect. 9.1, we showed how a simple Java program can be automatically generated from a Bellman net. Execution of that Java program will result in the computation of the goal, i.e., the value of the optimal solution, but *reconstruction* of the optimal *policy*, i.e., the series of optimal decisions leading to the goal, is not automated. Therefore, rather than generating the foregoing Java program, we generate a Java program that invokes specially designed routines, called by `CalculationObjects`, that perform the same operations but save sufficient information so that when the goal is reached, the reconstruction process automatically occurs and is recorded in a solution file.

The PN2Java module of DP2PN2Solver takes a Bellman Net as its input, and produces Java code as its output. The Java code efficiently produces the optimal solution to the underlying DP problem instance, along with optimal policy. Note the use of the term "efficient" in this context. If the underlying DP problem is inefficient (or intractable) — as in the TSP example — then PN2Java will produce an output inefficiently.

The selection of Java as the output solver code is somewhat arbitrary. Any other solver code that can be made to execute on a computer would serve as well. Our choice of Java takes advantage of the ubiquitous and free availability of the Java compile and run-time system. PN2Java also ensures that the produced code is well commented and human-readable, so it is possible to verify its correctness, or use the code as a building block for a larger software development project.

9.2.1 Java Solver Code Calculation Objects

The PN2Java module produces Java source code that solves the modeled DP problem instance. The automatically produced source code for some of the problem instances is listed in this section. In addition, the following rudimentary Java classes have been designed and implemented to support the automatically produced sources.

- `Out` merely facilitates simultaneous output to screen and file.
- `CalculationObject` represents the values associated with places encountered in the course of the computation. This class has the following subclasses.
  - `ECalculationObject` for computing the value of intermediate places
  - `MCalculationObject` for computing the value of state places that have a MTransition in their preset.

  Note that for base state places, we directly use `CalculationObject` instances.

For example, if state #36 is a base state with label "({})" and initial value 0.0, the following statement would be generated.

```
CalculationObject I36 = new CalculationObject("({})",0.0);
```

If a transition is an additive E-transition, whose (in this case, single) input is state #13, whose constant value is 0.3, whose output (successor) is state #11, and which is associated with a decision with label "d=2", the following statements would be generated.

```
operandList=new ArrayList();
operandList.add(I13);
ECalculationObject I11 = new ECalculationObject(
    "d=2", I13.getValue()+0.3, operandList);
```

If a transition is a minimizing M-transition, whose inputs are states #3, #7 and #11, whose output (successor) is state #1, and which has the label "({0,1,2})", the following statements would be generated.

```
argumentList=new ArrayList();
argumentList.add(I3);
argumentList.add(I7);
argumentList.add(I11);
MCalculationObject I1 = new MCalculationObject("({0,1,2})");
I1.performMinimization(argumentList);
```

For the nonserial MCM problem (Sect. 2.27), there are multiple next-states. The following excerpt from the automatically produced Java code illustrates this generalization.

```
List argumentList; //reusable temporary variable
                   //used for min/max
List operandList; //reusable temporary variable
                  //for eval. (Etransitions)
//code to initialize base state place variables
CalculationObject I5 = new CalculationObject("(1,1)",0.0);
CalculationObject I20 = new CalculationObject("(4,4)",0.0);
CalculationObject I25 = new CalculationObject("(2,2)",0.0);
CalculationObject I36 = new CalculationObject("(3,3)",0.0);

//code that simulates the transitions firing
operandList=new ArrayList();
operandList.add(I5);
operandList.add(I25);
ECalculationObject I32 = new ECalculationObject(
    "k=1",
    I5.getValue()+I25.getValue()+60.0,
    operandList);

argumentList=new ArrayList();
argumentList.add(I32);
MCalculationObject I12 = new MCalculationObject("(1,2)");
I12.performMinimization(argumentList);
```

For the INVEST problem (section 2.18), transition weights appear as factors in the construction of `ECalculationObjects`. The following excerpt from the automatically produced Java code illustrates this generalization.

```
operandList=new ArrayList();
operandList.add(I27);
ECalculationObject I25 = new ECalculationObject(
    "null",
    I27.getValue()*0.6,
    operandList);

operandList=new ArrayList();
operandList.add(I21);
operandList.add(I25);
ECalculationObject I19 = new ECalculationObject(
    "d=2",
    I21.getValue()+I25.getValue()+0.0,
    operandList);
```

9.2.2 Java Solver Code for LINSRCS

For the linear search example, for which the topologically sorted Bellman net matrix was given in Sect. 9.1, the following Java solver code would be generated and saved in a file with name `LINSRCSJavaSolver.java`.

```
//This file is automatically produced
//using the method toJava() from class BellmanNet
import pn2java.*;
import java.io.*;
import java.util.*;
public class linsrcsJavaSolver {
  public static void main(String[] args) throws IOException {
    final String subDirName="LINSRCSSolverCode";
    String currentWorkingDir=System.getProperty("user.dir");
    if(!currentWorkingDir.endsWith(subDirName)) {
          currentWorkingDir=currentWorkingDir+"/"+subDirName;
    }
    Out.pw=new PrintWriter(new FileWriter(
        new File(currentWorkingDir+"/"
              +"LINSRCSSolutionTree.txt")));
    List argumentList; //reusable temporary variable
                       //used for min/max
    List operandList; //reusable temporary variable
                      //for eval. (Etransitions)
    //code to initialize base state place variables
    CalculationObject I35 = new CalculationObject("({})",0.0);

    //code that simulates the transitions firing
    operandList=new ArrayList();
    operandList.add(I35);
    ECalculationObject I33 = new ECalculationObject(
        "d=2", I35.getValue()+0.8999999999999999, operandList);

    argumentList=new ArrayList();
    argumentList.add(I33);
    MCalculationObject I17 = new MCalculationObject("({2})");
    I17.performMinimization(argumentList);

    operandList=new ArrayList();
    operandList.add(I17);
    ECalculationObject I15 = new ECalculationObject(
        "d=1", I17.getValue()+1.0, operandList);

    operandList=new ArrayList();
    operandList.add(I17);
    ECalculationObject I23 = new ECalculationObject(
        "d=0", I17.getValue()+0.4, operandList);

    operandList=new ArrayList();
    operandList.add(I35);
    ECalculationObject I38 = new ECalculationObject(
        "d=1", I35.getValue()+1.5, operandList);

    argumentList=new ArrayList();
    argumentList.add(I38);
    MCalculationObject I21 = new MCalculationObject("({1})");
    I21.performMinimization(argumentList);

    operandList=new ArrayList();
    operandList.add(I21);
    ECalculationObject I19 = new ECalculationObject(
        "d=2", I21.getValue()+0.6, operandList);

    operandList=new ArrayList();
    operandList.add(I21);
    ECalculationObject I29 = new ECalculationObject(
        "d=0", I21.getValue()+0.4, operandList);

    argumentList=new ArrayList();
    argumentList.add(I15);
    argumentList.add(I19);
    MCalculationObject I5 = new MCalculationObject("({1,2})");
    I5.performMinimization(argumentList);

    operandList=new ArrayList();
    operandList.add(I5);
    ECalculationObject I3 = new ECalculationObject(
        "d=0", I5.getValue()+0.2, operandList);

    operandList=new ArrayList();
    operandList.add(I35);
    ECalculationObject I40 = new ECalculationObject(
        "d=0", I35.getValue()+0.6000000000000001, operandList);
```

```
    argumentList=new ArrayList();
    argumentList.add(I40);
    MCalculationObject I27 = new MCalculationObject("({0})");
    I27.performMinimization(argumentList);

    operandList=new ArrayList();
    operandList.add(I27);
    ECalculationObject I25 = new ECalculationObject(
        "d=2", I27.getValue()+0.6, operandList);

    operandList=new ArrayList();
    operandList.add(I27);
    ECalculationObject I31 = new ECalculationObject(
        "d=1", I27.getValue()+1.0, operandList);

    argumentList=new ArrayList();
    argumentList.add(I23);
    argumentList.add(I25);
    MCalculationObject I9 = new MCalculationObject("({0,2})");
    I9.performMinimization(argumentList);

    operandList=new ArrayList();
    operandList.add(I9);
    ECalculationObject I7 = new ECalculationObject(
        "d=1", I9.getValue()+0.5, operandList);

    argumentList=new ArrayList();
    argumentList.add(I29);
    argumentList.add(I31);
    MCalculationObject I13 = new MCalculationObject("({0,1})");
    I13.performMinimization(argumentList);

    operandList=new ArrayList();
    operandList.add(I13);
    ECalculationObject I11 = new ECalculationObject(
        "d=2", I13.getValue()+0.3, operandList);

    argumentList=new ArrayList();
    argumentList.add(I3);
    argumentList.add(I7);
    argumentList.add(I11);
    MCalculationObject I1 = new MCalculationObject("({0,1,2})");
    I1.performMinimization(argumentList);

    //code that gives us the final answer
    Out.put("The optimal value is: ");
    Out.putln(I1.getValue());
    Out.putln("The solution tree is: ");
    Out.putln(I1.predecessorSubtree(0));

    Out.pw.close(); //close the output stream
  } //end of main()
} //end of class
```

When the program `LINSRCSJavaSolver.java` is compiled to `LINSRCSJavaSolver.class` and executed, the output is placed in the output file `LINSRCSSolutionTree.txt`, which is as follows.

```
The optimal value is: 1.7000000000000002
The solution tree is:
State ({0,1,2}) has optimal value: 1.7000000000000002
 Decision d=1
  State ({0,2}) has optimal value: 1.2000000000000002
   Decision d=2
    State ({0}) has optimal value: 0.6000000000000001
     Decision d=0
      Base state ({}) has initial value: 0.0
```

The optimal solution is 1.7 obtained by making the sequence of decisions 1, 2, and 0, with costs 0.5, 0.6, and 0.6, respectively.

9.2.3 Java Solver Code for LSP

```
//This file is automatically produced.
//using the method toJava() from class BellmanNet
import pn2java.*;
import java.io.*;
import java.util.*;
public class LSPJavaSolver {
  public static void main(String[] args) throws IOException {
    //Out writes to screen and file at the same time
    Out.pw=new PrintWriter(new FileWriter(new File("LSPSolutionTree.txt")));
    List argumentList; //reusable temporary variable used for min/max
    List operandList; //reusable temporary variable for eval. (Etransitions)
    //code to initialize base state place variables
    CalculationObject I9 = new CalculationObject("({0,3},3)",0.0);
    CalculationObject I18 = new CalculationObject("({0,1,2,3},3)",0.0);

    //code that simulates the transitions firing
    operandList=new ArrayList();
    operandList.add(I9);
    ECalculationObject I7 = new ECalculationObject(
        "alpha=3",
        I9.getValue()+1.0,
        operandList);

    operandList=new ArrayList();
    operandList.add(I18);
    ECalculationObject I16 = new ECalculationObject(
        "alpha=3",
        I18.getValue()+1.0,
        operandList);

    argumentList=new ArrayList();
    argumentList.add(I16);
    MCalculationObject I14 = new MCalculationObject("({0,1,2},2)");
    I14.performMaximization(argumentList);

    operandList=new ArrayList();
    operandList.add(I14);
    ECalculationObject I12 = new ECalculationObject(
        "alpha=2",
        I14.getValue()+1.0,
        operandList);

    argumentList=new ArrayList();
    argumentList.add(I12);
    MCalculationObject I5 = new MCalculationObject("({0,1},1)");
    I5.performMaximization(argumentList);

    operandList=new ArrayList();
    operandList.add(I5);
    ECalculationObject I3 = new ECalculationObject(
        "alpha=1",
        I5.getValue()+1.0,
        operandList);

    argumentList=new ArrayList();
    argumentList.add(I3);
    argumentList.add(I7);
    MCalculationObject I1 = new MCalculationObject("({0},0)");
    I1.performMaximization(argumentList);

    //code that gives us the final answer
    Out.put("The optimal value is: ");
    Out.putln(I1.getValue());
    Out.putln("The solution tree is: ");
    Out.putln(I1.predecessorSubtree(0));

    Out.pw.close(); //close the output stream
  } //end of main()
} //end of class
```

9.2.4 Java Solver Code for MCM

```
//This file is automatically produced.
//using the method toJava() from class BellmanNet
import pn2java.*;
import java.io.*;
import java.util.*;
public class MCMJavaSolver {
  public static void main(String[] args) throws IOException {
    //Out writes to screen and file at the same time
    Out.pw=new PrintWriter(new FileWriter(new File("MCMSolutionTree.txt")));
    List argumentList; //reusable temporary variable used for min/max
```

```
List operandList; //reusable temporary variable for eval. (Etransitions)
//code to initialize base state place variables
CalculationObject I5 = new CalculationObject("(1,1)",0.0);
CalculationObject I20 = new CalculationObject("(4,4)",0.0);
CalculationObject I25 = new CalculationObject("(2,2)",0.0);
CalculationObject I36 = new CalculationObject("(3,3)",0.0);

//code that simulates the transitions firing
operandList=new ArrayList();
operandList.add(I5);
operandList.add(I25);
ECalculationObject I32 = new ECalculationObject(
    "k=1",
    I5.getValue()+I25.getValue()+60.0,
    operandList);

argumentList=new ArrayList();
argumentList.add(I32);
MCalculationObject I12 = new MCalculationObject("(1,2)");
I12.performMinimization(argumentList);

operandList=new ArrayList();
operandList.add(I20);
operandList.add(I36);
ECalculationObject I34 = new ECalculationObject(
    "k=3",
    I20.getValue()+I36.getValue()+20.0,
    operandList);

argumentList=new ArrayList();
argumentList.add(I34);
MCalculationObject I14 = new MCalculationObject("(3,4)");
I14.performMinimization(argumentList);

operandList=new ArrayList();
operandList.add(I12);
operandList.add(I14);
ECalculationObject I10 = new ECalculationObject(
    "k=2",
    I12.getValue()+I14.getValue()+30.0,
    operandList);

operandList=new ArrayList();
operandList.add(I14);
operandList.add(I25);
ECalculationObject I23 = new ECalculationObject(
    "k=2",
    I14.getValue()+I25.getValue()+40.0,
    operandList);

operandList=new ArrayList();
operandList.add(I12);
operandList.add(I36);
ECalculationObject I41 = new ECalculationObject(
    "k=2",
    I12.getValue()+I36.getValue()+30.0,
    operandList);

operandList=new ArrayList();
operandList.add(I25);
operandList.add(I36);
ECalculationObject I43 = new ECalculationObject(
    "k=2",
    I25.getValue()+I36.getValue()+40.0,
    operandList);

argumentList=new ArrayList();
argumentList.add(I43);
MCalculationObject I30 = new MCalculationObject("(2,3)");
I30.performMinimization(argumentList);

operandList=new ArrayList();
operandList.add(I20);
operandList.add(I30);
ECalculationObject I28 = new ECalculationObject(
    "k=3",
    I20.getValue()+I30.getValue()+16.0,
    operandList);

operandList=new ArrayList();
operandList.add(I5);
operandList.add(I30);
ECalculationObject I39 = new ECalculationObject(
    "k=1",
    I5.getValue()+I30.getValue()+24.0,
    operandList);

argumentList=new ArrayList();
argumentList.add(I23);
```

```
    argumentList.add(I28);
    MCalculationObject I8 = new MCalculationObject("(2,4)");
    I8.performMinimization(argumentList);

    operandList=new ArrayList();
    operandList.add(I5);
    operandList.add(I8);
    ECalculationObject I3 = new ECalculationObject(
        "k=1",
        I5.getValue()+I8.getValue()+24.0,
        operandList);

    argumentList=new ArrayList();
    argumentList.add(I39);
    argumentList.add(I41);
    MCalculationObject I18 = new MCalculationObject("(1,3)");
    I18.performMinimization(argumentList);

    operandList=new ArrayList();
    operandList.add(I18);
    operandList.add(I20);
    ECalculationObject I16 = new ECalculationObject(
        "k=3",
        I18.getValue()+I20.getValue()+12.0,
        operandList);

    argumentList=new ArrayList();
    argumentList.add(I3);
    argumentList.add(I10);
    argumentList.add(I16);
    MCalculationObject I1 = new MCalculationObject("(1,4)");
    I1.performMinimization(argumentList);

    //code that gives us the final answer
    Out.put("The optimal value is: ");
    Out.putln(I1.getValue());
    Out.putln("The solution tree is: ");
    Out.putln(I1.predecessorSubtree(0));

    Out.pw.close(); //close the output stream
  } //end of main()
} //end of class
```

9.2.5 Java Solver Code for SPA

```
//This file is automatically produced.
//using the method toJava() from class BellmanNet
import pn2java.*;
import java.io.*;
import java.util.*;
public class SPAJavaSolver {
  public static void main(String[] args) throws IOException {
    //Out writes to screen and file at the same time
    Out.pw=new PrintWriter(new FileWriter(new File("SPASolutionTree.txt")));
    List argumentList; //reusable temporary variable used for min/max
    List operandList; //reusable temporary variable for eval. (Etransitions)
    //code to initialize base state place variables
    CalculationObject I15 = new CalculationObject("(3)",0.0);

    //code that simulates the transitions firing
    operandList=new ArrayList();
    operandList.add(I15);
    ECalculationObject I13 = new ECalculationObject(
        "d=3",
        I15.getValue()+8.0,
        operandList);

    operandList=new ArrayList();
    operandList.add(I15);
    ECalculationObject I18 = new ECalculationObject(
        "d=3",
        I15.getValue()+5.0,
        operandList);

    argumentList=new ArrayList();
    argumentList.add(I18);
    MCalculationObject I9 = new MCalculationObject("(2)");
    I9.performMinimization(argumentList);

    operandList=new ArrayList();
    operandList.add(I9);
    ECalculationObject I7 = new ECalculationObject(
        "d=2",
        I9.getValue()+5.0,
        operandList);
```

```
    operandList=new ArrayList();
    operandList.add(I9);
    ECalculationObject I11 = new ECalculationObject(
        "d=2",
        I9.getValue()+1.0,
        operandList);

    argumentList=new ArrayList();
    argumentList.add(I11);
    argumentList.add(I13);
    MCalculationObject I5 = new MCalculationObject("(1)");
    I5.performMinimization(argumentList);

    operandList=new ArrayList();
    operandList.add(I5);
    ECalculationObject I3 = new ECalculationObject(
        "d=1",
        I5.getValue()+3.0,
        operandList);

    argumentList=new ArrayList();
    argumentList.add(I3);
    argumentList.add(I7);
    MCalculationObject I1 = new MCalculationObject("(0)");
    I1.performMinimization(argumentList);

    //code that gives us the final answer
    Out.put("The optimal value is: ");
    Out.putln(I1.getValue());
    Out.putln("The solution tree is: ");
    Out.putln(I1.predecessorSubtree(0));

    Out.pw.close(); //close the output stream
  } //end of main()
} //end of class
```

9.3 The PN2Spreadsheet Module

Spreadsheet formats are quite popular in the OR community and in management science. In this section, we describe the code generator PN2Spreadsheet that produces solver code in a spreadsheet language. Numerical solutions can then be obtained by use of any spreadsheet system, such as Excel.

For a small problem of fixed size, for which a program consisting of a sequence of assignment statements that solve a DP problem can be given, it is easy to express this program in spreadsheet form instead. For example, for the linear search problem of size $N = 3$, as given in Sect. 2.24, the following spreadsheet solves the problem.

```
   | A                       | B                  | C            | D  | E
---+-------------------------+--------------------+--------------+----+----
1  |=min(A5+B1,A6+B2,A7+B3)  |=min(B5+C1,B6+C2)   |=min(C5+D1)   |0   |
2  |                         |=min(B7+C1,B8+C3)   |=min(C6+D1)   |    |
3  |                         |=min(B9+C2,B10+C3)  |=min(C7+D1)   |    |
4  |                         |                    |              |    |
5  |.2                       |1.                  |.9            |    |
6  |.5                       |.6                  |1.5           |    |
7  |.3                       |.4                  |.6            |    |
8  |                         |.6                  |              |    |
9  |                         |.4                  |              |    |
10 |                         |1.                  |              |    |
11 |                         |                    |              |    |
```

When N is a variable, or is just a large constant, a spreadsheet solution is much more complex than for this simple example. Examples of such spreadsheets are given in operations research textbooks. Essentially, so that the formulas in all of the cells do not have to be entered individually, copying of formulas from one set of cells to another, possibly modified, may be necessary.

Furthermore, use of table lookup facilities (HLOOKUP and VLOOKUP) may also be necessary. For example, the following formula appears in [21]:

```
=HLOOKUP(I$17,$B$1:$H$4,$A18+1)
+HLOOKUP(I$66-I$17,$B$10:$H$14,#A18+1)
```

More complicated examples are given in [21, 63]; see also [50].

Because composing spreadsheets with such formulas is a complicated task, and is an extremely error-prone process with no easy way to debug the result, using spreadsheets to solve DP problems is not likely to become very useful in practice unless the spreadsheet generation process can be automated. We discuss how our software tool based on Bellman nets can be used to achieve such automation.

Spreadsheet solver code can be generated by following the procedure to obtain the sequence of assignment statements given at the end of Sect. 9.1. The righthand sides of these assignment statements would become the second column of a spreadsheet. The first column may contain declarative information that is not used in the numerical solution, such as a textstring copy of the formulas in the second column, as shown here. (The formulas are shown quoted, but these quotes may be omitted.)

```
   | A                        | B
---+--------------------------+------
 1  "0"                        =0
 2  "B1+0.9"                   =B1+0.9
 3  "min(B2)"                  =min(B2)
 4  "B3+1.0"                   =B3+1.0
 5  "B3+0.4"                   =B3+0.4
 6  "B1+1.5"                   =B1+1.5
 7  "min(B6)"                  =min(B6)
 8  "B7+0.6"                   =B7+0.6
 9  "B7+0.4"                   =B7+0.4
10  "min(B4,B8)"               =min(B4,B8)
11  "B10+0.2"                  =B10+0.2
12  "B1+0.6"                   =B1+0.6
13  "min(B12)"                 =min(B12)
14  "B13+0.6"                  =B13+0.6
15  "B13+1.0"                  =B13+1.0
16  "min(B5,B14)"              =min(B5,B14)
17  "B16+0.5"                  =B16+0.5
18  "min(B9,B15)"              =min(B9,B15)
19  "B18+0.3"                  =B18+0.3
20  "min(B11,B17,B19)"         =min(B11,B17,B19)
```

In our implementation of a spreadsheet solver code generator, we essentially produce the second column as shown above, but produce in the first column some information that relates the spreadsheet to its associated Bellman net. This spreadsheet solver code is placed in a `.csv` file.

9.3.1 PN2Spreadsheet Solver Code for LINSRCS

The PN2Spreadsheet module produces a `.csv` spreadsheet file from the Bellman net representation. Column A contains the descriptive names of the places of the PN. Since descriptions may contain commas, which can lead to a confusion with the comma separator, the descriptions are quoted. Column B contains the initial values for base state places as constants and cell formulas for the other places. In particular, minimization/maximization expressions (quoted since they might contain commas) appear for non-base state places to compute the output of M-transitions, and evaluation expressions appear for intermediate places to compute the output of E-transitions.

The `.csv` output `LINSRCS.csv` for the LINSRCS instance from Sect. 2.24 is shown below, listed as a text file. Each line of this text file corresponds to one of the place nodes of the Bellman net, and becomes a row of the spreadsheet. It includes a declaration (including a place node label), followed by either a constant for base-state places or an expression for the output places of transitions. The righthand sides of the assignment statements generated as described in the prior section simply become the second column (Column B, hence our use of the variable names `Bi` in the above) of the spreadsheet. The declarative place node information is not used in the numerical solution. (It contains labels used in the Petri net solver code.)

```
"statePlaceI35 ({})" ,0.0
"intermediatePlaceI33 p10" ,=B1+0.8999999999999999
"statePlaceI17 ({2})" ,"=min(B2)"
"intermediatePlaceI15 p4" ,=B3+1.0
"intermediatePlaceI23 p6" ,=B3+0.4
"intermediatePlaceI38 p11" ,=B1+1.5
"statePlaceI21 ({1})" ,"=min(B6)"
"intermediatePlaceI19 p5" ,=B7+0.6
"intermediatePlaceI29 p8" ,=B7+0.4
"statePlaceI5 ({1,2})" ,"=min(B4,B8)"
"intermediatePlaceI3 p1" ,=B10+0.2
"intermediatePlaceI40 p12" ,=B1+0.6000000000000001
"statePlaceI27 ({0})" ,"=min(B12)"
"intermediatePlaceI25 p7" ,=B13+0.6
"intermediatePlaceI31 p9" ,=B13+1.0
"statePlaceI9 ({0,2})" ,"=min(B5,B14)"
"intermediatePlaceI7 p2" ,=B16+0.5
"statePlaceI13 ({0,1})" ,"=min(B9,B15)"
"intermediatePlaceI11 p3" ,=B18+0.3
"statePlaceI1 ({0,1,2})" ,"=min(B11,B17,B19)"
```

The first column contains (e.g., `statePlaceI35` or `intermediatePlaceI33`) followed by a node label (e.g., a state description like `({})` for state places or a name like `p10` for intermediate places). Place ID numbers (e.g., 35 and 33) are internal sequence numbers; place labels (e.g., `({})` or `p10`) correspond to those in the `LINSRCSBN.csv` file.

The second column contains the formulas associated with the operations of the M-transition and E-transition nodes of the Bellman net. Evaluation of the formulas in Column B of the `LINSRCS.csv` file, by importing the file into a spreadsheet application, yields the following numerical solution.

```
   | A                        | B
---+--------------------------+------
 1  statePlaceI35 ({})         0
 2  intermediatePlaceI33 p10   0.9
 3  statePlaceI17 ({2})        0.9
 4  intermediatePlaceI15 p4    1.9
 5  intermediatePlaceI23 p6    1.3
 6  intermediatePlaceI38 p11   1.5
 7  statePlaceI21 ({1})        1.5
 8  intermediatePlaceI19 p5    2.1
 9  intermediatePlaceI29 p8    1.9
10  statePlaceI5 ({1,2})       1.9
11  intermediatePlaceI3 p1     2.1
12  intermediatePlaceI40 p12   0.6
13  statePlaceI27 ({0})        0.6
14  intermediatePlaceI25 p7    1.2
15  intermediatePlaceI31 p9    1.6
16  statePlaceI9 ({0,2})       1.2
17  intermediatePlaceI7 p2     1.7
18  statePlaceI13 ({0,1})      1.6
19  intermediatePlaceI11 p3    1.9
20  statePlaceI1 ({0,1,2})     1.7
```

A screenshot of the execution of this spreadsheet is shown in Fig. 11.1. The final answer, i.e., the goal, is in cell B20 and has value 1.7. We discuss how the optimal decisions can be found in Chap. 11.

In addition to the LINSRCS example, we provide the PN2Spreadsheet solver code for some other sample problems in the following sections.

9.3.2 PN2Spreadsheet Solver Code for Other Examples

The output file for the LSP instance from section 2.26 is as follows.

```
"statePlaceI9 ({0,3},3)" ,0.0
"statePlaceI18 ({0,1,2,3},3)" ,0.0
"intermediatePlaceI7 p2" ,=B1+1.0
"intermediatePlaceI16 p4" ,=B2+1.0
"statePlaceI14 ({0,1,2},2)" ,"=max(B4)"
"intermediatePlaceI12 p3" ,=B5+1.0
"statePlaceI5 ({0,1},1)" ,"=max(B6)"
"intermediatePlaceI3 p1" ,=B7+1.0
"statePlaceI1 ({0},0)" ,"=max(B8,B3)"
```

The output file for the MCM instance from section 2.27 is as follows.

```
"statePlaceI5 (1,1)" ,0.0
"statePlaceI20 (4,4)" ,0.0
"statePlaceI25 (2,2)" ,0.0
"statePlaceI36 (3,3)" ,0.0
"intermediatePlaceI32 p6" ,=B1+B3+60.0
"statePlaceI12 (1,2)" ,"=min(B5)"
"intermediatePlaceI34 p7" ,=B2+B4+20.0
"statePlaceI14 (3,4)" ,"=min(B7)"
"intermediatePlaceI10 p2" ,=B6+B8+30.0
"intermediatePlaceI23 p4" ,=B8+B3+40.0
"intermediatePlaceI41 p9" ,=B6+B4+30.0
"intermediatePlaceI43 p10" ,=B3+B4+40.0
"statePlaceI30 (2,3)" ,"=min(B12)"
"intermediatePlaceI28 p5" ,=B2+B13+16.0
"intermediatePlaceI39 p8" ,=B1+B13+24.0
"statePlaceI8 (2,4)" ,"=min(B10,B14)"
"intermediatePlaceI3 p1" ,=B1+B16+24.0
"statePlaceI18 (1,3)" ,"=min(B15,B11)"
"intermediatePlaceI16 p3" ,=B18+B2+12.0
"statePlaceI1 (1,4)" ,"=min(B17,B9,B19)"
```

The output file for the SPA instance from section 2.43 is as follows.

```
"statePlaceI15 (3)" ,0.0
"intermediatePlaceI13 p4" ,=B1+8.0
"intermediatePlaceI18 p5" ,=B1+5.0
"statePlaceI9 (2)" ,"=min(B3)"
"intermediatePlaceI7 p2" ,=B4+5.0
"intermediatePlaceI11 p3" ,=B4+1.0
"statePlaceI5 (1)" ,"=min(B6,B2)"
"intermediatePlaceI3 p1" ,=B7+3.0
"statePlaceI1 (0)" ,"=min(B8,B5)"
```

9.4 The PN2XML Module

Current standardization efforts [6, 62] introduce the Petri Net Markup Language (PNML) as an XML-based interchange format for PNs (the PNML standard is still undergoing changes). DP2PN2Solver contains a module PN2XML that is capable of producing a standard file format from a Bellman net. By importing the standard file into a PN simulator like Renew (see Sect. 5.1.5) this opens up the possibility to simulate the Bellman net with external PN systems (see Sect. 11.2). The shortcomings of solving DP problems in this way are performance penalties due to the overhead of the PN simulation system, and the fact that only the optimal *value* of the solution is directly observable, but not the decisions leading to the optimum (i.e. the optimal policy). The PN2Java module of DP2PN2Solver discussed in Sect. 9.2 does not have these problems.

The PNML representation of a Bellman net is simply a list of its place nodes, transition nodes, and arcs, together with associated *labels.* Labels can be text labels, graphical display information (collectively known as *annotations*) or labels can be *attributes.* (We will not discuss the display details here.) Place nodes are objects, and as such contain an internal ID; place node annotations include an external label, and an optional initial marking. Transition nodes are objects, and as such contain an internal ID; transition node annotations include a textual annotation, and an expression.

An arc has exactly one source and exactly one target node associated with it; these two references are represented by the source and target node ID. Arcs are considered Objects, which may have labels associated with them. In our implementation, labels to be displayed are given their own individual internal ID numbers.

Details are provided in the next section.

9.4.1 Petri Net Solver Code for LINSRCS

To illustrate the solver code generation process, we consider the linear search example. We assume we are given the following matrix representation of the Bellman net `LINSRCSBN.csv`.

The solver code consists of a prologue

```
<?xml version="1.0"?>
<!DOCTYPE net SYSTEM "http://www.renew.de/xrn1.dtd">
<net id="N" type="hlnet">
```

and an epilogue

```
  <annotation id="A1" type="name">
    <text>LINSRCS</text>
  </annotation>
</net>
```

in between which declarations for place nodes, transition nodes, and arcs are given. We note that the name of the Petri net (LINSRCS) appears in the epilogue. For the linear search examples, the generated solver code would have the following skeletal form.

```
<?xml version="1.0"?>
<!DOCTYPE net SYSTEM "http://www.renew.de/xrn1.dtd">
<net id="N" type="hlnet">

  <place id="I1">
    <annotation id="I2" type="name">      <text>({0,1,2})</text>    </annotation>
```

```
</place>
<place id="I3">
  <annotation id="I4" type="name">     <text>p1</text>   </annotation>
</place>
<place id="I5">
  <annotation id="I6" type="name">     <text>({1,2})</text>    </annotation>
</place>
...
<place id="I35">
  <annotation id="I36" type="name">     <text>({})</text>    </annotation>
  <annotation id="I37" type="initialmarking">     <text>0.0</text>    </annotation>
</place>
...
<place id="I40">
  <annotation id="I41" type="name">     <text>p12</text>    </annotation>
</place>
<place id="I48">
  <annotation id="I49" type="name">     <text>ep1</text>    </annotation>
  <annotation id="I50" type="initialmarking">     <text>[]</text>    </annotation>
</place>
...
<transition id="I42">
  <annotation id="I43" type="name">     <text>mt1</text>    </annotation>
  <annotation id="I44" type="expression">     <text>y=Math.min(x1,Math.min(x2,x3))</text>     </annotation>
</transition>
<transition id="I45">
  <annotation id="I46" type="name">     <text>st1</text>    </annotation>
  <annotation id="I47" type="expression">     <text>y=x1+0.2</text>    </annotation>
</transition>
...
<transition id="I140">
  <annotation id="I141" type="name">     <text>st12</text>    </annotation>
  <annotation id="I142" type="expression">     <text>y=x1+0.6000000000000001</text>     </annotation>
</transition>
...
<arc id="I51" source="I48" target="I45" type="ordinary">  </arc>
...
<arc id="I146" source="I143" target="I140" type="ordinary">  </arc>
<arc id="I147" source="I42" target="I1" type="ordinary">
  <annotation id="I148" type="expression">     <text>y</text>    </annotation>
</arc>
<arc id="I149" source="I3" target="I42" type="ordinary">
  <annotation id="I150" type="expression">     <text>x1</text>    </annotation>
</arc>
<arc id="I151" source="I7" target="I42" type="ordinary">
  <annotation id="I152" type="expression">     <text>x2</text>    </annotation>
</arc>
...
<arc id="I155" source="I45" target="I3" type="ordinary">
  <annotation id="I156" type="expression">     <text>y</text>    </annotation>
</arc>
<arc id="I157" source="I5" target="I45" type="double">
  <annotation id="I158" type="expression">     <text>x1</text>    </annotation>
</arc>
...
<arc id="I229" source="I35" target="I140" type="double">
  <annotation id="I230" type="expression">     <text>x1</text>    </annotation>
</arc>
<arc id="I231" source="I140" target="I40" type="ordinary">
  <annotation id="I232" type="expression">     <text>y</text>    </annotation>
</arc>

 <annotation id="A1" type="name">
   <text>linsrcs</text>
 </annotation>
</net>
```

The place nodes are specified first, in the row-order in which they appear in the `LINSRCSBN.csv` input file. Each place node is given an internal ID number and its textual label is attached as an *annotation*. These annotations are given internal ID numbers as well. The text for a place label annotation is obtained from the second column of the input file. Place nodes that are initially marked, which correspond to base-states, have a second annotation giving these base-condition values. Following these place nodes are the enabling place nodes (ep) for each transition; these nodes are initially marked with the value "`[]`" in a second annotation. (The token "`[]`" represents a single black token in Renew.) The transition nodes are then specified, in the column-order in which they appear in `LINSRCSBN.csv`. Each transition node is given an internal ID number and its textual label is attached as an annotation. These annotations are given internal ID numbers as well. The text for a transition label annotation is obtained from the top row of the input file. Each transition also has a second annotation giving the minimization or addition expression that the transition is to evaluate. This expression, given as a text string annotation, is of the form `y=Math.min(x1,...)` for M-transitions (`y=x1` in the special case where the minimization is over only a single value), or `y=x1+..+const` for E-expressions. The constant is obtained from the bottom row of the input file. Finally, the arcs are specified. Each arc identifies a source and a target, an arc type (ordinary or double, the latter for arcs from state places to E-transitions, so that a state place maintains its numerical token — once it has one — throughout the course of the simulation), and a label that is attached as an annotation. (In our current implementation, labels associated with decision arcs are of the form `xi`; the labels given in the last column of the input file are not used.)

For example, consider the excerpt given above.

- The goal-state place has ID=`I1`.
- The base-state place has ID=`I35`; note its initial marking.
- An example of an enabling place node has ID=`I48`.
- An example of an M-transition has ID=`I42`.
- An example of an E-transition has ID=`I45`.
- Examples of arcs have ID=`I229` and ID=`I231`.

To graphically display the Petri net, graphical information must also be included. The latter give screen size, location, and color attributes for the nodes, arcs, and labels. Examples of this are shown in the following excerpt.

```
<place id="I1">
  <graphics>
    <size w="20" h="20"/>
    <offset x="30" y="150"/>
    <fillcolor><RGBcolor r="112" g="219" b="147"/></fillcolor>
    <pencolor><RGBcolor r="0" g="0" b="0"/></pencolor>
    <textcolor><RGBcolor r="0" g="0" b="0"/></textcolor>
  </graphics>
  <annotation id="I2" type="name">
    <text>({0,1,2})</text>
    <graphics>
      <size w="20" h="16"/>
      <textsize size="12"/>
      <offset x="0" y="15"/>
      <fillcolor><transparent/></fillcolor>
      <pencolor><transparent/></pencolor>
      <textcolor><RGBcolor r="0" g="0" b="0"/></textcolor>
    </graphics>
  </annotation>
</place>

<arc id="I231" source="I140" target="I40" type="ordinary">
  <graphics>
    <fillcolor><RGBcolor r="112" g="219" b="147"/></fillcolor>
    <pencolor><RGBcolor r="0" g="0" b="0"/></pencolor>
    <textcolor><RGBcolor r="0" g="0" b="0"/></textcolor>
  </graphics>
  <annotation id="I232" type="expression">
    <text>y</text>
  </annotation>
</arc>

<transition id="I42">
  <graphics>
    <size w="20" h="20"/>
    <offset x="120" y="60"/>
    <fillcolor><RGBcolor r="112" g="219" b="147"/></fillcolor>
```

```
    <pencolor><RGBcolor r="0" g="0" b="0"/></pencolor>
    <textcolor><RGBcolor r="0" g="0" b="0"/></textcolor>
  </graphics>
  <annotation id="I43" type="name">
    <text>mt1</text>
    <graphics>
      <size w="20" h="16"/>
      <textsize size="12"/>
      <offset x="0" y="15"/>
      <fillcolor><transparent/></fillcolor>
      <pencolor><transparent/></pencolor>
      <textcolor><RGBcolor r="0" g="0" b="0"/></textcolor>
    </graphics>
  </annotation>
  <annotation id="I44" type="expression">
    <text>y=Math.min(x1,Math.min(x2,x3))</text>
  </annotation>
</transition>
```

The complete PNML code for `LINSRCS.xrn` is very long, so it will be omitted. For completeness, however, we show the PNML code for `SPA1.xrn`, a small version of SPA. It is listed in App. A.3 and described in the next section.

9.4.2 Petri Net Solver Code for SPA

For illustrative purposes, we show the Petri net solver code for a small example, SPA, for a single branch graph.

For the shortest path in an acyclic *single-branch* graph problem (SPA1), suppose there is a single branch connecting two nodes, a single source 0 (the goal state) and a single target 1, where this branch has weight $d(0,1) = 3$. The DPFE is $f(0) = \min\{d(0,1)+f(1)\}$, with base case $f(1) = 0$. This DPFE can be represented by a HLBN having two transitions: an E-transition that computes $p_1 = d(0,1) + f(1)$, or $p_1 = 3 + f(1)$; and an M-transition that computes $p_0 = \min\{p_1\}$, or more simply $p_0 = p_1$ (since the minimum is over a set having only one member). The E-transition `st1` has input place `(1)` and output place `p1`; the M-transition `mt1` has input place `p1` and output place `(0)`. The E-transition `st1` also has an enabling place `ep1`. A HLBN can be specified by listing the places p, transitions t, and their connecting arcs as pairs of nodes (p,t) or (t,p). For example, the HLBN for the single-branch SPA example can be represented by the following text file `SPA1.xrn` (shown here reformatted, with graphics information omitted):

```
<?xml version="1.0"?>
<!DOCTYPE net SYSTEM "http://www.renew.de/xrn1.dtd">
<net id="N" type="hlnet">
  <place id="I1">
    <annotation id="I2" type="name">             <text>(0)</text>       </annotation>
  </place>
  <place id="I3">
    <annotation id="I4" type="name">             <text>p1</text>        </annotation>
  </place>
  <place id="I5">
    <annotation id="I6" type="name">             <text>(1)</text>       </annotation>
    <annotation id="I7" type="initialmarking">   <text>0.0</text>       </annotation>
  </place>
  <place id="I14">
    <annotation id="I15" type="name">            <text>ep1</text>       </annotation>
    <annotation id="I16" type="initialmarking">  <text>[]</text>        </annotation>
  </place>
  <transition id="I8">
    <annotation id="I9" type="name">             <text>mt1</text>       </annotation>
    <annotation id="I10" type="expression">      <text>y=x1</text>      </annotation>
  </transition>
  <transition id="I11">
```

```
    <annotation id="I12" type="name">             <text>st1</text>       </annotation>
    <annotation id="I13" type="expression">       <text>y=x1+3.0</text>  </annotation>
  </transition>
  <arc id="I17" source="I14" target="I11" type="ordinary">
  </arc>
  <arc id="I18" source="I8" target="I1" type="ordinary">
    <annotation id="I19" type="expression">       <text>y</text>         </annotation>
  </arc>
  <arc id="I20" source="I3" target="I8" type="ordinary">
    <annotation id="I21" type="expression">       <text>x1</text>        </annotation>
  </arc>
  <arc id="I22" source="I11" target="I3" type="ordinary">
    <annotation id="I23" type="expression">       <text>y</text>         </annotation>
  </arc>
  <arc id="I24" source="I5" target="I11" type="double">
    <annotation id="I25" type="expression">       <text>x1</text>        </annotation>
  </arc>
  <annotation id="A1" type="name">
    <text>spa1</text>
  </annotation>
</net>
```

In this representation of the HLBN, the four places and two transitions, and their five connecting arcs are given internal ID numbers as follows: `(0)=I1`, `p1=I3`, `(1)=I5`, `ep1=I14`, `mt1=I8`, `st1=I11`, `(I14,I11)=I17`, `(I8,I1)=I18`, `(I3,I8)=I20`, `(I11,I3)=I22`, `(I5,I11)=I24`; the last of these arcs is a bidirectional "double" arc, i.e. it includes `(I11,I5)`. The base state place `I5` is initially marked with a token having value 0.0, and the enabling place `I14` is initially marked with a black token (denoted "`[]`"). E-transition `I11` computes the expression `I3=I5+3.0`, and M-transition `I8` computes the expression `I1=I3`.

To draw this HLBN (using circles for places, rectangles for transitions, and lines for arcs), "graphics" information, including size, position, and color, must be provided for each of these elements. In addition, any of the places, transitions, and arcs may have an "annotation" if these elements are to be drawn with a label (given in the specified "text"); "graphics" information for these labels must then also be provided. (In the example, only arc `(ep1,st1)` has no label.) For completeness, we provide the entire `SPA1.xrn` file in App. A.3.

9.5 Conclusion

In this book, we included Petri nets as a third type of solver code (in addition to java programs and spreadsheets). For the purpose of obtaining numerical solutions, the other two types are likely to be more useful. However, Petri-net solver code has one significant advantage: it is in a form for which many software tools exist that in principle permits analysis of theoretical properties DP formulations. There may also be circumstances where it may be useful to visualize solutions graphically. This is the subject of future research.

Part IV

Computational Results

10

Java Solver Results of DP Problems

The Java code obtained from the PN2Java module (see Chap. 9) can be used as the base building block for a solver function within a larger software component. Just executing the bare Java code without any further modifications leads to the output of the following solution trees for the examples introduced in Chap. 2. Each output gives both the optimal objective function value and also the optimal decision policy for every problem.

10.1 ALLOT Java Solver Output

The solution to the ALLOTt problem where allotment decisions and their costs are defined in separate tables:

```
The optimal value is: 0.06
The solution tree is:
State (0,0) has optimal value: 0.06
 Decision d=1
  State (1,1) has optimal value: 0.3
   Decision d=0
    State (2,1) has optimal value: 0.5
     Decision d=1
      Base state (3,2) has initial value: 1.0
```

The solution to the ALLOTf problem where the costs are defined by general functions:

```
The optimal value is: 49.0
The solution tree is:
State (1,6) has optimal value: 49.0
 Decision d=4
  State (2,2) has optimal value: 19.0
```

A. Lew and H. Mauch: *Java Solver Results of DP Problems*, Studies in Computational Intelligence (SCI) **38**, 293–320 (2007)
www.springerlink.com

```
   Decision d=1
    State (3,1) has optimal value: 9.0
     Decision d=1
      Base state (4,0) has initial value: 0.0
```

The solution to the ALLOTm problem where the costs are multiplicative rather than additive:

```
The optimal value is: 0.23099999999999998
The solution tree is:
State (1,2) has optimal value: 0.23099999999999998
 Decision d=0
  State (2,2) has optimal value: 0.385
   Decision d=1
    State (3,1) has optimal value: 0.55
     Decision d=1
      Base state (4,0) has initial value: 1.0
```

10.2 APSP Java Solver Output

For APSP we used two different DP models. The solution using the relaxation DP functional equation (2.2) is:

```
The optimal value is: 9.0
The solution tree is:
State (3,0,3) has optimal value: 9.0
 Decision d=1
  State (2,1,3) has optimal value: 6.0
   Decision d=2
    State (1,2,3) has optimal value: 5.0
     Decision d=3
      Base state (0,3,3) has initial value: 0.0
```

The solution using the Floyd-Warshall DP functional equation (2.4) is:

```
The optimal value is: 9.0
The solution tree is:
State (3,0,3) has optimal value: 9.0
 Decision d=0
   State (2,0,3) has optimal value: 9.0
    Decision d=1
      State (1,0,3) has optimal value: 11.0
       Decision d=1
```

```
      Base state (0,0,3) has initial value: 2.147483647E9
      Base state (0,0,1) has initial value: 3.0
      Base state (0,1,3) has initial value: 8.0
   State (1,0,2) has optimal value: 4.0
    Decision d=1
      Base state (0,0,2) has initial value: 5.0
      Base state (0,0,1) has initial value: 3.0
      Base state (0,1,2) has initial value: 1.0
   State (1,2,3) has optimal value: 5.0
    Decision d=0
      Base state (0,2,3) has initial value: 5.0
      Base state (0,2,1) has initial value: 2.0
      Base state (0,1,3) has initial value: 8.0
State (2,0,3) has optimal value: 9.0
 Decision d=1
   State (1,0,3) has optimal value: 11.0
    Decision d=1
      Base state (0,0,3) has initial value: 2.147483647E9
      Base state (0,0,1) has initial value: 3.0
      Base state (0,1,3) has initial value: 8.0
   State (1,0,2) has optimal value: 4.0
    Decision d=1
      Base state (0,0,2) has initial value: 5.0
      Base state (0,0,1) has initial value: 3.0
      Base state (0,1,2) has initial value: 1.0
   State (1,2,3) has optimal value: 5.0
    Decision d=0
      Base state (0,2,3) has initial value: 5.0
      Base state (0,2,1) has initial value: 2.0
      Base state (0,1,3) has initial value: 8.0
State (2,3,3) has optimal value: 0.0
 Decision d=0
   State (1,3,3) has optimal value: 0.0
    Decision d=0
      Base state (0,3,3) has initial value: 0.0
      Base state (0,3,1) has initial value: 2.147483647E9
      Base state (0,1,3) has initial value: 8.0
   State (1,3,2) has optimal value: 2.147483647E9
    Decision d=0
      Base state (0,3,2) has initial value: 2.147483647E9
      Base state (0,3,1) has initial value: 2.147483647E9
      Base state (0,1,2) has initial value: 1.0
   State (1,2,3) has optimal value: 5.0
    Decision d=0
      Base state (0,2,3) has initial value: 5.0
```

```
    Base state (0,2,1) has initial value: 2.0
    Base state (0,1,3) has initial value: 8.0
```

10.3 ARC Java Solver Output

For $S = (2,3,3,4)$, the solution is:

```
The optimal value is: 24.0
The solution tree is:
State (0,3) has optimal value: 24.0
 Decision d=1
  State (0,1) has optimal value: 5.0
   Decision d=0
    Base state (0,0) has initial value: 0.0
    Base state (1,1) has initial value: 0.0
  State (2,3) has optimal value: 7.0
   Decision d=2
    Base state (3,3) has initial value: 0.0
    Base state (2,2) has initial value: 0.0
```

10.4 ASMBAL Java Solver Output

For ASMBAL we came up with two different DP models. The solution using the staged DP functional equation (2.6) is:

```
The optimal value is: 38.0
The solution tree is:
State (0,0) has optimal value: 38.0
 Decision d=0
  State (1,0) has optimal value: 36.0
   Decision d=1
    State (2,1) has optimal value: 27.0
     Decision d=0
      State (3,0) has optimal value: 21.0
       Decision d=1
        State (4,1) has optimal value: 17.0
         Decision d=1
          State (5,1) has optimal value: 13.0
           Decision d=0
            State (6,0) has optimal value: 7.0
             Decision d=0
```

```
            Base state (7,0) has initial value: 0.0
```

The solution using the DP functional equation (2.7) is:

```
The optimal value is: 38.0
The solution tree is:
State (0) has optimal value: 38.0
 Decision d=1
  State (1) has optimal value: 36.0
   Decision d=4
    State (4) has optimal value: 27.0
     Decision d=5
      State (5) has optimal value: 21.0
       Decision d=8
        State (8) has optimal value: 17.0
         Decision d=10
          State (10) has optimal value: 13.0
           Decision d=11
            State (11) has optimal value: 7.0
             Decision d=13
              Base state (13) has initial value: 0.0
```

10.5 ASSIGN Java Solver Output

```
The optimal value is: 17.0
The solution tree is:
State ({0,1,2},1) has optimal value: 17.0
 Decision d=2
  State ({0,1},2) has optimal value: 15.0
   Decision d=0
    State ({1},3) has optimal value: 10.0
     Decision d=1
      Base state ({},4) has initial value: 0.0
```

10.6 BST Java Solver Output

```
The optimal value is: 1.9
The solution tree is:
State ({0,1,2,3,4}) has optimal value: 1.9
 Decision splitAtAlpha=3
```

```
  State ({0,1,2}) has optimal value: 0.8
   Decision splitAtAlpha=0
    Base state ({}) has initial value: 0.0
    State ({1,2}) has optimal value: 0.3
     Decision splitAtAlpha=2
      Base state ({}) has initial value: 0.0
      State ({1}) has optimal value: 0.05
       Decision splitAtAlpha=1
        Base state ({}) has initial value: 0.0
  State ({4}) has optimal value: 0.1
   Decision splitAtAlpha=4
    Base state ({}) has initial value: 0.0
```

10.7 COV Java Solver Output

```
The optimal value is: 129.0
The solution tree is:
State (3,9) has optimal value: 129.0
 Decision nextCoverSize=6
  State (2,6) has optimal value: 66.0
   Decision nextCoverSize=4
    Base state (1,4) has initial value: 40.0
```

10.8 DEADLINE Java Solver Output

```
The optimal value is: 40.0
The solution tree is:
State ({0,1,2,3,4},1) has optimal value: 40.0
 Decision d=1
  State ({0,2,3,4},2) has optimal value: 25.0
   Decision d=2
    State ({0,3,4},3) has optimal value: 5.0
     Decision d=4
      State ({0,3},4) has optimal value: 0.0
       Decision d=0
        State ({3},5) has optimal value: 0.0
         Decision d=3
          Base state ({},6) has initial value: 0.0
```

10.9 DPP Java Solver Output

```
The optimal value is: 19.047619047619047
The solution tree is:
State (1,10) has optimal value: 19.047619047619047
 Decision xt=0
   State (2,20) has optimal value: 20.0
    Decision xt=20
      Base state (3,0) has initial value: 0.0
```

10.10 EDP Java Solver Output

```
The optimal value is: 2.0
The solution tree is:
State (3,3) has optimal value: 2.0
 Decision dec=2
  State (3,2) has optimal value: 1.0
   Decision dec=12
    State (2,1) has optimal value: 1.0
     Decision dec=12
      Base state (1,0) has initial value: 1.0
```

10.11 FIB Java Solver Output

```
The optimal value is: 13.0
The solution tree is:
State (7) has optimal value: 13.0
 Decision dummy=777
  State (6) has optimal value: 8.0
   Decision dummy=777
    State (5) has optimal value: 5.0
     Decision dummy=777
      State (4) has optimal value: 3.0
       Decision dummy=777
        State (3) has optimal value: 2.0
         Decision dummy=777
          Base state (2) has initial value: 1.0
          Base state (1) has initial value: 1.0
        Base state (2) has initial value: 1.0
      State (3) has optimal value: 2.0
       Decision dummy=777
```

```
        Base state (2) has initial value: 1.0
        Base state (1) has initial value: 1.0
    State (4) has optimal value: 3.0
     Decision dummy=777
      State (3) has optimal value: 2.0
       Decision dummy=777
        Base state (2) has initial value: 1.0
        Base state (1) has initial value: 1.0
      Base state (2) has initial value: 1.0
  State (5) has optimal value: 5.0
   Decision dummy=777
    State (4) has optimal value: 3.0
     Decision dummy=777
      State (3) has optimal value: 2.0
       Decision dummy=777
        Base state (2) has initial value: 1.0
        Base state (1) has initial value: 1.0
      Base state (2) has initial value: 1.0
    State (3) has optimal value: 2.0
     Decision dummy=777
      Base state (2) has initial value: 1.0
      Base state (1) has initial value: 1.0
```

10.12 FLOWSHOP Java Solver Output

```
The optimal value is: 38.0
The solution tree is:
State ({0,1,2,3},0) has optimal value: 38.0
 Decision d=0
  State ({1,2,3},6) has optimal value: 35.0
   Decision d=2
    State ({1,3},9) has optimal value: 27.0
     Decision d=3
      State ({1},15) has optimal value: 17.0
       Decision d=1
        Base state ({},13) has initial value: 13.0
```

10.13 HANOI Java Solver Output

The number of moves is calculated, but not the sequence of moves. However, the set of moves is given in the base states; more work is needed to reconstruct the correct sequence, if desired.

```
The optimal value is: 7.0
The solution tree is:
State (3,1,2,3) has optimal value: 7.0
 Decision dummy=-1
  State (2,1,3,2) has optimal value: 3.0
   Decision dummy=-1
    Base state (1,1,2,3) has initial value: 1.0
    Base state (1,1,3,2) has initial value: 1.0
    Base state (1,2,3,1) has initial value: 1.0
  Base state (1,1,2,3) has initial value: 1.0
  State (2,3,2,1) has optimal value: 3.0
   Decision dummy=-1
    Base state (1,1,2,3) has initial value: 1.0
    Base state (1,3,1,2) has initial value: 1.0
    Base state (1,3,2,1) has initial value: 1.0
```

10.14 ILP Java Solver Output

```
The optimal value is: 36.0
The solution tree is:
State (0,4,12,18) has optimal value: 36.0
 Decision d=2
  State (1,2,12,12) has optimal value: 30.0
   Decision d=6
    Base state (2,2,0,0) has initial value: 0.0
```

10.15 ILPKNAP Java Solver Output

```
The optimal value is: 25.0
The solution tree is:
State (0,22,1,1,1) has optimal value: 25.0
 Decision d=0
  State (1,22,1,1,1) has optimal value: 25.0
   Decision d=1
    State (2,4,1,0,1) has optimal value: 0.0
     Decision d=0
      Base state (3,4,1,0,1) has initial value: 0.0
```

10.16 INTVL Java Solver Output

The solution tree based on intvl1.dp, which uses DPFE (2.21) is as follows.

```
The optimal value is: 8.0
The solution tree is:
State (6) has optimal value: 8.0
 Decision d=0
  State (5) has optimal value: 8.0
   Decision d=1
    State (3) has optimal value: 6.0
     Decision d=1
      State (1) has optimal value: 2.0
       Decision d=1
        Base state (0) has initial value: 0.0
```

The solution tree based on intvl3.dp, which uses DPFE (2.22) is as follows.

```
The optimal value is: 8.0
The solution tree is:
State (6) has optimal value: 8.0
 Decision d=0
   State (3) has optimal value: 6.0
    Decision d=1
      State (1) has optimal value: 2.0
       Decision d=1
         Base state (0) has initial value: 0.0
         Base state (0) has initial value: 0.0
      State (2) has optimal value: 4.0
       Decision d=1
         Base state (0) has initial value: 0.0
         State (1) has optimal value: 2.0
          Decision d=1
            Base state (0) has initial value: 0.0
            Base state (0) has initial value: 0.0
   State (5) has optimal value: 8.0
    Decision d=1
      State (3) has optimal value: 6.0
       Decision d=1
         State (1) has optimal value: 2.0
          Decision d=1
            Base state (0) has initial value: 0.0
            Base state (0) has initial value: 0.0
         State (2) has optimal value: 4.0
```

```
        Decision d=1
          Base state (0) has initial value: 0.0
          State (1) has optimal value: 2.0
           Decision d=1
             Base state (0) has initial value: 0.0
             Base state (0) has initial value: 0.0
    State (4) has optimal value: 7.0
     Decision d=1
       Base state (0) has initial value: 0.0
       State (3) has optimal value: 6.0
        Decision d=1
          State (1) has optimal value: 2.0
           Decision d=1
             Base state (0) has initial value: 0.0
             Base state (0) has initial value: 0.0
          State (2) has optimal value: 4.0
           Decision d=1
             Base state (0) has initial value: 0.0
             State (1) has optimal value: 2.0
              Decision d=1
                Base state (0) has initial value: 0.0
                Base state (0) has initial value: 0.0
```

The solution tree based on intvl2.dp, which uses DPFE (2.20) is as follows.

```
The optimal value is: 8.0
The solution tree is:
State ({0,1,2,3,4,5},0,20) has optimal value: 8.0
 Decision d=1
  State ({3,4,5},0,8) has optimal value: 6.0
   Decision d=3
    State ({5},0,5) has optimal value: 2.0
     Decision d=5
      Base state ({},0,1) has initial value: 0.0
      Base state ({},4,5) has initial value: 0.0
    Base state ({},7,8) has initial value: 0.0
  Base state ({},11,20) has initial value: 0.0
```

10.17 INVENT Java Solver Output

```
The optimal value is: 20.0
The solution tree is:
```

```
State (0,0) has optimal value: 20.0
 Decision x=1
  State (1,0) has optimal value: 16.0
   Decision x=5
    State (2,2) has optimal value: 7.0
     Decision x=0
      State (3,0) has optimal value: 7.0
       Decision x=4
        Base state (4,0) has initial value: 0.0
```

10.18 INVEST Java Solver Output

```
The optimal value is: 2.6000000000000005
The solution tree is:
State (1,2) has optimal value: 2.6000000000000005
 Decision d=1
   State (2,3) has optimal value: 3.4000000000000004
    Decision d=1
      State (3,4) has optimal value: 4.2
       Decision d=1
         Base state (4,5) has initial value: 5.0
         Base state (4,4) has initial value: 4.0
      State (3,3) has optimal value: 3.2
       Decision d=1
         Base state (4,4) has initial value: 4.0
         Base state (4,3) has initial value: 3.0
   State (2,2) has optimal value: 2.4000000000000004
    Decision d=1
      State (3,3) has optimal value: 3.2
       Decision d=1
         Base state (4,4) has initial value: 4.0
         Base state (4,3) has initial value: 3.0
      State (3,2) has optimal value: 2.2
       Decision d=1
         Base state (4,3) has initial value: 3.0
         Base state (4,2) has initial value: 2.0
```

10.19 INVESTWLV Java Solver Output

Since INVESTWLV is a probabilistic DP problem, the solution tree should be read as a decision tree where after each decision, chance plays a role in which state we enter next. Values associated with a state must be interpreted as *expected* values.

```
The optimal value is: 0.7407407407407407
The solution tree is:
State (1,3) has optimal value: 0.7407407407407407
 Decision xn=1
   State (2,2) has optimal value: 0.4444444444444444
    Decision xn=1
      State (3,1) has optimal value: 0.0
       Decision xn=0
         Base state (4,1) has initial value: 0.0
         Base state (4,1) has initial value: 0.0
      State (3,3) has optimal value: 0.6666666666666666
       Decision xn=2
         Base state (4,1) has initial value: 0.0
         Base state (4,5) has initial value: 1.0
   State (2,4) has optimal value: 0.8888888888888888
    Decision xn=1
      State (3,3) has optimal value: 0.6666666666666666
       Decision xn=2
         Base state (4,1) has initial value: 0.0
         Base state (4,5) has initial value: 1.0
      State (3,5) has optimal value: 1.0
       Decision xn=0
         Base state (4,5) has initial value: 1.0
         Base state (4,5) has initial value: 1.0
```

10.20 KS01 Java Solver Output

```
The optimal value is: 25.0
The solution tree is:
State (2,22) has optimal value: 25.0
 Decision d=0
  State (1,22) has optimal value: 25.0
   Decision d=0
    State (0,22) has optimal value: 25.0
     Decision d=1
      Base state (-1,4) has initial value: 0.0
```

10.21 KSCOV Java Solver Output

```
The optimal value is: 129.0
The solution tree is:
State (1,10) has optimal value: 129.0
 Decision d=3
  State (2,7) has optimal value: 66.0
   Decision d=2
    State (3,5) has optimal value: 40.0
     Decision d=5
      Base state (4,0) has initial value: 0.0
```

10.22 KSINT Java Solver Output

```
The optimal value is: 30.0
The solution tree is:
State (2,22) has optimal value: 30.0
 Decision d=0
  State (1,22) has optimal value: 30.0
   Decision d=2
    State (0,2) has optimal value: 0.0
     Decision d=0
      Base state (-1,2) has initial value: 0.0
```

10.23 LCS Java Solver Output

For LCS we came up with two different DP models. One uses the DP functional equation (2.30), the other uses the improved DP functional equation (2.29) that produces fewer states. The fact that the DP2PN2JavaSolver produces two identical solution trees for both approaches shows the versatility of our software and increases the confidence in the correctness of our models (and our software!). The solution for both model reads as follows.

```
The optimal value is: 4.0
The solution tree is:
State (7,6) has optimal value: 4.0
 Decision pruneD=1
  State (6,6) has optimal value: 4.0
   Decision pruneD=12
    State (5,5) has optimal value: 3.0
     Decision pruneD=1
```

```
State (4,5) has optimal value: 3.0
 Decision pruneD=12
  State (3,4) has optimal value: 2.0
   Decision pruneD=2
    State (3,3) has optimal value: 2.0
     Decision pruneD=12
      State (2,2) has optimal value: 1.0
       Decision pruneD=2
        State (2,1) has optimal value: 1.0
         Decision pruneD=12
          Base state (1,0) has initial value: 0.0
```

10.24 LINSRC Java Solver Output

For LINSRC we came up with two different DP models. The first one uses method W as a cost function and produces the following output.

```
The optimal value is: 1.7
The solution tree is:
State ({0,1,2}) has optimal value: 1.7
 Decision d=1
  State ({0,2}) has optimal value: 0.7
   Decision d=2
    State ({0}) has optimal value: 0.2
     Decision d=0
      Base state ({}) has initial value: 0.0
```

The second one uses method W as a cost function and produces the following output.

```
The optimal value is: 1.7000000000000002
The solution tree is:
State ({0,1,2}) has optimal value: 1.7000000000000002
 Decision d=1
  State ({0,2}) has optimal value: 1.2000000000000002
   Decision d=2
    State ({0}) has optimal value: 0.6000000000000001
     Decision d=0
      Base state ({}) has initial value: 0.0
```

10.25 LOT Java Solver Output

```
The optimal value is: 3680.0
The solution tree is:
State (1) has optimal value: 3680.0
 Decision x=0
  State (2) has optimal value: 2990.0
   Decision x=0
    State (3) has optimal value: 2180.0
     Decision x=1
      State (5) has optimal value: 790.0
       Decision x=0
        Base state (6) has initial value: 0.0
```

10.26 LSP Java Solver Output

```
The optimal value is: 3.0
The solution tree is:
State ({0},0) has optimal value: 3.0
 Decision alpha=1
  State ({0,1},1) has optimal value: 2.0
   Decision alpha=2
    State ({0,1,2},2) has optimal value: 1.0
     Decision alpha=3
      Base state ({0,1,2,3},3) has initial value: 0.0
```

10.27 MCM Java Solver Output

```
The optimal value is: 76.0
The solution tree is:
State (1,4) has optimal value: 76.0
 Decision k=3
  State (1,3) has optimal value: 64.0
   Decision k=1
    Base state (1,1) has initial value: 0.0
    State (2,3) has optimal value: 40.0
     Decision k=2
      Base state (2,2) has initial value: 0.0
      Base state (3,3) has initial value: 0.0
  Base state (4,4) has initial value: 0.0
```

10.28 MINMAX Java Solver Output

```
The optimal value is: 8.0
The solution tree is:
State (1,{0},0) has optimal value: 8.0
 Decision alpha=2
  State (2,{0,2},2) has optimal value: 8.0
   Decision alpha=4
    State (3,{0,2,4},4) has optimal value: 8.0
     Decision alpha=7
      State (4,{0,2,4,7},7) has optimal value: 8.0
       Decision alpha=9
        Base state (5,{0,2,4,7,9},9) has initial value: 8.0
```

10.29 MWST Java Solver Output

```
The optimal value is: 7.0
The solution tree is:
State ({0,1,2,3,4},0) has optimal value: 7.0
 Decision d=1
  State ({0,2,3,4},1) has optimal value: 3.0
   Decision d=3
    State ({0,2,4},2) has optimal value: 1.0
     Decision d=4
      Base state ({0,2},3) has initial value: 0.0
```

10.30 NIM Java Solver Output

The following output gives the strategy if we are in a winning state.

```
The optimal value is: 1.0
The solution tree is:
State (10) has optimal value: 1.0
 Decision d=1
  State (8) has optimal value: 1.0
   Decision d=3
    State (4) has optimal value: 1.0
     Decision d=3
      State (0) has optimal value: 1.0
       Decision d=1
        Base state (-2) has initial value: 1.0
```

```
        Base state (-3) has initial value: 1.0
        Base state (-4) has initial value: 1.0
      Base state (-1) has initial value: 1.0
      Base state (-2) has initial value: 1.0
    State (3) has optimal value: 1.0
     Decision d=2
      State (0) has optimal value: 1.0
       Decision d=1
        Base state (-2) has initial value: 1.0
        Base state (-3) has initial value: 1.0
        Base state (-4) has initial value: 1.0
      Base state (-1) has initial value: 1.0
      Base state (-2) has initial value: 1.0
    State (2) has optimal value: 1.0
     Decision d=1
      State (0) has optimal value: 1.0
       Decision d=1
        Base state (-2) has initial value: 1.0
        Base state (-3) has initial value: 1.0
        Base state (-4) has initial value: 1.0
      Base state (-1) has initial value: 1.0
      Base state (-2) has initial value: 1.0
  State (7) has optimal value: 1.0
   Decision d=2
    State (4) has optimal value: 1.0
     Decision d=3
      State (0) has optimal value: 1.0
       Decision d=1
        Base state (-2) has initial value: 1.0
        Base state (-3) has initial value: 1.0
        Base state (-4) has initial value: 1.0
      Base state (-1) has initial value: 1.0
      Base state (-2) has initial value: 1.0
    State (3) has optimal value: 1.0
     Decision d=2
      State (0) has optimal value: 1.0
       Decision d=1
        Base state (-2) has initial value: 1.0
        Base state (-3) has initial value: 1.0
        Base state (-4) has initial value: 1.0
      Base state (-1) has initial value: 1.0
      Base state (-2) has initial value: 1.0
    State (2) has optimal value: 1.0
     Decision d=1
      State (0) has optimal value: 1.0
```

```
      Decision d=1
        Base state (-2) has initial value: 1.0
        Base state (-3) has initial value: 1.0
        Base state (-4) has initial value: 1.0
      Base state (-1) has initial value: 1.0
      Base state (-2) has initial value: 1.0
  State (6) has optimal value: 1.0
   Decision d=1
    State (4) has optimal value: 1.0
     Decision d=3
      State (0) has optimal value: 1.0
       Decision d=1
        Base state (-2) has initial value: 1.0
        Base state (-3) has initial value: 1.0
        Base state (-4) has initial value: 1.0
      Base state (-1) has initial value: 1.0
      Base state (-2) has initial value: 1.0
    State (3) has optimal value: 1.0
     Decision d=2
      State (0) has optimal value: 1.0
       Decision d=1
        Base state (-2) has initial value: 1.0
        Base state (-3) has initial value: 1.0
        Base state (-4) has initial value: 1.0
      Base state (-1) has initial value: 1.0
      Base state (-2) has initial value: 1.0
    State (2) has optimal value: 1.0
     Decision d=1
      State (0) has optimal value: 1.0
       Decision d=1
        Base state (-2) has initial value: 1.0
        Base state (-3) has initial value: 1.0
        Base state (-4) has initial value: 1.0
      Base state (-1) has initial value: 1.0
      Base state (-2) has initial value: 1.0
```

The output for $m = 10$ is shown. The optimal value of 1.0 indicates that a win is guaranteed with optimal play; the optimal next play is that of removing $d = 1$ matchsticks. If the optimal value is 0.0, such as for $m = 9$, our adversary has a guaranteed win provided it plays optimally regardless of our next play (chosen arbitrarily to be $d = 1$).

The solution tree is very large for larger values of m since, as is also the case for the recursive solution of Fibonacci numbers (FIB). Although overlapping substates are only calculated once, common subtrees reappear over and over again in the output tree.

10.31 ODP Java Solver Output

```
The optimal value is: 31.0
The solution tree is:
State (0,0) has optimal value: 31.0
 Decision d=1
  State (1,1) has optimal value: 27.0
   Decision d=2
    State (2,3) has optimal value: 16.0
     Decision d=1
      Base state (3,6) has initial value: 0.0
```

10.32 PERM Java Solver Output

```
The optimal value is: 17.0
The solution tree is:
State ({0,1,2}) has optimal value: 17.0
 Decision d=2
  State ({0,1}) has optimal value: 11.0
   Decision d=1
    State ({0}) has optimal value: 5.0
     Decision d=0
      Base state ({}) has initial value: 0.0
```

10.33 POUR Java Solver Output

```
The optimal value is: 8.0
The solution tree is:
State (1,0,0,13) has optimal value: 8.0
 Decision d=1
  State (2,9,0,4) has optimal value: 7.0
   Decision d=5
    State (3,5,4,4) has optimal value: 6.0
     Decision d=4
      State (4,5,0,8) has optimal value: 5.0
       Decision d=5
        State (5,1,4,8) has optimal value: 4.0
```

```
Decision d=4
 State (6,1,0,12) has optimal value: 3.0
  Decision d=5
   State (7,0,1,12) has optimal value: 2.0
    Decision d=1
     State (8,9,1,3) has optimal value: 1.0
      Decision d=5
       Base state (9,6,4,3) has initial value: 0.0
```

10.34 PROD Java Solver Output

```
The optimal value is: 42.244
The solution tree is:
State (1,0) has optimal value: 42.244
 Decision d=2
   State (2,1) has optimal value: 19.900000000000002
    Decision d=0
      State (3,0) has optimal value: 14.600000000000001
       Decision d=1
         State (4,0) has optimal value: 0.0
          Decision d=0
            Base state (5,-1) has initial value: 0.0
            Base state (5,-2) has initial value: 0.0
         State (4,-1) has optimal value: 11.5
          Decision d=1
            Base state (5,-1) has initial value: 0.0
            Base state (5,-2) has initial value: 0.0
      State (3,-1) has optimal value: 25.1
       Decision d=2
         State (4,0) has optimal value: 0.0
          Decision d=0
            Base state (5,-1) has initial value: 0.0
            Base state (5,-2) has initial value: 0.0
         State (4,-1) has optimal value: 11.5
          Decision d=1
            Base state (5,-1) has initial value: 0.0
            Base state (5,-2) has initial value: 0.0
   State (2,0) has optimal value: 28.26
    Decision d=2
      State (3,1) has optimal value: 5.700000000000001
       Decision d=0
         State (4,0) has optimal value: 0.0
          Decision d=0
```

```
          Base state (5,-1) has initial value: 0.0
          Base state (5,-2) has initial value: 0.0
       State (4,-1) has optimal value: 11.5
        Decision d=1
          Base state (5,-1) has initial value: 0.0
          Base state (5,-2) has initial value: 0.0
    State (3,0) has optimal value: 14.600000000000001
     Decision d=1
       State (4,0) has optimal value: 0.0
        Decision d=0
          Base state (5,-1) has initial value: 0.0
          Base state (5,-2) has initial value: 0.0
       State (4,-1) has optimal value: 11.5
        Decision d=1
          Base state (5,-1) has initial value: 0.0
          Base state (5,-2) has initial value: 0.0
```

10.35 PRODRAP Java Solver Output

Since PRODRAP is a probabilistic DP problem, the solution tree should be read as a decision tree where after each decision, chance plays a role in whether the decision process terminates or not. Values associated with a state must be interpreted as *expected* values.

```
The optimal value is: 6.75
The solution tree is:
State (1) has optimal value: 6.75
 Decision xn=2
   State (2) has optimal value: 7.0
    Decision xn=2
      State (3) has optimal value: 8.0
       Decision xn=3
         Base state (4) has initial value: 16.0
```

10.36 RDP Java Solver Output

```
The optimal value is: 0.648
The solution tree is:
State (2,105) has optimal value: 0.648
 Decision m=2
  State (1,65) has optimal value: 0.864
```

```
    Decision m=2
     State (0,35) has optimal value: 0.9
      Decision m=1
       Base state (-1,5) has initial value: 1.0
```

10.37 REPLACE Java Solver Output

```
The optimal value is: 1280.0
The solution tree is:
State (0) has optimal value: 1280.0
 Decision d=1
  State (1) has optimal value: 1020.0
   Decision d=1
    State (2) has optimal value: 760.0
     Decision d=3
      Base state (5) has initial value: 0.0
```

10.38 SCP Java Solver Output

```
The optimal value is: 2870.0
The solution tree is:
State (0,0) has optimal value: 2870.0
 Decision d=1
  State (1,1) has optimal value: 2320.0
   Decision d=4
    State (2,4) has optimal value: 1640.0
     Decision d=7
      State (3,7) has optimal value: 1030.0
       Decision d=9
        Base state (4,9) has initial value: 0.0
```

10.39 SEEK Java Solver Output

```
The optimal value is: 190.0
The solution tree is:
State ({0,1,2},140) has optimal value: 190.0
 Decision d=2
  State ({0,1},190) has optimal value: 140.0
   Decision d=0
```

```
    State ({1},100) has optimal value: 50.0
     Decision d=1
      Base state ({},50) has initial value: 0.0
```

10.40 SEGLINE Java Solver Output

For SEGLINE, we gave two different DP models. One used the DP functional equation (2.40), the other used the DP functional equation (2.41).

For the former, no limit on the number of segments was assumed. For a segment cost $K = 1$, we obtain the following solution.

```
The optimal value is: 2.0556
The solution tree is:
State (0) has optimal value: 2.0556
 Decision d=2
  State (2) has optimal value: 1.0
   Decision d=3
    Base state (3) has initial value: 0.0
```

In an alternate model, an upper or exact limit LIM on the number of segments was assumed. For $K = 10$ and an exact $LIM = 2$, we obtain the following solution.

```
The optimal value is: 20.0556
The solution tree is:
State (2,0) has optimal value: 20.0556
 Decision d=2
  State (1,2) has optimal value: 10.0
   Decision d=3
    Base state (0,3) has initial value: 0.0
```

10.41 SEGPAGE Java Solver Output

```
The optimal value is: 87.0
The solution tree is:
State (0) has optimal value: 87.0
 Decision d=1
  State (1) has optimal value: 87.0
   Decision d=5
    State (5) has optimal value: 5.0
     Decision d=7
```

```
    State (7) has optimal value: 4.0
     Decision d=10
      State (10) has optimal value: 2.0
       Decision d=13
        State (13) has optimal value: 0.0
         Decision d=15
          State (15) has optimal value: 0.0
           Decision d=16
            Base state (16) has initial value: 0.0
```

10.42 SELECT Java Solver Output

```
The optimal value is: 21.0
The solution tree is:
State (1,4,1,10) has optimal value: 21.0
 Decision k=2
  State (1,1,1,5) has optimal value: 5.0
   Decision k=1
    Base state (1,0,1,2) has initial value: 0.0
    Base state (2,1,4,5) has initial value: 0.0
  State (3,4,7,10) has optimal value: 6.0
   Decision k=3
    State (4,4,9,10) has optimal value: 2.0
     Decision k=4
      Base state (5,4,11,10) has initial value: 0.0
      Base state (4,3,9,9) has initial value: 0.0
    Base state (3,2,7,7) has initial value: 0.0
```

10.43 SPA Java Solver Output

```
The optimal value is: 9.0
The solution tree is:
State (0) has optimal value: 9.0
 Decision d=1
  State (1) has optimal value: 6.0
   Decision d=2
    State (2) has optimal value: 5.0
     Decision d=3
      Base state (3) has initial value: 0.0
```

10.44 SPC Java Solver Output

For SPC we came up with two different DP models. One used the DP functional equation (2.45), the other used the DP functional equation (2.46). For the former, we get the following solution.

```
The optimal value is: 9.0
The solution tree is:
State (0,{0}) has optimal value: 9.0
 Decision alpha=1
  State (1,{0,1}) has optimal value: 6.0
   Decision alpha=2
    State (2,{0,1,2}) has optimal value: 5.0
     Decision alpha=3
      Base state (3,{0,1,2,3}) has initial value: 0.0
```

For the latter, we get the following solution.

```
The optimal value is: 9.0
The solution tree is:
State (0,3) has optimal value: 9.0
 Decision d=1
  State (1,2) has optimal value: 6.0
   Decision d=2
    State (2,1) has optimal value: 5.0
     Decision d=3
      Base state (3,0) has initial value: 0.0
```

One can easily verify that the two decision trees are equivalent, they only differ in the way the states are named.

10.45 SPT Java Solver Output

```
The optimal value is: 17.0
The solution tree is:
State (0,{0,1,2}) has optimal value: 17.0
 Decision d=2
  State (2,{0,1}) has optimal value: 15.0
   Decision d=0
    State (5,{1}) has optimal value: 10.0
     Decision d=1
      Base state (10,{}) has initial value: 0.0
```

10.46 TRANSPO Java Solver Output

```
The optimal value is: 239.0
The solution tree is:
State (0,0) has optimal value: 239.0
 Decision x=3
  State (1,1) has optimal value: 126.0
   Decision x=1
    State (2,0) has optimal value: 80.0
     Decision x=2
      Base state (3,0) has initial value: 0.0
```

10.47 TSP Java Solver Output

For TSP we came up with two different DP models. The first model keeps the set of nodes visited as a part of the state. The second model is an alternative formulation that keeps the set of nodes not yet visited as a part of the state. For the former, we get the following solution.

```
The optimal value is: 39.0
The solution tree is:
State (0,{0}) has optimal value: 39.0
 Decision alpha=1
  State (1,{0,1}) has optimal value: 38.0
   Decision alpha=3
    State (3,{0,1,3}) has optimal value: 35.0
     Decision alpha=4
      State (4,{0,1,3,4}) has optimal value: 20.0
       Decision alpha=2
        Base state (2,{0,1,2,3,4}) has initial value: 7.0
```

For the latter, we get the following solution.

```
The optimal value is: 39.0
The solution tree is:
State (0,{1,2,3,4}) has optimal value: 39.0
 Decision alpha=1
  State (1,{2,3,4}) has optimal value: 38.0
   Decision alpha=3
    State (3,{2,4}) has optimal value: 35.0
     Decision alpha=4
      State (4,{2}) has optimal value: 20.0
       Decision alpha=2
```

```
Base state (2,{}) has initial value: 7.0
```

One can easily verify that the two decision trees are equivalent, they only differ in the way the states are named.

11

Other Solver Results

This chapter shows numerical output for the spreadsheet and PNML solver codes, as generated by PN2Solver in the fashion described in Chap. 9. Recall that this solver code can be automatically obtained from the internal Bellman net representation. In this exposition, we focus on the linear search example.

11.1 PN2Spreadsheet Solver Code Output

11.1.1 PN2Spreadsheet Solver Code for LINSRCS

The spreadsheet solver code for LINSRCS is in the `LINSRCS.csv` file, as given in Sect. 9.3.1. We show this file again here, adding in an adjoining third column the formulas that are in Column B. When this `LINSRCS.csv` file is imported into a spreadsheet application, the formulas in Column B are evaluated; a screenshot of this is shown in Fig. 11.1. The result looks like this:

```
   | A                        | B
---+--------------------------+------
 1  statePlaceI35 ({})          0
 2  intermediatePlaceI33 p10    0.9    B1+0.8999999999999999
 3  statePlaceI17 ({2})         0.9    min(B2)
 4  intermediatePlaceI15 p4     1.9    B3+1.0
 5  intermediatePlaceI23 p6     1.3    B3+0.4
 6  intermediatePlaceI38 p11    1.5    B1+1.5
 7  statePlaceI21 ({1})         1.5    min(B6)
 8  intermediatePlaceI19 p5     2.1    B7+0.6
 9  intermediatePlaceI29 p8     1.9    B7+0.4
10  statePlaceI5 ({1,2})        1.9    min(B4,B8)
11  intermediatePlaceI3 p1      2.1    B10+0.2
12  intermediatePlaceI40 p12    0.6    B1+0.6000000000000001
13  statePlaceI27 ({0})         0.6    min(B12)
14  intermediatePlaceI25 p7     1.2    B13+0.6
15  intermediatePlaceI31 p9     1.6    B13+1.0
16  statePlaceI9 ({0,2})        1.2    min(B5,B14)
17  intermediatePlaceI7 p2      1.7    B16+0.5
18  statePlaceI13 ({0,1})       1.6    min(B9,B15)
19  intermediatePlaceI11 p3     1.9    B18+0.3
20  statePlaceI1 ({0,1,2})      1.7    min(B11,B17,B19)
```

Columns 1 and 4 are not part of the file; column 4 has been added to the figure to explicitly show the formulas in column B for expositional purposes. This redundant information makes it easier to interpret the spreadsheet.

A. Lew and H. Mauch: *Other Solver Results*, Studies in Computational Intelligence (SCI) **38**, 321–327 (2007)
www.springerlink.com

In the given spreadsheet, the final answer, i.e., the goal, is in cell B20 and has value 1.7. To determine the decisions that result in this solution, we note that 1.7 is the minimum of B11, B17, and B19, with in turn have values 2.1, 1.7, and 1.9, respectively. Thus, the optimal initial decision is to choose the middle one B17, which equals B16+0.5; the cost of the decision is 0.5. The value 1.2 in cell B16 is the minimum of B5 and B14, which in turn have values 1.3 and 1.2, respectively. Thus, the optimal next decision is to choose the latter one B14, which equals B13+0.6; the cost of the decision is 0.6. The value 0.6 in cell B13 is the minimum of B12, which in turn has value B1+0.6; the cost of the decision is 0.6. Thus, the optimal last (and in this example the only possible) decision is to choose B1, which corresponds to a base-state with value 0. We conclude that the "solution tree" is as follows:

```
The optimal value is: 1.7
The solution tree is:
State ({0,1,2}) has optimal value: 1.7
 Decision d=1          {choose middle one, at cost 0.5}
  State ({0,2}) has optimal value: 1.2
   Decision d=2         {choose latter one, at cost 0.6}
    State ({0}) has optimal value: 0.6
     Decision d=0       {make only choice, at cost 0.6}
      Base state ({}) has initial value: 0.0
```

In summary, the output produced by the PN2Spreadsheet module (see section 9.3) can be imported into a spreadsheet application. After the spreadsheet application has updated all the cells to its correct value, the user can obtain the optimal objective function value from the cell representing the goal state place. The reconstruction of the optimal policy within the spreadsheet application is not automated, and must be done "by hand". For future improvements, a graphically appealing output of the optimal decisions within a spreadsheet should be a valuable tool for decision makers.

In addition to the LINSRCS example given in the preceding section, we provide the PN2Spreadsheet solver code for some other sample problems in the following sections.

11.1.2 PN2Spreadsheet Solver Code for LSP

The output file for the LSP instance from section 2.26 is as follows.

```
"statePlaceI9 ({0,3},3)" ,0.0
"statePlaceI18 ({0,1,2,3},3)" ,0.0
"intermediatePlaceI7 p2" ,=B1+1.0
"intermediatePlaceI16 p4" ,=B2+1.0
"statePlaceI14 ({0,1,2},2)" ,"=max(B4)"
"intermediatePlaceI12 p3" ,=B5+1.0
"statePlaceI5 ({0,1},1)" ,"=max(B6)"
"intermediatePlaceI3 p1" ,=B7+1.0
"statePlaceI1 ({0},0)" ,"=max(B8,B3)"
```

11.1.3 PN2Spreadsheet Solver Code for MCM

The output file for the MCM instance from section 2.27 is as follows.

```
"statePlaceI5 (1,1)" ,0.0
"statePlaceI20 (4,4)" ,0.0
"statePlaceI25 (2,2)" ,0.0
"statePlaceI36 (3,3)" ,0.0
"intermediatePlaceI32 p6" ,=B1+B3+60.0
"statePlaceI12 (1,2)" ,"=min(B5)"
"intermediatePlaceI34 p7" ,=B2+B4+20.0
"statePlaceI14 (3,4)" ,"=min(B7)"
"intermediatePlaceI10 p2" ,=B6+B8+30.0
"intermediatePlaceI23 p4" ,=B8+B3+40.0
"intermediatePlaceI41 p9" ,=B6+B4+30.0
"intermediatePlaceI43 p10" ,=B3+B4+40.0
"statePlaceI30 (2,3)" ,"=min(B12)"
"intermediatePlaceI28 p5" ,=B2+B13+16.0
"intermediatePlaceI39 p8" ,=B1+B13+24.0
"statePlaceI8 (2,4)" ,"=min(B10,B14)"
"intermediatePlaceI3 p1" ,=B1+B16+24.0
"statePlaceI18 (1,3)" ,"=min(B15,B11)"
"intermediatePlaceI16 p3" ,=B18+B2+12.0
"statePlaceI1 (1,4)" ,"=min(B17,B9,B19)"
```

11.1.4 PN2Spreadsheet Solver Code for SPA

The output file for the SPA instance from section 2.43 is as follows.

```
"statePlaceI15 (3)" ,0.0
"intermediatePlaceI13 p4" ,=B1+8.0
"intermediatePlaceI18 p5" ,=B1+5.0
"statePlaceI9 (2)" ,"=min(B3)"
"intermediatePlaceI7 p2" ,=B4+5.0
"intermediatePlaceI11 p3" ,=B4+1.0
"statePlaceI5 (1)" ,"=min(B6,B2)"
"intermediatePlaceI3 p1" ,=B7+3.0
"statePlaceI1 (0)" ,"=min(B8,B5)"
```

11.1.5 Spreadsheet Output

The solver codes given above are shown in text format. When imported into a spreadsheet application, the formulas are evaluated. Screenshots of some examples are given here.

Spreadsheet Output for LINSRCS

For the LINSRCS problem, the optimal objective function value of $f(\{0,1,2\}) = 17.0$ appears in cell B20 in the spreadsheet in Fig. 11.1.

Spreadsheet Output for LSP

The optimal objective function value of $f(\{0\},0) = 3.0$ appears in cell B9 in the spreadsheet in Fig. 11.2.

Spreadsheet Output for MCM

The optimal objective function value of $f(1,4) = 76.0$ appears in cell B20 in the spreadsheet in Fig. 11.3.

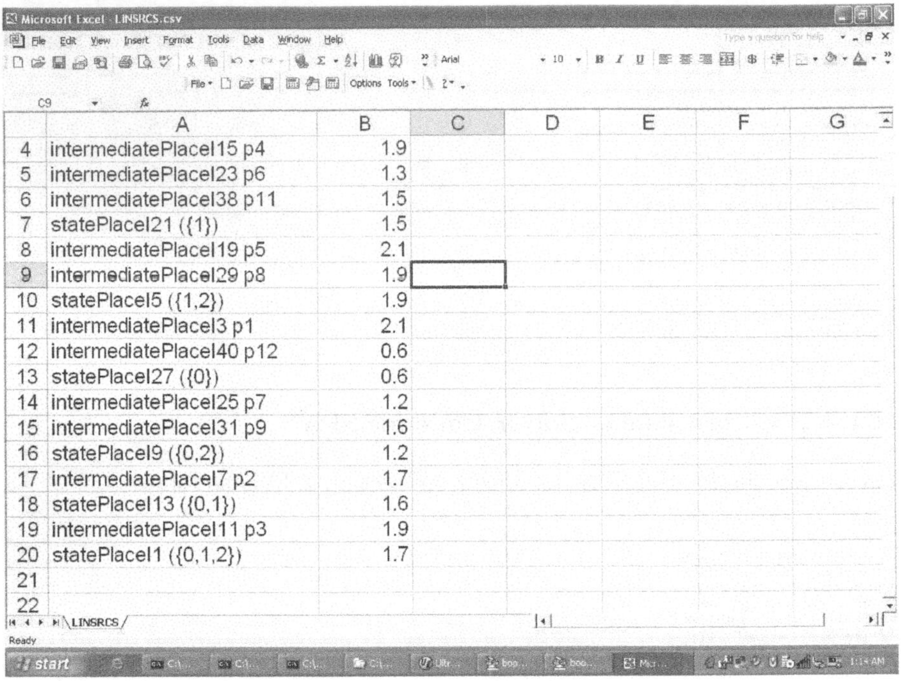
Microsoft Excel - LINSRCS.csv
File Edit View Insert Format Tools Data Window Help
C9
A B C D E F G
4 intermediatePlacel15 p4 1.9
5 intermediatePlacel23 p6 1.3
6 intermediatePlacel38 p11 1.5
7 statePlacel21 ({1}) 1.5
8 intermediatePlacel19 p5 2.1
9 intermediatePlacel29 p8 1.9
10 statePlacel5 ({1,2}) 1.9
11 intermediatePlacel3 p1 2.1
12 intermediatePlacel40 p12 0.6
13 statePlacel27 ({0}) 0.6
14 intermediatePlacel25 p7 1.2
15 intermediatePlacel31 p9 1.6
16 statePlacel9 ({0,2}) 1.2
17 intermediatePlacel7 p2 1.7
18 statePlacel13 ({0,1}) 1.6
19 intermediatePlacel11 p3 1.9
20 statePlacel1 ({0,1,2}) 1.7
21
22
LINSRCS
Ready
start

Fig. 11.1. Screenshot after importing LINSRCS.csv into Microsoft Excel

Spreadsheet Output for SPA

The optimal objective function value of $f(0) = 9.0$ appears in cell B9 in the spreadsheet in Fig. 11.4.

11.2 PN2XML Solver Code Output

The Petri net markup language solver code for LINSRCS was excerpted in Sect. 9.4.1. Because of its size, much of which is associated with specifying how to draw the graph, the complete PNML code for a much smaller example (`SPA.xrn`) is listed in App. A.3.

Since the PN2XML module produces standardized XML code that can readily be imported into PN simulators like Renew (see Sect. 5.1.5), one can use the simulator as a solver. The user can watch how a DP problem instance is solved in an illustrative graphical fashion by observing the simulation progress. Starting with the initial marking we fire transitions until the net is dead. Such a net is said to be in its final marking. The solution to the problem instance can be obtained by inspecting the net in its final marking. The graphical results of this simulation approach are documented for various examples in Sect. 6.2. For the LINSRCS example, the initial and final nets are shown in

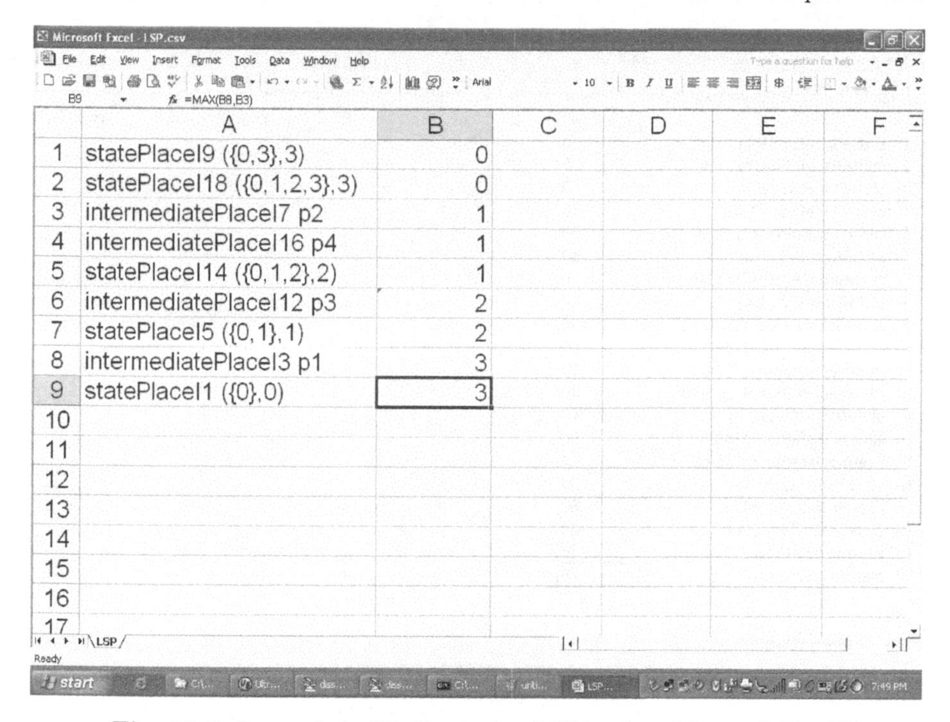

	A	B
1	statePlace l9 ({0,3},3)	0
2	statePlace l18 ({0,1,2,3},3)	0
3	intermediatePlace l7 p2	1
4	intermediatePlace l16 p4	1
5	statePlace l14 ({0,1,2},2)	1
6	intermediatePlace l12 p3	2
7	statePlace l5 ({0,1},1)	2
8	intermediatePlace l3 p1	3
9	statePlace l1 ({0},0)	3

Fig. 11.2. Screenshot after importing LSP.csv into Microsoft Excel

Figs. 6.21 and 6.22. These figures depict what appears on the screen using Renew. For legibility reasons, we show in this book (in Fig. 5.1) only a single screenshot, for a smaller example (`LSP.xrn`); Figure 5.1 should be compared with Fig. 6.23.

11.2.1 PN2XML Simulation Output for LINSRCS

For the initial net shown in Fig. 6.21, the only initially marked place (ignoring the enabling places) is the base-state place, marked with the value 0.0. For the final net shown in Fig. 6.22, the goal-state place is marked with the result 1.7. The final markings in the intermediate places can be used to reconstruct the optimal policy in a fashion similar to that used in the spreadsheet case. However, use of a Petri net *simulation* tool (such as Renew) which permits animation that can be paused, the optimal policy can also be determined by single stepping through the simulation process. We will not describe all of the details here since we are constrained to the printed page. To illustrate the basic idea, we do so tabularly rather than graphically.

We show in the following table a sequence of changes in markings, starting from the initial one in column 0, and ending with the final marking in column 6. The rows of the table correspond to the place nodes of the Bellman net, and the columns specify their markings as a function of time. The

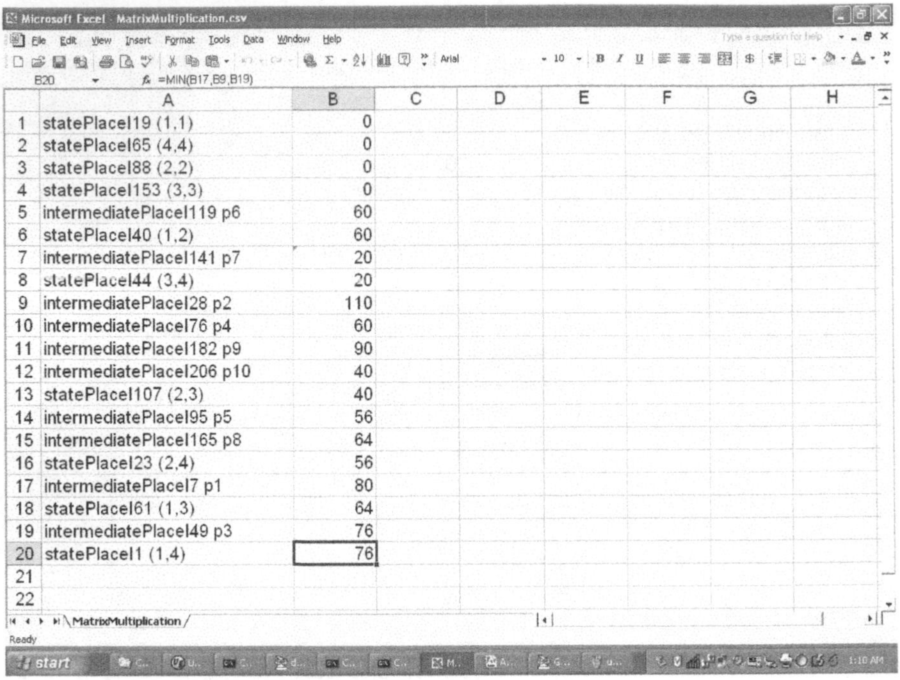

Microsoft Excel - MatrixMultiplication.csv

B20 =MIN(B17,B9,B19)

	A	B
1	statePlace119 (1,1)	0
2	statePlace165 (4,4)	0
3	statePlace188 (2,2)	0
4	statePlace1153 (3,3)	0
5	intermediatePlace1119 p6	60
6	statePlace140 (1,2)	60
7	intermediatePlace1141 p7	20
8	statePlace144 (3,4)	20
9	intermediatePlace128 p2	110
10	intermediatePlace176 p4	60
11	intermediatePlace1182 p9	90
12	intermediatePlace1206 p10	40
13	statePlace1107 (2,3)	40
14	intermediatePlace195 p5	56
15	intermediatePlace1165 p8	64
16	statePlace123 (2,4)	56
17	intermediatePlace17 p1	80
18	statePlace161 (1,3)	64
19	intermediatePlace149 p3	76
20	statePlace11 (1,4)	76
21		
22		

MatrixMultiplication

Fig. 11.3. Screenshot after importing MCM.csv into Microsoft Excel

leftmost column gives the place node id and label for each of the state and intermediate place nodes in the net, whose marking is determined by evaluating the transition expression given in the rightmost column. Column 1 shows that places I33, I38, and I40 can change their markings after the firing of their associated E-transitions in any order; each of these E-transitions are initially enabled since they depend only on the base-state place I35 which is initially marked. Column 6 shows that the goal place I1 can change its marking after the firing of its associated M-transition, which is only enabled after place nodes I3, I6, and I11 have all had their markings changed from their initial empty state.

```
place node |  0  |  1  |  2  |  3  |  4  |  5  |  6
-----------+-----+-----+-----+-----+-----+-----+-----+
I35 ({})    :  0
I33 p10     :       0.9                                    I35+0.9
I38 p11     :       1.5                                    I35+1.5
I40 p12     :       0.6                                    I35+0.6
I17 ({2})   :             0.9                              min(I33)
I21 ({1})   :             1.5                              min(I38)
I27 ({0})   :             0.6                              min(I40)
I15 p4      :                   1.9                        I17+1.0
I23 p6      :                   1.3                        I17+0.4
I19 p5      :                   2.1                        I21+0.6
I29 p8      :                   1.9                        I21+0.4
I25 p7      :                   1.2                        I27+0.6
I31 p9      :                   1.6                        I27+1.0
I5 ({1,2})  :                         1.9                  min(I15,I19)
I9 ({0,2})  :                         1.2                  min(I23,I25)
I13 ({0,1}) :                         1.6                  min(I29,I31)
I3 p1       :                               2.1            I5+0.2
I7 p2       :                               1.7            I9+0.5
I11 p3      :                               1.9            I13+0.3
```

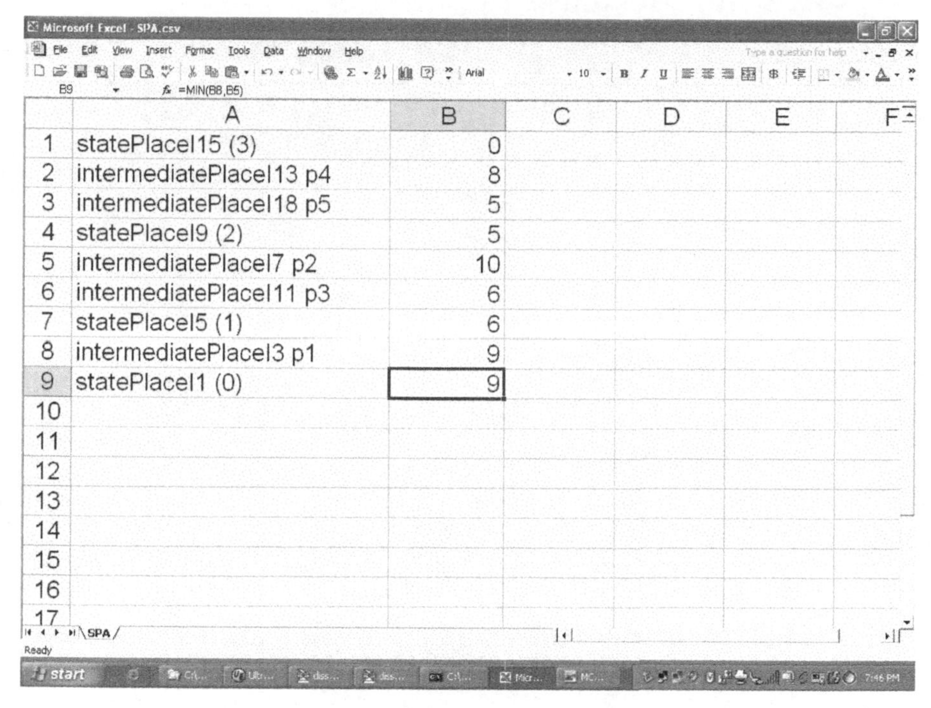

Fig. 11.4. Screenshot after importing SPA.csv into Microsoft Excel

```
I1 ({0,1,2}):                           1.7    min(I3,I7,I11)
```

12

Conclusions

In this book, we discussed dynamic programming as a general method for solving discrete optimization problems. This book is distinguished from many others, such as [1, 12, 13, 57], by its restriction to discrete problems and by its pragmatic approach. Here, we emphasize computational aspects rather than, say, theoretical aspects such as conditions under which problems are solvable by DP. The key to the use of DP is the formulation of a problem as a sequential decision process, where separable costs can be attached to each decision. For a multitude of problems, we derive such formulations, in the form of a dynamic programming functional equation that solves the problem. We assume these DPFEs are "proper" (in the sense that their solutions exist in theory), and focus on the computational problem of numerically obtaining these solutions.

It would be ideal if there existed a *mechanical* procedure to derive a DPFE from a description of a problem. Unfortunately, no such procedure exists for arbitrary problems. Thus, in the first part of this book, we provide an extensive catalogue of DP formulations. Numerous examples from a wide range of application areas are given. Exposure to these examples should make it easier to apply DP to new problems. The remainder of the book is devoted to the problem of obtaining the numerical solutions of these and other DPFEs. We specifically discuss the design and implementation of a computational tool (DP2PN2Solver) that yields these solutions.

12.1 Applicability of DP and DP2PN2Solver

In order to demonstrate the generality and flexibility of DP as a method to solve optimization problems and of the DP2PN2Solver tool for numerically solving these DP problems, we included numerous applications in Chap. 2. In addition to choosing examples from an exceptionally large variety of sources, we also chose to implement all (with a few exceptions, as noted below) of the examples found in the DP chapters of several major textbooks in the fields of operations research and computer science.

A. Lew and H. Mauch: *Conclusions*, Studies in Computational Intelligence (SCI) **38**, 329–337 (2007)
www.springerlink.com

- The following examples appear in the operations research book by Hillier and Liebermann [21, Chap. 11]: Ex.1: SCP; Ex.2: ALLOT; Ex.3: ALLOTm; Ex.5: ILP; Ex.6: PRODRAP; Ex.7: INVESTWLV. (Example 4 was skipped since it involves a continuous decision set. In principle, discretization could be used for such examples.)
- The following examples appear in the operations research book by Winston [63, Chap. 18]: Ex.1: NIM; Ex.2: POUR; Ex.3: SCP; Ex.4: INVENT; Ex.5: ALLOTf; Ex.6: KSINT; Ex.7: REPLACE; Ex.8: DDP; Ex.12: TSP; Ex.13: MINMAX; Ex.14: ALLOTm; Ex.15: LOT. (Examples 9–11 were skipped since they were only sketched in the book, and thus lack specificity.)
- The following examples appear in the computer science algorithms book by Cormen, et al. [10, Chap. 15]: Sect. 1: ASMBAL; Sect. 2: MCM; Sect. 3: LSP; Sect. 4: LCS; Sect. 5: BST.
- The following examples appear in the computer science algorithms book by Horowitz, et al. [22, Chap. 5]: Sect. 1: KS01; Sect. 2: SCP; Sect. 3: APSPFW; Sect. 4: SPC; Sect. 5: BST; Sect. 6: EDP; Sect. 7: KS01; Sect. 8: RDP; Sect. 9: TSP; Sect. 10: FLOWSHOP. (KS01 was treated twice, the second time in greater depth.)

The two operations research books also contain a chapter illustrating the application of DP to Markovian decision processes. Iterative methods to solve such problems are not discussed in this book. The two computer science books also discuss "all-pairs shortest-path" algorithms, where a matrix of values rather than a single value is computed, such as for APSPFW. For these and some related algorithms, the order in which calculations are performed affect efficiency. In this book, we have neglected these concerns, but of course the utility of DP and of our DP tool depends very much on efficiency. We address this issue later in this chapter.

12.2 The DP2PN2Solver Tool

A problem that has existed for over fifty years is the development of general-purpose computer software for solving large classes of DP problems, comparable to what has long existed for linear programming (LP). The problem for LP is much simpler, in part because the linearity assumption for objective functions and constraints can easily be represented in a standard form. For example, a large class of LP problems can be represent in a tabular or matrix form known as a "simplex tableaux", and early linear programming software were matrix based. It was natural for more recently developed spreadsheet systems to incorporate simplex algorithms in "add-in" solvers for LP problems. However, an analogous general-purpose solver for DP has been sorely lacking. Some attempts have been made over the years, such as by Hastings [19] and by Sarmiento [55], but have not been successful, at least as measured by their recognization in the community of scholars and practitioners. For example, in operations research textbooks, such as those of Hillier and Lieberman [21] or Winston [63], there are many examples of computer solutions of LP problems, generally using Excel or Lingo, but there are no comparable examples for DP. This is largely because other optimization software systems were not designed expressly for, hence are not especially well suited for, solving DP problems in general.

A software system for *automating* the computational solution of an optimization problem having a given DP formulation is described in this book. This system, called DP2PN2Solver, was designed expressly for solving DP problems. A DP specification source language (gDPS) was first designed, general enough to accommodate a very large class of problems, but simple enough for nonprogrammers to learn with a minimum of effort. In essence, this DP specification language is a transliteration of the DPFE. For flexibility, the DP2PN2Solver system permits other specification languages to be used as an alternative, such as a basic language for simpler problems, or perhaps more complex mathematical programming languages. The design of translators from any specification language to a common intermediate representation is then discussed. Rather than implementing different compilers for different source languages, it may be advantageous for specifications in other source languages to instead be translated (say, using a macro processor) into gDPS.

A key decision was the adoption of an intermediate representation, called a Bellman net, based upon Petri net models. A solver code generator was then designed to translate a Bellman net into computer object code (in, say, a programming language such as Java) from which an object code compiler can be used to produce the numerical solution of the original DP problem. This task is complicated by the desire to produce as answers not only optimal values but also "reconstructed" optimal policies. The Bellman net representation of a DP problem facilitates this solver code generation process for a variety of different object languages. As an illustration of the flexibility of this system, it is also shown how a spreadsheet language or a Petri net simulator can be used an alternative to Java as an object language.

Our Bellman net model was designed to suit the needs of both the translator and the solver code generator. Furthermore, to allow the use of Petri net theory in the analysis of DP problems, the Bellman net model was defined in a manner consistent with high-level Petri net conventions. Certain DP problems can also be modeled using a subclass of low-level Petri nets. High-level colored Petri nets allow for more generality, and low-level place/transition nets allow for easier automated analyses of structural or behavioral properties of the model. Consistency with Petri net conventions also allows Bellman nets to be processed using standardized Petri net software, such as Renew.

DP2PN2Solver guarantees that the optimal solution to a properly formulated DP instance will be found, provided of course that the DPFE is correct and its solution does exist. Many other optimization software systems incorporate search mechanisms. This can in the best case be very efficient, but in the worst case not lead to the optimal solution at all.

In summary, this book describes the design and implementation of a software system DP2PN2Solver that includes a DP specification language, the Bellman net representation, a translator between the two, and a solver code generator that produces executable code which in turn yields numerical solutions. Numerous examples are given to demonstrate the general applicability of this software system to the solution of a wide variety of DP problems.

The DP2PN2Solver system addresses the main shortcomings of other DP software tools. It can be used for a myriad of problems, it does not require programming expertise, and it does not require proprietary commercial systems (neither special-purpose ones, nor office packages such as Excel). Furthermore, the modular structure of the DP2PN2Solver system allows for changes in the input source specification languages and for alternative output solution solvers. In the worst case, it is conceivable that if a DP problem involves features unanticipated by our choice of source syntax, the associated Bellman net can be manually constructed. Furthermore, the production of an intermediate Bellman net representation as a text file also allows for analysis or other processing of DP formulations by independently developed tools.

12.3 Research Directions

There are numerous extensions of the work reported in this book that are currently the subject of current research. Some extensions relate to extending the applicability of our software tool DP2PN2Solver, especially so as to:

- solve continuous problems (by automating discretization procedures, for example).
- solve problems with more complex state types, e.g., matrices, strings, multisets (sets with nonunique elements).
- solve iterative problems associated with Markovian decision processes.

While any restriction is one too many, the variety of applications that can be solved using our current tool, as presented in Chapter 2, is quite large indeed. Nevertheless, we continue to explore methods of extending the class of DP problems that can be automatically solved. These extensions relate to:

- methods of improving the user interfaces to make the tool easier and more productive to use.
- methods of solving large-scale problems, including those with larger state spaces, and with larger solution trees.
- methods of modeling DP problems using other classes of Petri nets, that may have better diagnostic or efficiency advantages, for example.

We mention here a few areas in which we have a special interest.

12.3.1 User Functionality

The utility of a software tool for solving DP problems, as with other problems, depends largely upon its functionality, which loosely speaking is what it outputs (the more the better) for various inputs (the less the better). Ideally, users should be able to provide a DPFE as input easily and naturally, and the system should automatically produce the computational solution to the DPFE in a form that is easy to interpret.

User Interfaces

As noted previously, changes to gDPS to improve its utility are relatively easy to make, since we used a compiler-compiler to parse gDPS programs. We discuss language extensions below. In addition, alternate front-ends (source languages and compilers) and back-ends (solver code generators and execution systems) can also be incorporated as part of the DP2PN2Solver tool. Thus, alternatives that would make our tool more convenient to use are being explored.

- Provide a windows-based rather than command-line environment.
- Add a graphical or tabular user interface to input DP specifications and extend ways to produce and output optimal policies.
- Reconstruct the solution in the spreadsheet solver output.
- Incorporate more complex layout algorithms to improve the graphical display of Bellman nets.
- Extend the error checking capabilities of the proposed system.

The latter is of course of special value, and will be discussed further below.

Language Extensions

Language extensions range from simple syntax refinements to changes that require significant redesign efforts.

- Only additive and multiplicative DPFEs, where decision costs or rewards are combined by addition or multiplication operations, are currently permitted in gDPS. Permitting the minimum and maximum operations as well is one extension. (This would simplify the solution of the MINMAX problem.) In general, any binary operation that is associative, commutative, and has an identity element, is a reasonable candidate for inclusion in gDPS.
- In our current implementation of gDPS, data is hard-coded into the `GENERAL_VARIABLES` section. Thus, to solve a problem with varying values of data, the gDPS source program must be repeatedly recompiled. Allowing for data to be input from a file (or as a command-line argument) would be a convenient extension.

- gDPS is basically a "specification" language rather than a "procedural" language in that, except for in the GENERAL_FUNCTIONS section, no (control) *sequence* of assignment, alternation, or iteration statements are allowed. One extension is to remove the procedural elements in the GENERAL_FUNCTIONS section, perhaps by having general functions expressed in, say, a *functional* language [17] or an abstract *relational* language [11].

Error Diagnostics

In Sect. 8.4, we described the error diagnostics capabilities of DP2PN2Solver. Diagnostic error messages are given in App. B.7. Improvements in diagnostics is one area of continuing research.

Of special theoretical interest is the investigation of methods of testing whether a DPFE is proper based upon Petri net analysis techniques. For example, as we noted previously (in Sect. 8.4), certain errors, such as the existence of circular definitions and incorrect or omitted base conditions, are detectable by an analysis of net properties such as cyclicness and deadlock. Much more along these lines remains to be done.

One important class of errors occurs when memory size limitations are exceeded. This may be due to the state space being too large. Even when the size of the state space is modest, since the DPFE is solved recursively, the solution tree (such as for the Fibonacci example) may become too large. Methods of reducing size are discussed in the next section.

12.3.2 Reduction of Dimensionality

One limitation of our tool (and every tool) is on the size or dimensionality of the problems that can be solved. Thus, means to reduce dimensionality, of both the state space and the size of the solution tree, is an important objective. As suggested in the preceding section, one avenue of future research is the automation of means to convert recursion to iteration whenever possible.

Time and Space Complexity

The computational requirements of DP2PN2Solver are asymptotically the same as the requirements for other DP codes implementing the DP functional equation directly, since the intermediate PN model may be regarded as just another representation of a DP functional equation. For example, since a DP solution of the traveling-salesman problem is inefficient (exponential-time), so is a PN modeled solution. Nevertheless, small instances of such "intractable" problems can still be handled using DP, and methods that expand these instances are being investigated.

The space requirements of DP2PN2Solver are dominated by the space requirements needed to store the intermediate PN representation. This could become a problem for larger instance sizes that deplete the space resources. One technique that alleviates this problem is to reallocate those parts of the memory that are no longer needed in the course of the computation of the current instance being solved. Applying this technique to DP2PN2Solver means that a state-decision-reward-transformation-table like Table 8.1 is no longer constructed in its entire size. Rather, parts of the intermediate PN representation are put together in a bottom-up fashion starting with the base place states (as opposed to the current top-down build starting with the goal place state). Whenever a state place has served all transitions in its postset, it and its remaining predecessor subtree can be deleted. That the Bellman net is in essence an acyclic graph ensures that deleted portions will not be needed again. The space freed in this way can then be recycled. Implementing this technique into DP2PN2Solver is one avenue of future research.

Potential for Parallelization

DP2PN2Solver's use of a PN representation has the potential for parallelization. Computations that can take place concurrently are identified by transitions that are enabled concurrently. In theory, the PN's intrinsic concurrency allows the distribution of the computational load encountered when solving a single DP problem instance to several computational units. This opens up various possibilities for machines with multiple parallel processors.

Rather than use existing parallel processing systems, a special-purpose system may be designed expressly to solve DP problems. That DP problems can be represented by Bellman nets suggests the adaptation of *dataflow* computer architectures long associated with Petri nets. For example, a dataflow architecture proposed for spreadsheets ([39]) may prove especially useful for the execution of DP problems.

12.3.3 Petri Net Modeling

The adoption of Petri nets in our tool has the advantage that there is an abundance of theoretical results in the PN literature, some of which may be useful in analyzing DP models. Consistency checks such as a test for circularity, a test whether the net is dead before a proper final marking is reached and tests whether there are unmarked source state places are obvious examples. It will be a part of future research to find additional examples.

Additional research into our underlying Petri net models may also prove helpful:

- Develop software for automated translation between LLBN and HLBN model.
- Extend the LLBN model to allow more general DP functional equations, e.g. involving multiplication.
- Extend the HLBN model to allow more general E-transitions.

- In the HLBN drawings, maintain the preset of M-transitions (using self-loops), so that the optimal solution can be reconstructed from the drawing.
- Examine whether it is worthwhile to simulate PNs directly, i.e. solve the problem on the PN level, instead of transforming it to a high-level general purpose programming language such as JAVA? The advantage is that the graphical representation of a PN allows the user to visually follow the solution process. The disadvantage is that the PN simulator might cause additional overhead with respect to time complexity. If HLBNs are used, then there are two more disadvantages. First a specialized PN simulator capable of simulating HLBNs needs to be used. Second, the reconstruction of the optimal policy is difficult, since only the optimal objective function value is computed directly. (This poses no problem for LLBNs, the tree obtained by following empty min-in places represents the optimal policy.)
- Improve the suggested Bellman net models. Develop more theory about Bellman nets. How do they compare to the standard PN model with respect to static and dynamic properties?
- Consider ways to model other DP solution methodologies, including iterative ones. This may ultimately lead to ways to handle Markovian decision processes.

With respect to iterative approaches, we are investigating an "iterative Bellman net" model in which a Petri net automatically restarts with final markings used to determine new initial markings.

12.4 Summary

In this book, we described the solution of discrete optimization problems using dynamic programming, emphasizing the design and use of a computational tool named DP2PN2Solver for numerically solving these problems. We formulated and numerically solved about 50 different problems from a wide variety of application areas. We also described:

- the design and implementation of specification source languages (gDPS, bDPS, ilpDPS) that express given DPFEs for solving a myriad of DP problems, all of which were implemented in gDPS;
- Bellman-net models of DPFEs and a suitable internal computer representation for these Bellman nets;
- a compiler (gDPS2PN) that translates from gDPS source to Bellman nets;
- code-generators (PN2Java, PN2SS, PN2XML) that translate from Bellman nets to object code that can be executed by "solver" systems so as to obtain the numerical solutions to the given DPFEs (in Java or Excel or Renew);
- and the overall structure of the DP2PN2Solver system.

Central to the design of this system is the Bellman net model, whose definition had to satisfy the requirements of generality, so that large classes of

DPFEs can be so modeled, and of standardization, i.e. adherence to Petri net conventions, so that Petri net theories can be applied to the analysis of the Bellman nets. Standardization also permits the option of using Petri tools for the graphical display of Bellman net models as well as for object-code execution. The choice of internal representation for Bellman nets was also of great practical importance because of how it affects the design of compilers and code-generators, certainly with respect to efficiency.

DP2PN2Solver should be useful as an educational tool for teaching DP in a practical hands on approach where students can work on case studies, model the underlying DP problem, and obtain an actual solution to the problem without concern about the solution process itself. Practitioners should also find DP2PN2Solver useful, certainly for problems (perhaps scaled down) similar to those solved in this book. Researchers should find the task of extending this work in various ways a rich source of interesting problems.

A

Supplementary Material

A.1 Pseudocode of the DP2PN Module

This section shows the individual components of the DP2PN module in a Java inspired pseudocode style notation. The problem specific `Main` and `State` classes are generated automatically by JavaCC. They are shown here for the LINSRCS problem. All other classes are problem independent; they represent general data structures and functions needed across all DP problems. Classes that are rather trivial extensions of basic structures (`StateList` extends `List`, `StateSet` extends `Set`, and `NodeSet` extends `Set`, `StateValueMap` extends `Map`) are not shown here.

A.1.1 Main Class for LINSRCS

```
//This Main class has been
//automatically generated by the DPspecificationParser
//for the problem: LINSRCS

public class LINSRCSMain {
  String problemName="LINSRCS";
  private static double[] prob= {.2,.5,.3};
  private static int N = prob.length;

  void main() {

    State goalState = new State(setOfAllItems);
    StateValueMap baseStatesWithValues = new StateValueMap();
    baseStatesWithValues.put(new State(emptySet),
                             new Double(0.0));

    DPInstance theInstance
      = new DPInstance(goalState,baseStatesWithValues);
```

```
theInstance.setTransitionOperator("+");
theInstance.setMinOrMax("min");

//make table entries automatically
StateSet statesSeen =new StateSet(); //keep track for which
                                     //states table entries
                                     //have been made
//add goal state to statesSeen
statesSeen.add(goalState);
LinkedList stateQueue = new LinkedList();
//initialize queue with the goal state
stateQueue.addLast(goalState);
//loop as long as there is something in the queue
while (stateQueue.size()>0) {
  //dequeue the first state and make it the current state
  State currentState=(State) stateQueue.removeFirst();
  //check whether the current state is a base state as
  //defined in DPFE_BASE_CONDITIONS section; if so, add
  //it and its value to baseStatesWithValues
  if (determineWhetherBaseState(currentState)==true) {
    baseStatesWithValues.put(currentState,
      new Double(determineBaseStateValue(currentState)));
  }

  //if current state is not a base state...
  if (!baseStatesWithValues.containsKey(currentState)) {
    //create the decision set
    NodeSet _setInDDN0=new NodeSet();
    { //extra block so _i reusable
      for (int _i=((Integer)
               currentState.getCoordinate(0)).intValue();
           _i<=((Integer)
               currentState.getCoordinate(1)).intValue()-1;
           _i++) {
        _setInDDN0.add(new Integer(_i));
      }
    } //close extra block
    NodeSet decisionSet=_setInDDN0;
    //loop over all decisions now
    for (Iterator it=decisionSet.iterator();it.hasNext();) {
      Integer currentDecisionCandidate = (Integer)it.next();
      Decision d
        = new Decision(currentDecisionCandidate.intValue());
      d.setDecisionPrefix("k=");
      //determine the successor states of current state
```

```
        StateList successorStates
            = calculateTransformation(currentState,d);
        //determine the transition weights given the
        //current state
        ArrayList transitionWeights
            = calculateTransitionWeights(currentState,d);
        //generate table entry
        DPInstanceTableEntry entry = new DPInstanceTableEntry(
            currentState,
            d,
            calculateReward(currentState,d),
            successorStates,
            transitionWeights);
        theInstance.addTableEntry(entry);
        //enqueue unseen destination states by iterating
        //over successorStates
        for(Iterator it2=successorStates.iterator();
            it2.hasNext();) {
          State currentSuccessor = (State) it2.next();
          if(!statesSeen.contains(currentSuccessor)) {
            stateQueue.addLast(currentSuccessor);
            //mark state as seen
            statesSeen.add(currentSuccessor);
          }
        }
      } //end of for loop over the decisions
    } //end of if
  } //end of loop once queue is empty

  //Build the BellmanNet from the instance.
  BellmanNet bn
      = theInstance.buildBellmanPNTopDown(problemName);
  //Output the BellmanNet as incidence matrix.
  bn.toIncidenceMatrix();

}//end of main() method


double calculateReward(State s, Decision d) {
  double result;
  result = (N+1-size(((NodeSet)
      ((NodeSet) s.getCoordinate(0)).clone())))
      *prob[d.getDecision()];
  return result;
}
```

```
  StateList calculateTransformation(State s, Decision d) {
    StateList result=new StateList();
    NodeSet items=((NodeSet)
        ((NodeSet) s.getCoordinate(0)).clone());
    NodeSet _setExplicit2=new NodeSet();
    _setExplicit2.add(new Integer(d.getDecision()));
    items.removeAll(_setExplicit2);
    result.add(new State(items));
    return result;
  }

  double determineBaseStateValue(State s) {
    if(s.getCoordinate(0)==s.getCoordinate(1)) return 0.0;
    return Double.NaN; //NaN denotes: not a base state
  }

  boolean determineWhetherBaseState(State s) {
    if (Double.isNaN(determineBaseStateValue(s))) {
      return false;
    }
    return true;
  }

}//end of MCMMain class
```

A.1.2 State Class for LINSRCS

```
//This State class has been
//automatically generated by the DPspecificationParser
//for the problem: LINSRCS

class State {

  List theState;

  //Constructor for State class
  State(NodeSet items) {
    theState=new ArrayList();
    theState.add(items);
  } //end of constructor

  //Detemine the lexicographical order of two states.
  //Iterate over instance variable theState
  //and compare each coordinate with state2.
```

```
  int compareTo(Object state2) {
    if states are equal return 0;
    if theState<state2 return -1;
    if theState>state2 return 1;
  }

  //Returns the i-th coordinate of this state
  Object getCoordinate(int i) {
    return theState.get(i);
  }

  //Set i-th coordinate in the state to object o
  public void setCoordinate(int i, Object o) {
    theState.set(i,o);
  }

}//end of State class
```

A.1.3 Decision Class

```
//For most DP problems a decision is essentially an int
//plus a String prefix like "k=" or "goto"

public class Decision {

  private String decisionPrefix;
  private int decision;

  //constructor
  Decision(int d) {
    decision=d;
    decisionPrefix="";
  }

  //accessor method for the int core of the decision
  //without the prefix
  int getDecision() {
    return decision;
  }

  //accessor method to set the decisionPrefix
  void setDecisionPrefix(String s) {
    decisionPrefix=s;
  }
```

```
}//end of class
```

A.1.4 DPInstanceTableEntry Class

```
//Represents info for the DPInstance class
//format:
// state--decision--reward--List of nextStates--List of weights

class DPInstanceTableEntry {

  State state;
  Decision decision;
  double reward;
  StateList nextStates;
  ArrayList transitionWeights;

  //constructor and basic accessor methods not shown here

}//end of class
```

A.1.5 DPInstance Class

```
//DP instance represents all the information necessary to
//build a complete state diagram for the DP instance

public class DPInstance {

  ArrayList stateDecisionRewardTransformationTable;
  private String operatorForTransition;
      //e.g. "+" for sum transition, or "*" for mult transition
  private String minOrMax; //min or max problem?
  private State goalState;
  private StateValueMap baseStatesWithValues;//store base states
                                             //along with values

  //The constructor creates a DP instance with no table entries.
  //Add entries with add() method.
  public DPInstance(State goal, StateValueMap bases) {
  ...
  }

  //Accessor methods not shown
  ...

  void addTableEntry(DPInstanceTableEntry e){
```

```
  stateDecisionRewardTransformationTable.add(e);
}

//The following is the central method where the BellmanNet
//is built and returned.
BellmanNet buildBellmanPNTopDown(String netName) {
  //initialize place and transition counters
  int sp=0; //for state places (not used for naming,
            //only for graphical layout)
  int ip=0; //for intermediate places (i.e. places
            //not associated with a state)
  int ep=0; //for enabling places(allow eval
            //transitions to only fire once)
  int st=0; //for evaluation transitions
            //the type of operator ("+","*") is specified
            //in the instance variable operatorForTransition
  int mt=0; //for min transitions

  StateSet statesBuilt = new StateSet();
      //to keep track which states are already built
      //in the Bellman Net

  //create an empty Bellman net
  BellmanNet bn=new BellmanNet(netName);

  //build the goal state
  Place goalPlace=new StatePlace(goalState.toString());
  bn.addPlace(goalPlace);
  statesBuilt.add(goalState);

  LinkedList stateQueue = new LinkedList();
  //initialize queue with the goal state
  stateQueue.addLast(goalState);

  //loop as long as there is something in the queue
  while (stateQueue.size()>0) {
    //dequeue the first state and make it the current state
    //(it has already been built when it was discovered)
    State currentState=(State) stateQueue.removeFirst();
    if (!baseStatesWithValues.containsKey(currentState)) {
        //regular state?
     //count how many times currentState appears on left
     //hand side of the table now
     int stateAppearanceInTableCounter=0;
     for(Iterator i
```

```
            =stateDecisionRewardTransformationTable.iterator();
            i.hasNext();) {
          DPInstanceTableEntry anEntry
            = (DPInstanceTableEntry) i.next();
          if (anEntry.getState().equals(currentState)) {
            stateAppearanceInTableCounter++;
          }
        }//end of for
        if (stateAppearanceInTableCounter>0) {
            // Making sure that current state
            // has successors. Flawed BellmanNets might have
            // states that do not have successor states,
            // but are not base states either.
         mt++; //update the counter for min/max transitions
         MTransition newMinMaxTransition
            = new MTransition("mt"+mt,minOrMax);
         newMinMaxTransition.setNumberOfArguments
             (stateAppearanceInTableCounter);
         newMinMaxTransition.addInscription();
         bn.addTransition(newMinMaxTransition);

         Arc arcFromMinMaxTransToState
           =new Arc(newMinMaxTransition,
                    bn.getPlace(currentState.toString()),
                    "ordinary");
         arcFromMinMaxTransToState.addInscription("y");
            //y is output var. of min/max transition
         bn.addArc(arcFromMinMaxTransToState);
        }//end of if block making sure that currentState
         //has successor states
       }//end of if block making sure that currentState
        //is not a base state
       else { //currentState is a base state
         double tokenValue
             = (baseStatesWithValues.get(currentState));
         //get a reference for the place:
         Place basePlace=bn.getPlace(currentState.toString());
         basePlace.addToInitialMarking(tokenValue);
       }
       //loop through through the left column of the table
       int inputArcForMinMaxTransCounter=0;
          //counter to label the arcs x1,x2,etc.
       for(Iterator i
            =stateDecisionRewardTransformationTable.iterator();
          i.hasNext();) {
```

```
DPInstanceTableEntry currentEntry
  =(DPInstanceTableEntry) i.next();
if (currentEntry.getState().equals(currentState)) {
  //found the current state in the left column
  ip++; //update counter for intermediate places
  Place intermedPlace = new IntermediatePlace("p"+ip,
      currentEntry.getDecision().toString());
  bn.addPlace(intermedPlace);

  Arc arcFromIntermedPlaceToMinMaxTrans
    =new Arc(intermedPlace, //bn.getPlace("p"+ip),
             bn.getTransition("mt"+mt),
             "ordinary");
  inputArcForMinMaxTransCounter++; //update arc counter
  arcFromIntermedPlaceToMinMaxTrans.addInscription(
      "x"+inputArcForMinMaxTransCounter);
  bn.addArc(arcFromIntermedPlaceToMinMaxTrans);

  st++; //update counter for E(val)-transition
  ETransition evalTrans
      =new ETransition("st"+st,
                       operatorForTransition,
                       currentEntry.getReward());
  bn.addTransition(evalTrans);

  Arc arcFromETransToIntermedPlace
    =new Arc(evalTrans, //bn.getTransition("st"+st),
             intermedPlace, //bn.getPlace("p"+ip),
             "ordinary");
  arcFromETransToIntermedPlace.addInscription("y");
      //y is output var. of eval. transition
  bn.addArc(arcFromETransToIntermedPlace);

  //now make enabling places with a single black token
  //as initial marking to allow eval transitions to
  //only fire once
  ep++; //update counter for enabling places
  Place enablingPlace=new EnablingPlace("ep"+ep);
  bn.addPlace(enablingPlace);

  Arc arcFromEnablingPlaceToEvalTrans
    =new Arc(enablingPlace,
             evalTrans,
             "ordinary");
  bn.addArc(arcFromEnablingPlaceToEvalTrans);
```

```
        //make placeS for the newly found stateS and enqueue
        //THEM, if THEY do not already exist
        StateList destinationStates
            = currentEntry.getNextStates();
        State currentDestination;
        //loop over all destination states now
        int inputArcForEtransCounter=0;
            //counter to label the arcs x1,x2,etc.
        for(Iterator i2=destinationStates.iterator();
            i2.hasNext(); ) {
          currentDestination=(State) i2.next();
          if(!statesBuilt.contains(currentDestination)) {
            sp++; //update counter for state places
            Place statePlace
               =new StatePlace(currentDestination.toString());
            bn.addPlace(statePlace);
            statesBuilt.add(currentDestination);
            //now enqueue it
            stateQueue.addLast(currentDestination);
          }
          //make an arc (regardless whether state
          //is new or not)
          Arc doubleArcBetweenStateAndEtrans
              =new Arc(
                bn.getPlace(currentDestination.toString()),
                evalTrans,
                "double");
          inputArcForEtransCounter++;
          doubleArcBetweenStateAndEtrans.addInscription(
              "x"+inputArcForEtransCounter);
          bn.addArc(doubleArcBetweenStateAndEtrans);
        }//end of for loop over destination states
        //At the end of the above for-loop we know how many
        //arcs go into the ETransition, so we can build its
        //inscription now
        evalTrans.setNumberOfVariables(
            inputArcForEtransCounter);
        evalTrans.addInscription();
      }//end of if
    }//end of for-loop over table entries
  }//end of while

  return bn;
}//end of buildBellmanPNTopDown() method
```

```
}//end of class
```

A.1.6 BellmanNet Class

```
class BellmanNet {

  String title; //holds the title of the net
  ArrayList placeList; //a list of Place objects
  ArrayList transitionList; //a list of Transition objects
  ArrayList arcList;  //a list of Arc objects

  //useful to have is an adjacency matrix. It can be calculated
  //using the method calculateAdjacencyMatrix()
  //the indices run from:
  // 0,..,|transitionList|-1,|transitionList|,..,
  //   |transitionList|+|placeList|-1
  //adjacencyMatrix[x][y]=1 means that there is an arc
  //from x to y
  int[][] adjacencyMatrix;

  //An incidence matrix is used for the file I/O of Bellman nets
  //(for output via the method toIncidenceMatrix())
  //We use the instance variable incidenceMatrix to store the
  //mere numbers.  It can be calculated using the method
  //calculateIncidenceMatrix()
  //Columns are labeled with transitions, rows with places
  int[][] incidenceMatrix;

  ArrayList topologicallySortedTransitionList;

  //constructor creates an initially empty BellmanNet
  BellmanNet(String title) {}

  //add a place
  void addPlace(Place p) {
    placeList.add(p);
  }

  //add a transition
  void addTransition(Transition t) {
    transitionList.add(t);
  }

  //add an arc
```

```
void addArc(Arc a) {
  arcList.add(a);
}

//get a place by index
Place getPlace(int index) {
  return placeList.get(index);
}

//get a transition by index
Transition getTransition(int index) {
  return transitionList.get(index);
}

//get a place by name
Place getPlace(String placeName) {
  //iterate through placeList
  for(Iterator i=placeList.iterator(); i.hasNext();){
    Place currentPlace=(Place) i.next();
    if (currentPlace.getName().equals(placeName)) {
      return currentPlace;
    }
  }
  return null; //not found
}

//get a transition by name
Transition getTransition(String transitionName) {
  //iterate through transitionList
  for(Iterator i=transitionList.iterator(); i.hasNext();){
    Transition currentTransition=(Transition) i.next();
    if (currentTransition.getName().equals(transitionName)) {
      return currentTransition;
    }
  }
  return null; //not found
}

//this helper method topologically sorts all the transitions
//from transitionList and puts the sorted list into
//topologicallySortedTransitionList.
void topologicallySortTransitions() {...}

void calculateAdjacencyMatrix() {...}
```

```
void calculateIncidenceMatrix() {...}

ArrayList getPresetIgnoreSelfLoopsAndEnablingPlaces
    (NetNode node) {...}

ArrayList getPostsetIgnoreSelfLoops(NetNode node) {...}

//Convert the BellmanNet to a comma separated
//exportable format based on the incidence matrix.
String toIncidenceMatrix() {
  calculateIncidenceMatrix(); //update the instance variable
                              //incidenceMatrix
  String result;
  //make the first line with transition names
  for (Iterator j=transitionList.iterator(); j.hasNext();) {
    result.append(j.next().getName());
  }
  //outer loop iterates over the places
  for (Iterator i=placeList.iterator(); i.hasNext();) {
    Place currentPlace= i.next();
    if(!(currentPlace instanceof EnablingPlace)) {
          //ignore enabling places
      if(currentPlace instanceof StatePlace) {
        result.append("s,");
      }
      else { //currentPlace is intermediate place
        result.append("i,");
      }
      result.append(currentPlace.getName()); //append place
                                             //name
      //inner loop iterates over the transitions
      for(int j=0; j<transitionList.size(); j++) {
        result.append(incidenceMatrix
            [placeList.indexOf(currentPlace)][j]);
      }////end of inner for loop over the transitions
      if(currentPlace instanceof StatePlace) {
        //make the init entry for INIT/DEC column
        //if currentPlace is a base state
        if(currentPlace.getInitialMarking()!=null) {
          result.append(currentPlace.getInitialMarking());
        }
      }
      else { //currentPlace is intermediate place
        //make the DEC entry for INIT/DEC column
        //if cip has a decision entry
```

```
          if(currentPlace.getDecision()!=null) {
            result.append(currentPlace.getDecision());
          }
        }
      }//end of if
    }//end of outer for loop over the places
    //second to last line produces the transition types
    for (Iterator j=transitionList.iterator(); j.hasNext();) {
      Transition currentTransition=j.next();
      if(currentTransition instanceof MTransition) {
       result.append(currentTransition.getMinOrMaxDesignator());
      }
      else { //ETransition
       result.append(currentTransition.getArithmeticOperator());
      }
    }
    //last line produces the constants for ETransitions
    for (Iterator j=transitionList.iterator(); j.hasNext();) {
      Transition currentTransition=j.next();
      if(currentTransition instanceof MTransition) {
        //MTransitions do not have constants, leave blank
      }
      else { //ETransition
        result.append(currentTransition.getConstantValue());
      }
    }
    return result;
  }

}//end of class
```

A.2 DP2PN System Files

The system file `Main1.txt` is as follows.

```
//////////////////////////////////////////////////////
//beginning of the fixed code Main1.txt for Main.java
//////////////////////////////////////////////////////

public static void main(String[] args) throws IOException{

  //set default value for name of logfile that records details
  //about how the Bellman net is built
  String outputFileName="buildBNlog.txt";

  //determine the correct current working directory,
  //so that it works for
  //case 1: if this class is launched directly with "java" from within
  //this directory (and therefore currentWorkingDir==user.dir)
  //case 2: if this class is launched with
  //          'java -cp thisDir classname' from its parent directory
  //        or with
  //          'rt.exec(java -cp thisDir classname)' from a class located in
  //          the parent directory
  final String subDirName="DP2PN"+problemName;
                                  //needs to be hardcoded, no way to find out
                                  //this name here, if called from parent dir
  String currentWorkingDir=System.getProperty("user.dir");
  //now append subDirName, if launched from parent dir
  //which is (most likely) the case if currentWorkingDir does not end in
  //subDirName
  if(!currentWorkingDir.endsWith(subDirName)) {
    currentWorkingDir=currentWorkingDir+"/"+subDirName;
  }

  //create an output directory for the Bellman net
  File newDir=new File(currentWorkingDir+"/"+problemName+"BN");
  newDir.mkdir(); //make the directory

  //Out writes to screen and file at the same time
  //to create a log file
  Out.pw=new PrintWriter(new FileWriter(
             new File(currentWorkingDir+"/"+problemName+"BN"+"/"
                 +outputFileName)));

  //Out5 writes to file
  //to create the incidence matrix of the Bellman net in .CSV format
  //the file is written into the subdirectory "PN2Java"+problemName
  Out5.pw=new PrintWriter(new FileWriter(
              new File(currentWorkingDir+"/"+problemName+"BN"+"/"
                  +problemName+"BN.csv")));

  Out.putln("Starting...");

//////////////////////////////////////////////////////
//end of the fixed code Main1.txt for Main.java
//////////////////////////////////////////////////////
```

The system file `Main2.txt` is as follows.

```
//////////////////////////////////////////////////////
//beginning of the fixed code Main3.txt for Main.java
//////////////////////////////////////////////////////

        //determine the successor states of current state
        StateList successorStates=calculateTransformation(currentState,d);
        //determine the transition weights given the current state
        ArrayList transitionWeights
            =calculateTransitionWeights(currentState,d);
        //generate table entry
        DPInstanceTableEntry entry = new DPInstanceTableEntry(
                                  currentState,
                                  d,
                                  calculateReward(currentState,d),
                                  successorStates,
                                  transitionWeights);
        theInstance.addTableEntry(entry);
        //enqueue unseen destination states by iterating over successorStates
        //for TSP only one iteration in the following loop
        //for MatMul two iterations
        for(Iterator it2=successorStates.iterator(); it2.hasNext();) {
          State currentSuccessor = (State) it2.next();
          if(!statesSeen.contains(currentSuccessor)) {
```

```
          stateQueue.addLast(currentSuccessor);
          statesSeen.add(currentSuccessor);//mark state as seen
        }
      }
    } //end of for loop over the decisions
  } //end of if
  else { //base state
    //do nothing
  }
} //end of loop once queue is empty

//print the instance
theInstance.print();
Out.putln();

//build the Bellman PN from the instance
//if there are no transition weights, use the ordinary method
if(!(theInstance.hasTransitionWeights())) {
  BellmanNet bn=theInstance.buildBellmanPNTopDown(problemName);
  Out5.putln(bn.toIncidenceMatrix()); //write the BellmanNet as incidence
                                      //matrix.
}
else { //there are transition weights, use the newly crafted method
  BellmanNet bn=theInstance.buildBellmanPNTopDownWithWeights(problemName);
  Out5.putln(bn.toIncidenceMatrix()); //write the BellmanNet as incidence
                                      //matrix.
}

//finish the output files
Out.putln("End.");
Out.pw.close(); //close the output stream (the log file)
Out5.pw.close(); //close the 5th output stream (the BellmanNet
                 //incidence matrix in .CSV)

}//end of main() method

//////////////////////////////////////////////////
//end of the fixed code Main3.txt for Main.java
//////////////////////////////////////////////////
```

The system file `Main3.txt` is as follows.

```
/////////////////////////////////////////////////////
//beginning of the fixed code Main3.txt for Main.java
/////////////////////////////////////////////////////

        //determine the successor states of current state
        StateList successorStates=calculateTransformation(currentState,d);
        //determine the transition weights given the current state
        ArrayList transitionWeights
            =calculateTransitionWeights(currentState,d);
        //generate table entry
        DPInstanceTableEntry entry = new DPInstanceTableEntry(
                              currentState,
                              d,
                              calculateReward(currentState,d),
                              successorStates,
                              transitionWeights);
        theInstance.addTableEntry(entry);
        //enqueue unseen destination states by iterating over successorStates
        //for TSP only one iteration in the following loop
        //for MatMul two iterations
        for(Iterator it2=successorStates.iterator(); it2.hasNext();) {
          State currentSuccessor = (State) it2.next();
          if(!statesSeen.contains(currentSuccessor)) {
            stateQueue.addLast(currentSuccessor);
            statesSeen.add(currentSuccessor);//mark state as seen
          }
        }
      } //end of for loop over the decisions
    } //end of if
    else { //base state
      //do nothing
    }
  } //end of loop once queue is empty

  //print the instance
  theInstance.print();
  Out.putln();

  //build the Bellman PN from the instance
  //if there are no transition weights, use the ordinary method
  if(!(theInstance.hasTransitionWeights())) {
    BellmanNet bn=theInstance.buildBellmanPNTopDown(problemName);
    Out5.putln(bn.toIncidenceMatrix()); //write the BellmanNet as incidence
                                        //matrix.
```

```
    }
    else { //there are transition weights, use the newly crafted method
      BellmanNet bn=theInstance.buildBellmanPNTopDownWithWeights(problemName);
      Out5.putln(bn.toIncidenceMatrix()); //write the BellmanNet as incidence
                                           //matrix.
    }

    //finish the output files
    Out.putln("End.");
    Out.pw.close(); //close the output stream (the log file)
    Out5.pw.close(); //close the 5th output stream (the BellmanNet
                     //incidence matrix in .CSV)

  }//end of main() method

  /////////////////////////////////////////////////////
  //end of the fixed code Main3.txt for Main.java
  /////////////////////////////////////////////////////
```

A.3 Output from PN2XML

In this section, the XML output from the PN2XML module is shown. In the form presented, such files (with the suffix `.xrn`) can be imported into the PN software system Renew version 1.6 [32]. Renew has the capabilities to display PNs (and thus the Bellman nets produced) graphically and also to simulate PNs. Unfortunately, the graphical layout is not optimized and looks rather crude for large PNs. Very often however, Renew's automatic net layout feature (when choosing the *Automatic Net Layout* option of the *Layout* menu) rearranges the nodes of a PN nicely, especially for smaller nets. For space reasons, we include only the XML file for SPA1 (see Sect. 9.4.2).

A.3.1 High-Level Bellman Net XML file for SPA1


```
<?xml version="1.0"?>
<!DOCTYPE net SYSTEM "http://www.renew.de/xrn1.dtd">
<net id="N" type="hlnet">
  <place id="I1">
    <graphics>
      <size w="20" h="20"/>
      <offset x="30" y="150"/>
      <fillcolor><RGBcolor r="112" g="219" b="147"/></fillcolor>
      <pencolor><RGBcolor r="0" g="0" b="0"/></pencolor>
      <textcolor><RGBcolor r="0" g="0" b="0"/></textcolor>
    </graphics>
    <annotation id="I2" type="name">
      <text>(0)</text>
      <graphics>
        <size w="20" h="16"/>
        <textsize size="12"/>
        <offset x="0" y="15"/>
        <fillcolor><transparent/></fillcolor>
        <pencolor><transparent/></pencolor>
        <textcolor><RGBcolor r="0" g="0" b="0"/></textcolor>
      </graphics>
    </annotation>
  </place>
  <place id="I3">
    <graphics>
      <size w="20" h="20"/>
      <offset x="30" y="240"/>
      <fillcolor><RGBcolor r="112" g="219" b="147"/></fillcolor>
      <pencolor><RGBcolor r="0" g="0" b="0"/></pencolor>
      <textcolor><RGBcolor r="0" g="0" b="0"/></textcolor>
    </graphics>
    <annotation id="I4" type="name">
      <text>p1</text>
      <graphics>
        <size w="20" h="16"/>
        <textsize size="12"/>
        <offset x="0" y="15"/>
        <fillcolor><transparent/></fillcolor>
        <pencolor><transparent/></pencolor>
        <textcolor><RGBcolor r="0" g="0" b="0"/></textcolor>
      </graphics>
    </annotation>
  </place>
  <place id="I5">
    <graphics>
      <size w="20" h="20"/>
      <offset x="30" y="330"/>
      <fillcolor><RGBcolor r="112" g="219" b="147"/></fillcolor>
      <pencolor><RGBcolor r="0" g="0" b="0"/></pencolor>
      <textcolor><RGBcolor r="0" g="0" b="0"/></textcolor>
    </graphics>
    <annotation id="I6" type="name">
      <text>(1)</text>
      <graphics>
        <size w="20" h="16"/>
        <textsize size="12"/>
        <offset x="0" y="15"/>
        <fillcolor><transparent/></fillcolor>
        <pencolor><transparent/></pencolor>
        <textcolor><RGBcolor r="0" g="0" b="0"/></textcolor>
      </graphics>
    </annotation>
```

```
      <annotation id="I7" type="initialmarking">
        <text>0.0</text>
      </annotation>
    </place>
    <place id="I14">
      <graphics>
        <size w="15" h="15"/>
        <offset x="210" y="30"/>
        <fillcolor><RGBcolor r="112" g="219" b="147"/></fillcolor>
        <pencolor><RGBcolor r="0" g="0" b="0"/></pencolor>
        <textcolor><RGBcolor r="0" g="0" b="0"/></textcolor>
      </graphics>
      <annotation id="I15" type="name">
        <text>ep1</text>
        <graphics>
          <size w="20" h="16"/>
          <textsize size="12"/>
          <offset x="0" y="15"/>
          <fillcolor><transparent/></fillcolor>
          <pencolor><transparent/></pencolor>
          <textcolor><RGBcolor r="0" g="0" b="0"/></textcolor>
        </graphics>
      </annotation>
      <annotation id="I16" type="initialmarking">
        <text>[]</text>
      </annotation>
    </place>
    <transition id="I8">
      <graphics>
        <size w="20" h="20"/>
        <offset x="120" y="60"/>
        <fillcolor><RGBcolor r="112" g="219" b="147"/></fillcolor>
        <pencolor><RGBcolor r="0" g="0" b="0"/></pencolor>
        <textcolor><RGBcolor r="0" g="0" b="0"/></textcolor>
      </graphics>
      <annotation id="I9" type="name">
        <text>mt1</text>
        <graphics>
          <size w="20" h="16"/>
          <textsize size="12"/>
          <offset x="0" y="15"/>
          <fillcolor><transparent/></fillcolor>
          <pencolor><transparent/></pencolor>
          <textcolor><RGBcolor r="0" g="0" b="0"/></textcolor>
        </graphics>
      </annotation>
      <annotation id="I10" type="expression">
        <text>y=x1</text>
      </annotation>
    </transition>
    <transition id="I11">
      <graphics>
        <size w="20" h="20"/>
        <offset x="210" y="60"/>
        <fillcolor><RGBcolor r="112" g="219" b="147"/></fillcolor>
        <pencolor><RGBcolor r="0" g="0" b="0"/></pencolor>
        <textcolor><RGBcolor r="0" g="0" b="0"/></textcolor>
      </graphics>
      <annotation id="I12" type="name">
        <text>st1</text>
        <graphics>
          <size w="20" h="16"/>
          <textsize size="12"/>
          <offset x="0" y="15"/>
          <fillcolor><transparent/></fillcolor>
          <pencolor><transparent/></pencolor>
          <textcolor><RGBcolor r="0" g="0" b="0"/></textcolor>
        </graphics>
      </annotation>
      <annotation id="I13" type="expression">
        <text>y=x1+3.0</text>
      </annotation>
    </transition>
    <arc id="I17" source="I14" target="I11" type="ordinary">
      <graphics>
        <fillcolor><RGBcolor r="112" g="219" b="147"/></fillcolor>
        <pencolor><RGBcolor r="0" g="0" b="0"/></pencolor>
        <textcolor><RGBcolor r="0" g="0" b="0"/></textcolor>
      </graphics>
    </arc>
    <arc id="I18" source="I8" target="I1" type="ordinary">
      <graphics>
        <fillcolor><RGBcolor r="112" g="219" b="147"/></fillcolor>
        <pencolor><RGBcolor r="0" g="0" b="0"/></pencolor>
        <textcolor><RGBcolor r="0" g="0" b="0"/></textcolor>
      </graphics>
      <annotation id="I19" type="expression">
        <text>y</text>
      </annotation>
```

```
  </arc>
  <arc id="I20" source="I3" target="I8" type="ordinary">
    <graphics>
      <fillcolor><RGBcolor r="112" g="219" b="147"/></fillcolor>
      <pencolor><RGBcolor r="0" g="0" b="0"/></pencolor>
      <textcolor><RGBcolor r="0" g="0" b="0"/></textcolor>
    </graphics>
    <annotation id="I21" type="expression">
      <text>x1</text>
    </annotation>
  </arc>
  <arc id="I22" source="I11" target="I3" type="ordinary">
    <graphics>
      <fillcolor><RGBcolor r="112" g="219" b="147"/></fillcolor>
      <pencolor><RGBcolor r="0" g="0" b="0"/></pencolor>
      <textcolor><RGBcolor r="0" g="0" b="0"/></textcolor>
    </graphics>
    <annotation id="I23" type="expression">
      <text>y</text>
    </annotation>
  </arc>
  <arc id="I24" source="I5" target="I11" type="double">
    <graphics>
      <fillcolor><RGBcolor r="112" g="219" b="147"/></fillcolor>
      <pencolor><RGBcolor r="0" g="0" b="0"/></pencolor>
      <textcolor><RGBcolor r="0" g="0" b="0"/></textcolor>
    </graphics>
    <annotation id="I25" type="expression">
      <text>x1</text>
    </annotation>
  </arc>
  <annotation id="A1" type="name">
    <text>spa1</text>
  </annotation>
</net>
```

B

User Guide for DP2PN2Solver

B.1 System Requirements for DP2PN2Solver

The DP2PN2Solver tool should run on any computer system on which a java compiler and runtime systems (javac and java) is installed. (We have only tested the tool using JDK 1.4.2 and JDK 1.5.) To obtain numerical solutions in a spreadsheet or graphically, a spreadsheet system (such as Excel) or a Petri net tool (such as Renew) is also required.

B.1.1 Java Environment

Make sure that JDK 1.4.2 or 1.5 is installed on your system. (DP2PN2Solver might also work with earlier and later versions, but this has not been tested.) You may download it for free from `http://www.java.sun.com`

DP2PN2Solver needs the compiler "javac", so having merely a Java Runtime Environment (JRE) installed will not be sufficient. The compiler "javac" should be universally accessible, so make sure to include the directory where the binary of "javac" resides to your current path.

For example, on Windows XP you might add the following to your path by going through the sequence "Start—Control Panel—System—Advanced—Environment Variables" and then under "System variables" (or "User variables" if you are not the admin), edit the variable "path" by prepending the prefix `c:\j2sdk1.4.2_01\bin;` to the path. (We assumed that the binary `javac.exe` is located in `c:\j2sdk1.4.2_01\bin`.)

You can easily check whether the compiler "javac" is universally accessible by typing `javac` on the command line from any directory of your choice; if you always get a response that explains the usage of "javac" then the path is set correctly; if you get a response that "javac" is not recognized as a command, then the path is not set correctly.

B.2 Obtaining DP2PN2Solver

The DP2PN2Solver software can be downloaded from the following websites:

`http://natsci.eckerd.edu/~mauchh/Research/DP2PN2Solver`

`http://www2.hawaii.edu/~icl/DP2PN2Solver`

free of charge. Version 6 of the software comes as a zipped file named DP2NP2SolverV6.zip. (Over time, the version numbers may have progressed.)

When unzipped, the folder will contain:

1. four folders (named bellman_net, BN2SolverV6, DP2PNv6, and pn2java) that contain the software and sample programs,
2. a read me file that contains special notices (regarding rights, disclaimers, etc.), including a reference to the book, and
3. a UserGuide that provides usage and reference information, explaining how to install and use the DP2PN2Solver tool and giving some implementation details.

The UserGuide includes instructions for compiling gDPS source programs into Bellman nets and for obtaining numerical solutions form these Bellman nets. Suggestions for debugging gDPS programs using the tool, as well as a list of diagnostic error messages, are also included.

(Note: This Appendix duplicates what is in the UserGuide.)

B.3 Installation of DP2PN2Solver

This section describes how and where to deploy the files to your computer.

B.3.1 Deployment of the Files

Take the downloaded ZIP file and unzip it into a directory of your choice (the installation directory). Make sure that when you unzip and extract, you preserve the pathname information for the files (check your zip utility for that) — what you definitely do not want to happen is that after unzipping all files are in a single flat directory.

Throughout these instructions let us assume that the installation directory that contains the software is named `DP2PN2Solver`. (You may pick another name if you wish.) In Windows systems, the software might have been installed to `C:\Program Files\DP2PN2Solver`, for example. Note that this manual also applies to UNIX-like systems; the DOS/Windows specific separation character '\' (backslash) will have to be read as '/' (slash) instead, and the DOS/Windows typical path separation character ';' (semicolon) has to be read as ':' (colon).

After the unzip is completed the directory `DP2PN2Solver` contains the following four subdirectories

```
bellman_net
BN2SolverV5
DP2BNv5
pn2java
```

The directories `bellman_net` and `pn2java` are Java packages that contain parts of the implementation. While they might be of interest to developers, they need not concern ordinary users of DP2PN2Solver.

The directory `DP2BNv5` contains the software for the module DP2PN; this is also the directory where you have to place your gDPS source files.

The directory `BN2SolverV5` contains the software for the module PN2Solver; this is also the directory where you have to place your Bellman net source file, which you probably created using the DP2PN module.

B.4 Running DP2PN2Solver

This section describes how to invoke DP2PN2Solver's two major modules DP2PN and PN2Solver.

B.4.1 The DP2PN Module

Preparations

Switch to the directory

```
DP2PN2Solver\DP2BNv5
```

Make sure you have your gDPS source file ready to be parsed in this directory. Remember, the gDPS source file is the one that contains all the information of your DP problem instance. Usually it needs to be created manually by the DP modeler. (With the exception of integer linear programming problems, for which a preprocessor exists.) You will notice that there are already some example gDPS source files present in this directory, ending with the suffix ".dp", e.g.

```
act.dp
...
tspAlt.dp
```

For details on how to create a gDPS source file, please refer to Chap. 3 and Chap. 4.

The following other files in this directory (ending in `.class` and in `.txt`) make up the DP parser and should not be modified:

```
Main1.txt
Main2.txt
Main3.txt
State.txt
CodeCreator.class
DPFEdata.class
DPspecificationParser$1.class
DPspecificationParser$JJCalls.class
DPspecificationParser$LookaheadSuccess.class
DPspecificationParser.class
DPspecificationParserConstants.class
DPspecificationParserTokenManager.class
FileCopy.class
Out.class
Out2.class
Out3.class
ParseException.class
SimpleCharStream.class
StreamGobbler.class
SymbolTableEntry.class
Token.class
TokenMgrError.class
```

Invocation of DP2PN

The gDPS2BN parser is invoked as follows, where the name of the gDPS source specification file (e.g. `mcm.dp`) is provided as a command line argument; optionally, a name for the parser log file can be specified.

```
java DPspecificationParser mcm.dp [parserLogFileName]
```

As an alternative, on DOS/Windows systems the batch file `DP2BN.bat` can simplify the invocation:

```
DP2BN mcm.dp [parserLogFileName]
```

Consequences of Invocation

This will create a directory for your problem instance, named DP2PNXXX where XXX is the name given in the "NAME" section of the gDPS specification, e.g. BST, MCM, etc. The contents of this new directory are explained below; we will from now on assume that the name of the problem instance is MCM.

Note that the name of the gDPS source specification file does not necessarily need to match the name given in the "NAME" section of this gDPS specification (e.g. MCM), but it seems good practice to match them anyway.

The parser log file is stored by default in

```
DP2PN2Solver\DP2BNv5\dpParserLog.txt
```

but the log file name can be changed to something else by specifying `[parserLogFileName]` in the invocation launch.

Now look at the directory

```
DP2PN2Solver\DP2BNv5\DP2PNMCM
```

which contains some intermediate data files that might be useful for debugging, if you encountered a problem parsing or compiling your gDPS source file. Otherwise they are of no further interest to the normal user. If the gDPS source is parsed, compiled, and successfully translated into a Bellman net, then the subdirectory `MCMBN` (`XXXBN` in general) will hold the output, i.e. the Bellman net, and a log file of the Bellman net building process in:

```
DP2PN2Solver\DP2BNv5\DP2PNMCM\MCMBN
  MCMBN.csv
  DP2PNlog.txt
```

The first file `MCMBN.csv` is the desired Bellman net, which can be fed into the PN2Solver software module.

This concludes the use of the DP2PN module. Starting from a gDPS source, we have produced a Bellman net. The next section deals with the use of the PN2Solver module, that automatically produces solver code from a Bellman net.

B.4.2 The PN2Solver Module

Preparations

The output of DP2PN can and should be used as the input for PN2Solver, so now manually copy or move the Bellman net you just produced (e.g. `MCMBN.csv`) to the directory

```
DP2PN2Solver\BN2SolverV5
```

which already contains twelve `.class` files that make up the Bellman net parser. So before you invoke the Bellman net parser, the directory will look like this:

```
DP2PN2Solver\BN2SolverV5
  BNspecificationParser.class
  BNspecificationParserConstants.class
  BNspecificationParserTokenManager.class
  Out.class
  Out2.class
  Out3.class
  Out4.class
  Out6.class
  ParseException.class
  SimpleCharStream.class
  Token.class
  TokenMgrError.class
  MCMBN.csv
```

In our distribution, there may already be additional sample Bellman net files in the directory, all having a name of the form `XXXBN.csv`.

Invocation of PN2Solver

After establishing the Bellman net file `MCMBN.csv` in

`DP2PN2Solver\BN2SolverV5`

and changing to this directory we are ready to invoke the BN2Solver module; we provide the name of the Bellman net file `MCMBN.csv` as a command line argument as follows (do not omit the classpath information, otherwise you will get an error message):

```
java -classpath .;.. BNspecificationParser MCMBN.csv
```

As an alternative, on DOS/Windows systems the batch file `BN2Solver.bat` can simplify the invocation:

```
BN2Solver MCMBN.csv
```

Consequences of Invocation

This produces a directory named `MCMSolverCode` that contains three solvers where each solver uses a different technology to produce a final result. In

`DP2PN2Solver\BN2SolverV5\MCMSolverCode`

the first solver is

```
MCM.csv
```

This is the speadsheet solver; load this file into a spreadsheet application such as Microsoft Excel, update the cells, and you get the solution to the DP problem instance.

The second solver is

```
MCM.xrn
```

This is the Petri net solver; import this file (which is in PNML standard format) into a PN application such as Renew, simulate the net, and you get the solution to the DP problem instance.

The third solver is the Java solver

```
MCMJavaSolver.java
```

The Java Solver file is automatically compiled to

```
MCMJavaSolver.class
```

and executed by the invocation from Section B.4.2 and the resulting solution tree of the problem instance can be found in the file

```
MCMSolutionTree.txt
```

This file is the desired output providing not only the optimal function value but also the optimal decision policy for the DP problem instance.

The file

```
PN2SolverLog.txt
```

contains a log of the transformation process from a Bellman net to the solver files.

If for some reason you would like to trace the automated compilation and execution of the JavaSolver, see the file

```
runJavaSolver.bat
```

which contains the necessary steps to compile and launch the JavaSolver:

```
javac -classpath ..\.. MCMJavaSolver.java
rem for running, do not forget to include the current directory
java -classpath .;..\.. MCMJavaSolver
```

B.5 Creation of the gDPS Source File

A gDPS source file is a plain text file ending in the suffix `.dp`. It can be created and modified with a simple text editor. Details on how to create a gDPS source files are given in Chapter 3. Numerous gDPS examples are shown in Chapter 4. We refer the reader to these chapters.

B.6 Debugging gDPS Code

B.6.1 Omission of Base Cases

Suppose in INTKSSCA, instead of correctly specifying the base cases as

```
DPFE_BASE_CONDITIONS:
  f(k,s) = 0.0 WHEN (s==0);
  f(k,s) = 99990.0 WHEN ((k==M1)&&(s>0));
```

we accidentally omit the base cases $(2,0)$ and $(3,0)$ by specifying

```
DPFE_BASE_CONDITIONS:
  f(k,s) = 0.0 WHEN ((k==M1)&&(s==0));
  f(k,s) = 99990.0 WHEN ((k==M1)&&(s>0));
```

then there is no feasible decision to take from the states $(2,0)$ and $(3,0)$ and hence the recursive process terminates, yet $(2,0)$ and $(3,0)$ are not declared as base states.

The DP2PN module will not report an error in this case, and produce a PN. (This is not a bug of the DP2PN module, which has performed its job of translating the gDPS source into a PN — the error becomes apparent after performing consistency checks on the PN.) In the resulting PN, the states $(2,0)$ and $(3,0)$ are source places without a proper initial marking.

One way to detect this, would be to examine the intermediate PN produced by the DP2PN module.

Another way of getting valuable hints for debugging is to examine the file `buildBNlog.txt` which is located in the same directory as the resulting Bellman net. The file contains the base states and their initial values and also the state-decision-reward-transformation-table, which shows the states $(2,0)$ and $(3,0)$ appearing as successor states in the transformation column, but not as states from which transformations originate (i.e. they do not appear in the state-column). It also shows that the states $(2,0)$ and $(3,0)$ are not among the base states.

If it is attempted to invoke BN2Solver upon the flawed PN, for the Java solver we get a bunch of error messages, e.g.

```
errStream>intKSscaSolverCode\intKSscaJavaSolver.java:44:
          cannot resolve symbol
errStream>symbol  : variable I41
errStream>location: class intKSscaJavaSolver
errStream>    operandList.add(I41);
errStream>                    ^
errStream>intKSscaSolverCode\intKSscaJavaSolver.java:47:
          cannot resolve symbol
errStream>symbol  : variable I41
errStream>location: class intKSscaJavaSolver
errStream>         I41.getValue()+210.0,
errStream>         ^
```

These error messages do not provide a good starting point for debugging. The debugging process should be initiated one step earlier, as mentioned above.

B.6.2 Common Mistakes

Space Before Minus

The DP2PN module's lexicographical analysis scans negative numbers as tokens, which sometimes leads to an ambiguity when subtraction expressions are desired and there is no whitespace between the minus operator and the expression to be subtracted. In this case an error is reported. For example,

```
TRANSFORMATION_FUNCTION: t1(m,dummy)
                           =(m-1);
```

causes an error that looks something like:

```
Exception in thread "main" ParseException:
  Encountered "-1" at line 29, column 32.
Was expecting one of:
    ")" ...
    "[" ...
    "," ...
    "+" ...
    "-" ...
    "*" ...
    "/" ...
    "%" ...
    "(" ...
```

Avoid this error by adding a space after the minus sign, as in

```
TRANSFORMATION_FUNCTION: t1(m,dummy)
                           =(m- 1);
```

and now it works.

Forgotten Braces in Singleton

If a set expression involves a singleton, it is easy to forget the curly braces around the single element. But those are necessary to correctly identify the expression as a set. For example, with the integer decision variable d,

```
TRANSFORMATION_FUNCTION:
  tLeft(k,S,i,j,d)  = (k+1, S SETMINUS d, i, begintime[d]);
  tRight(k,S,i,j,d) = (k+1, S SETMINUS d, endtime[d], j  );
```

causes the following error when attempting to construct the Bellman net:

```
errStream>DP2PNact2\act2Main.java:239: cannot resolve symbol
errStream>symbol  : constructor NodeSet (Decision)
errStream>location: class NodeSet
errStream>    NodeSet _globalSet2=new NodeSet(d);
errStream>                        ^
errStream>DP2PNact2\act2Main.java:246: cannot resolve symbol
errStream>symbol  : constructor NodeSet (Decision)
errStream>location: class NodeSet
errStream>    NodeSet _globalSet3=new NodeSet(d);
errStream>                        ^
errStream>[total 1502ms]
errStream>2 errors
```

The error is not caught by the syntax parser, because if d were a set variable, syntactically there would be nothing wrong. The correct formulation would be:

```
TRANSFORMATION_FUNCTION:
  tLeft(k,S,i,j,d)  = (k+1, S SETMINUS {d}, i, begintime[d]);
  tRight(k,S,i,j,d) = (k+1, S SETMINUS {d}, endtime[d], j  );
```

B.7 Error Messages of DP2PN2Solver

In addition to obvious syntax errors, the following error messages are reported by the DP2PN module.

- *Illegal type in state section.* This error is reported if a component of the state is not of type `int` or `Set`.
- *Illegal type of decisionSetGlobalFunctionalArgument.* This error is reported if a variable used in the decision set is not of type `int` or `Set`.
- *A* `DPFE_BASE_CONDITION` *functional does not match the one declared in GOAL section.* This error is reported if there is a mismatch of the functional name used in the goal statement and the one used in a base condition.
- *Base section functional does not match the one declared in GOAL section.* This error is reported if there is a mismatch of the functional name used in the goal statement and the one used in a base statement.
- *DPFE functional does not match the one declared in GOAL section.* This error is reported if there is a mismatch of the functional name used in the goal statement and the one used in the DPFE.
- *Decision variable in DPFE does not match the one declared after* `DECISION_VARIABLE`. This error is reported if there is a mismatch of the decision variable name declared in the decision variable section and the one used in the DPFE.

- *Decision set identifier in DPFE does not match the one declared after* `DECISION_SPACE`. This error is reported if there is a mismatch of the decision set identifier used in the decision space section and the one used in the DPFE.
- *More than one reward functional in DPFE.* The current version of DP2PN2Solver requires exactly one reward functional to present in the DPFE. This error is reported if there is more than one reward functional.
- *Recursive functional mismatch in DPFE.* There must be exactly one name that is used as the functional in the recurrence relation. This error is reported if there is a mismatch of the functional names used within the DPFE.
- *In* `REWARD_FUNCTION` *section, functional identifier does not match the one in DPFE.* This error is reported if there is a mismatch of the functional name used for the reward function in the DPFE and the one used in the reward function section.
- *Illegal type of rewardFunctionGlobalFunctionalArgument.* This error is reported if a variable used in the reward function section is not of type `int` or `Set`.
- *In* `TRANSFORMATION_FUNCTION` *a functional appears that is not present in DPFE.* This error is reported if a functional name is used in the transformation function section, but not in the DPFE.
- *Illegal type of transformationFunctionGlobalFunctionalArgument.* This error is reported if a variable used in the transformation function section is not of type `int` or `Set`.
- *Illegal type of transformationFunctionSetGlobalFunctionalArgument.* This error is reported if a variable used in the transformation function section is a state coordinate of illegal type.
- *In* `TRANSITION_WEIGHTS` *a functional appears that is not present in DPFE.* All weight functions defined in the transition weight section must be used in the DPFE.

References

1. Richard E. Bellman. *Dynamic Programming.* Princeton University Press, Princeton, New Jersey, 1957.
2. Richard E. Bellman. On the approximation of curves by line segments using dynamic programming. *Communications of the ACM*, 4(6):284, 1961.
3. Richard E. Bellman. *An Introduction to Artificial Intelligence: Can Computers Think?* Boyd and Fraser, San Francisco, California, 1978.
4. Richard E. Bellman and Stuart E. Dreyfus. *Applied Dynamic Programming.* Princeton University Press, Princeton, New Jersey, 1962.
5. Alan W. Biermann and Dietolf Ramm. *Great Ideas in Computer Science with Java.* MIT Press, Cambridge, MA, 2001.
6. Jonathan Billington, Søren Christensen, Kees van Hee, Ekkart Kindler, Olaf Kummer, Laure Petrucci, Reinier Post, Christian Stehno, and Michael Weber. The Petri net markup language: Concepts, technology, and tools. In W.M.P. van der Aalst and E. Best, editors, *Proceedings of the 24th International Conference on Applications and Theory of Petri Nets (ICATPN 2003), Eindhoven, The Netherlands, June 23-27, 2003 — Volume 2679 of Lecture Notes in Computer Science / Wil M. P. van der Aalst and Eike Best (Eds.)*, volume 2679 of *LNCS*, pages 483–505. Springer-Verlag, June 2003.
7. Allan Borodin, Morten N. Nielsen, and Charles Rackoff. (Incremental) priority algorithms. In *Proc. 13th Annual ACM-SIAM Symp. on Discrete Algorithms, San Francisco*, pages 752–761, 2002.
8. Richard Bronson and Govindasami Naadimuthu. *Schaum's Outline of Theory and Problems of Operations Research.* McGraw-Hill, New York, New York, 2nd edition, 1997.
9. Kevin Q. Brown. Dynamic programming in computer science. Technical Report CMU-CS-79-106, Carnegie-Mellon University, February 1979.
10. Thomas H. Cormen, Charles E. Leiserson, Ronald L. Rivest, and Clifford Stein. *Introduction to Algorithms.* McGraw-Hill Book Company, Boston, 2nd edition, 2001.
11. Sharon Curtis. A relational approach to optimization problems. Technical monograph PRG-122, Oxford University Computing Laboratory, 1996.
12. Eric V. Denardo. *Dynamic Programming — Models and Applications.* Prentice Hall, Englewood Cliffs, New Jersey, 1982.

13. Stuart E. Dreyfus and Averill M. Law. *The Art and Theory of Dynamic Programming.* Academic Press, New York, 1977.
14. Hartmann J. Genrich. Predicate/transition nets. In W. Brauer, W. Reisig, and G. Rozenberg, editors, *Petri Nets: Central Models and Their Properties, Advances in Petri Nets 1986, Part I, Proceedings of an Advanced Course, Bad Honnef, Germany, September 8–19, 1986*, volume 254 of *LNCS*, pages 207–247, Bad Honnef, Germany, 1987. Springer-Verlag.
15. Hartmann J. Genrich and K. Lautenbach. System modelling with high-level Petri nets. *Theoretical Computer Science*, 13:109–136, 1981.
16. Dan Gusfield. *Algorithms on Strings, Trees, and Sequences: Computer Science and Computational Biology.* Cambridge University Press, 1st edition, 1997.
17. Rachel Harrison and Celia A. Glass. Dynamic programming in a pure functional language. In *Proceedings of the 1993 ACM/SIGAPP Symposium on Applied Computing*, pages 179–186. ACM Press, 1993.
18. N.A.J. Hastings. *Dynamic Programming with Management Applications.* Butterworths, London, England, 1973.
19. N.A.J. Hastings. *DYNACODE Dynamic Programming Systems Handbook.* Management Center, University of Bradford, Bradford, England, 1974.
20. Frederick S. Hillier and Gerald J. Lieberman. *Introduction to Operations Research.* McGraw-Hill Publishing Company, New York, 5th edition, 1990.
21. Frederick S. Hillier and Gerald J. Lieberman. *Introduction to Operations Research.* McGraw-Hill Publishing Company, Boston, 7th edition, 2001.
22. Ellis Horowitz, Sartaj Sahni, and Sanguthevar Rajasekaran. *Computer Algorithms/C++.* Computer Science Press, New York, New York, 1996.
23. T.C. Hu. *Combinatorial Algorithms.* Addison-Wesley, Reading, 1982.
24. Matthias Jantzen. Complexity of place/transition nets. In W. Brauer, W. Reisig, and G. Rozenberg, editors, *Petri Nets: Central Models and Their Properties, Advances in Petri Nets 1986, Part I, Proceedings of an Advanced Course, Bad Honnef, Germany, September 8–19, 1986*, volume 254 of *LNCS*, pages 413–434, Bad Honnef, Germany, 1987. Springer-Verlag.
25. Kurt Jensen. Coloured Petri nets and the invariant-method. *Theoretical Computer Science*, 14:317–336, 1981.
26. Kurt Jensen. Coloured Petri nets. In W. Brauer, W. Reisig, and G. Rozenberg, editors, *Petri Nets: Central Models and Their Properties, Advances in Petri Nets 1986, Part I, Proceedings of an Advanced Course, Bad Honnef, Germany, September 8–19, 1986*, volume 254 of *LNCS*, pages 248–299, Bad Honnef, Germany, 1987. Springer-Verlag.
27. Kurt Jensen. *Coloured Petri Nets, Vol. 1.* Springer-Verlag, Berlin, Germany, 1992.
28. Brian W. Kernighan. Optimal sequential partitions of graphs. *Journal of the ACM*, 18(1):34–40, 1971.
29. Jeffrey H. Kingston. *Algorithms and Data Structures: Design, Correctness, Analysis.* Addison Wesley Longman, Harlow, England, 2nd edition, 1998.
30. Jon Kleinberg and Éva Tardos. *Algorithm Design.* Pearson Addison-Wesley, Boston, 2006.
31. Olaf Kummer. Introduction to Petri nets and reference nets. *Sozionik aktuell*, 1:7–16, 2001.
32. Olaf Kummer, Frank Wienberg, and Michael Duvigneau. *Renew User Guide Release 1.6.* Department of Informatics, University of Hamburg, Hamburg, Germany, 2002.

33. Olaf Kummer, Frank Wienberg, Michael Duvigneau, Jörn Schumacher, Michael Köhler, Daniel Moldt, Heiko Rölke, and Rüdiger Valk. An Extensible Editor and Simulation Engine for Petri Nets: Renew. In Jordi Cortadella and Wolfgang Reisig, editors, *Applications and Theory of Petri Nets 2004. 25th International Conference, ICATPN 2004, Bologna, Italy, June 2004. Proceedings*, volume 3099 of *Lecture Notes in Computer Science*, pages 484–493, Heidelberg, June 2004. Springer.
34. K. Lautenbach and A. Pagnoni. Invariance and duality in predicate/transition nets. Arbeitspapier der GMD 132, Gesellschaft für Math. und Datenverarbeitung mbH, Bonn, Germany, February 1985.
35. Art Lew. *Computer Science: A Mathematical Introduction.* Prentice-Hall International, Englewood Cliffs, New Jersey, 1985.
36. Art Lew. *N* degrees of separation: Influences of dynamic programming on computer science. *J. Math. Analysis and Applications*, 249(1):232–242, 2000.
37. Art Lew. A Petri net model for discrete dynamic programming. In *Proceedings of the 9th Bellman Continuum: International Workshop on Uncertain Systems and Soft Computing, Beijing, China, July 24–27, 2002*, pages 16–21, 2002.
38. Art Lew. Canonical greedy algorithms and dynamic programming. *Journal of Control and Cybernetics*, 2006.
39. Art Lew and R. Halverson, Jr. A FCCM for dataflow (spreadsheet) programs. In *Proceedings. IEEE Symposium on FPGAs for Custom Computing Machines (FCCM '95).* IEEE Computer Society, 1995.
40. Art Lew and Holger Mauch. Solving integer dynamic programming using Petri nets. In *Proceedings of the Multiconference on Computational Engineering in Systems Applications (CESA), Lille, France, July 9–11, 2003*, 2003.
41. Art Lew and Holger Mauch. Bellman nets: A Petri net model and tool for dynamic programming. In Le Thi Hoai An and Pham Dinh Tao, editors, *Proceedings of Modelling, Computation and Optimization in Information Systems and Management Sciences (MCO), Metz, France, July 1–3, 2004*, pages 241–248, Metz, France, 2004. Hermes Science Publishing Limited.
42. William J. Masek and Michael S. Paterson. A faster algorithm computing string edit distances. *Journal of Computer and System Sciences*, 20:18–31, 1980.
43. Holger Mauch. A Petri net representation for dynamic programming problems in management applications. In *Proceedings of the 37th Hawaii International Conference on System Sciences (HICSS2004), Waikoloa, Hawaii, January 5–8, 2004*, pages 72–80. IEEE Computer Society, 2004.
44. Holger Mauch. *Automated Translation of Dynamic Programming Problems to Java code and their Solution via an Intermediate Petri Net Representation.* PhD thesis, University of Hawaii at Manoa, 2005.
45. Holger Mauch. DP2PN2Solver: A flexible dynamic programming solver software tool. *Journal of Control and Cybernetics*, 2006.
46. Boleslaw Mikolajczak and John T. Rumbut, Jr. Distributed dynamic programming using concurrent object-orientedness with actors visualized by high-level Petri nets. *Computers and Mathematics with Applications*, 37(11–12):23–34, 1999.
47. Tadao Murata. Petri nets: Properties, analysis and applications. *Proceedings of the IEEE*, 77(4):541–580, April 1989.
48. G. L. Nemhauser. *Introduction to Dynamic Programming.* Wiley, New York, 1966.

49. Christos H. Papadimitriou and Kenneth Steiglitz. *Combinatorial Optimization: Algorithms and Complexity.* Prentice-Hall, Englewood Cliffs, New Jersey, 1982.
50. John F. Raffensperger and Pascal Richard. Implementing dynamic programs in spreadsheets. *INFORMS Transactions on Education*, 5(2), January 2005.
51. Wolfgang Reisig. *Petri Nets: an Introduction.* Springer-Verlag, Berlin, Germany, 1985.
52. P. Richard. Modelling integer linear programs with Petri nets. *RAIRO - Recherche Opérationnelle - Operations Research*, 34(3):305–312, Jul-Sep 2000.
53. Kenneth H. Rosen. *Discrete Mathematics and Its Applications.* WCB/McGraw-Hill, Boston, MA, fourth edition, 1999.
54. William Sacco, Wayne Copes, Clifford Sloyer, and Robert Stark. *Dynamic Programming — An Elegant Problem Solver.* Janson Publications, Inc., Providence, Rhode Island, 1987.
55. Alfonso T. Sarmiento. P4 dynamic programming solver. http://www.geocities.com/p4software/.
56. Moshe Sniedovich. Use of APL in operations research: An interactive dynamic programming model. *APL Quote Quad*, 12(1):291–297, 1981.
57. Moshe Sniedovich. *Dynamic Programming.* Marcel Dekker, Inc., New York, New York, 1992.
58. Moshe Sniedovich. OR/MS games: 2. the tower of hanoi problem. *INFORMS Transactions on Education*, 3(1):34–51, 2002.
59. Moshe Sniedovich. Dijkstra's algorithm revisited: the DP connexion. *Journal of Control and Cybernetics*, 2006.
60. Harald Störrle. An evaluation of high-end tools for Petri nets. Technical Report 9802, Institute for Computer Science, Ludwig-Maximilians-University Munich, Munich, Germany, 1998.
61. Harvey M. Wagner and Thomson M. Whitin. Dynamic version of the economic lot size model. *Management Science*, 5(1):89–96, October 1958.
62. Michael Weber and Ekkart Kindler. The Petri net markup language. In Hartmut Ehrig, Wolfgang Reisig, Grzegorz Rozenberg, and Herbert Weber, editors, *Petri Net Technology for Communication Based Systems*, volume 2472 of *LNCS*, pages 124–144. Springer-Verlag, November 2003.
63. Wayne L. Winston. *Operations Research: Applications and Algorithms.* Brooks/Cole — Thomson Learning, Pacific Grove, CA, 4th edition, 2004.
64. Wayne L. Winston and Munirpallam Venkataramanan. *Introduction to Mathematical Programming: Applications and Algorithms.* Brooks/Cole – Thomson Learning, Pacific Grove, CA, 4th edition, 2002.

Index